THE ENCYCLOPEDIA OF
AVIATION

THE ENCYCLOPEDIA OF
AVIATION

GENERAL EDITOR: PAUL BEAVER

GALLERY BOOKS

First published in Gt Britain by
Octopus Books Ltd,
Michelin House, 81 Fulham Road,
London SW3 6RB

This edition published by Gallery Books
An imprint of W H Smith Publishers Inc.
112 Madison Avenue, New York, NY 10016

© Octopus Books, 1986, 1989

Reprinted 1989

ISBN 0-8317-2776-4

Produced by Mandarin Offset
Printed and bound in Hong Kong

CONTENTS

INTRODUCTION

Although aviation is not truly a child of the 20th century, the bulk of experience and endeavour in the subject has taken place within living memory; there are still those who remember the Wright Brothers' achievements being published in newspapers and reading about Blériot's courageous first flight across the English Channel. Then came the daring exploits of the military pilots of World War I, forced to fly and fight without parachutes because senior officers, on both sides, thought their young aviators might leap out of the frail machines at the first sign of danger. The significance of the aeroplane is underlined by its rapid development: by 1918, there were

Until the early days of World War II aviation implied in the main biplane aircraft of limited performance. The Airco DH 9A bomber carried out policing in various corners of the world for Britain's Royal Air Force.

22,500 flying machines in military colours. The aeroplane had superseded the balloon and was beginning to be accepted as necessary for success on the battlefield or at sea.

This growth was not only military. After the 'war to end all wars' there was a tremendous exploitation of air routes, using for the most part converted military surplus aircraft. This period saw the great feats of aviation in the 1920s and 1930s, pioneering work such as Alcock and Brown's transatlantic crossing from west to east in June 1919. In the same year the first passenger domestic and international air services started, in Germany and between France and Belgium respectively (five decades later the Franco-Belgian border saw the first international helicopter airline service).

The year 1919 was momentous for aviation. As well as the above achievements, the year saw the first scheduled daily international commercial airline flight anywhere, in August, when a de Havilland DH-16 flew from Hounslow (near the site of the current busiest international airport in the world, London

Heathrow) to Paris Le Bourget (now home of the Paris Air Show). And the oldest airline still flying under its original name – KLM – first began operations in that same year.

This was the great pioneering era of aviation. Setting and breaking records was all the rage. The first round-the-world flight, by Douglas World Cruisers, was in 1924; the first flight over the North Pole, by Lieutenant Commander Richard Byrd in a Fokker F VIIA tri-motor, was in 1926; the first non-stop solo crossing of the Atlantic, by Captain Charles Lindbergh in a Ryan monoplane, was in 1927; it was followed in the same year by Captain Dieudonne Costes of the French Navy in a Breguet XIX crossing the South Atlantic. In 1928, Captain Charles Kingford Smith and C T P Ulm made the first flight across the Pacific Ocean, again using the Fokker F VII tri-motor, flying from California to Queensland, via Hawaii and Fiji. By the late 1920s many famous company names had appeared on the scene, including Pan American Airways (1927) and the Australian Flying Doctor Service (1928); a Zeppelin airship had

Military needs spurred further aircraft development in the Second World War. The American B-24 Liberator bombers' role was to protect maritime power, especially in the mid-Atlantic 'black hole' during the battle against Hitler's U-Boats.

flown around the world (1929) and Commander Byrd had taken a Ford Tri-Motor over the South Pole (1929).

The 1930s saw the continuation of pioneering efforts in civil air transport and in the development of the military aeroplane, including autogyros and some of the first metal-winged monoplanes. Although its significance could not have been fully anticipated at the time, Douglas Aircraft first flew the DC-3 in 1935; 50 years later 2,500 of the type are still in service.

When war clouds again gathered over Europe and later the Far East, it was the aeroplane that was seen as the greatest threat, terrorising civilian populations from Poland to the Ukraine and Manchuria to Singapore. Many nations, particularly the Americans, British, Germans and Japanese, had developed air power in the 1930s, and it came to one of its periodic peaks in the mid-1940s. Aerial weapons were used with devastating effect on military targets as well as civilian, but with the coming of the atomic bomb, delivered by an aircraft on Japan in 1945, most people thought all war was ended. The proxy conflicts between what are now called superpowers, as well as the former imperialist world powers,

The nuclear world since 1945 has seen the dominance of missiles. A fairly recent addition to superpower weaponry is the cruise missile like the McDonnell Sub-Harpoon and other stand-off aerial devices, launched from ships, submarines, and aircraft.

led however, to the large number of 'brush fire' wars of the late 1940s and 1950s, especially during the Cold War period, which showed the aeroplane dominating many nations' high-technology industries.

Air power from the late 1960s onwards includes aerospace power, with the development of satellites and missiles of mass destruction. Recently there has been considerable debate about President Reagan's SDI programme – better known as 'Star Wars' – which could be a way to eliminate the threat of mass nuclear destruction. It could also add yet another dimension to the balance of terror as the Soviet Union tries to catch the American lead in technology.

The 1970s and 1980s have seen the growth of aviation industries in many of the developing countries of the world. Initially, these served local needs, especially in countries where air transport is considered a social necessity in binding together peoples of different regions and ethnic groupings, often in far-flung provinces. In the mid-1980s, many of these nations – Brazil, Indonesia, Chile and Argentina – have become exporters of airframes and systems. The same period has also seen the re-development of the barter and trade-off deal, whereby one nation will sell an

Of all the many airliner manufacturers who have helped span the world and brought communities closer together none has been more successful than Boeing; this is a 737 short-haul airliner in the colours of the American carrier Piedmont.

Helicopters have recently proved invaluable in many roles, both civil and military. This Bell 206L LongRanger of the US Park Police, Washington DC, took part in the heroic rescue of Air Florida crash survivors from the icy Potomac River in 1984.

aircraft to another, but many of the components will be manufactured or assembled in the buyer's country, thus developing aviation expertise. Collaboration is a new watchword in the industry.

Today, although European and North American manufacturers still control the bulk of aviation activity, their positions are not as strong as they were and to survive many projects have become multi-national. A classic example of this type of operation is the Airbus, a European airliner series which directly rivals the American Boeing company for business around the world.

Over the years aviation has played an important, even vital, role in the success of man on this planet. Despite the fact that the aeroplane has brought the concept of mass destruction to civilian populations previously little affected by wars, its social and political advantages have far outweighed its disadvantages. In terms of air travel alone, the advent of the long-range passenger aircraft, developed from World War II military aircraft such as the heavy bomber, has brought the population of the world closer together and allowed the rapid transportation of people and freight to almost every corner of the globe within 24 hours.

Many a shipwrecked sailor, pest-infested farmer or critically ill patient has had cause to thank the aircraft in the fixed-wing or rotary-winged forms for saving life and property. In military terms too, soldiers and sailors have been saved by aircraft, although the primary role of the military aircraft is still the destruction of the enemy, whether in defence or offensive operations. As aviation develops, certain technological trends create advances in other sectors of life: it is in this way that aviation has changed the world.

Typical of modern highly sophisticated fighters, the Grumman F-14A Tomcat is an advanced two-seat naval aircraft, armed with Phoenix guided missiles which can engage a target at very long range. It is capable of defending America's strike aircraft carrier forces against almost any aerial attack.

This encyclopedia aims to put current aviation trends into perspective, and at the same time to cover the history of a particular concept: the notion that aerospace is pushing back the barriers of knowledge, with research and development being carried out for the twenty-first century. No book of this size can be a catch-all, nor can it hope to be a 'text book' – it should be read and re-read as a source book, a reference work which allows those who wish to know something of what is behind, say, the development of transatlantic air routes, or why aircraft are fitted with guided weapons, to find out at least the basic facts that answer their questions. In many cases the authors have used their intimate knowledge of the aerospace industry to predict future patterns and trends – one look at the technical terms section will show the way in which the electronic revolution is altering the flight deck and cockpits of modern aircraft. The section on learning to fly and gliding will enable readers to identify ways in which they, too, can join in aviation as a sport and perhaps as a career.

The writers gathered together for this work have special expertise in the subjects that form the basis of their contributions, and the emphasis throughout the book is on a balanced appraisal of past and present, informed by the latest future perspectives on this essential part of 20th century life.

1

THE PIONEER YEARS

Man has always entertained dreams of flying, and from the days of the ancients, the study of aviation has led to great technological achievement. However, when aviation did eventually triumph, it was slow to gain official recognition, and even the Wright brothers were unable to convince the people of the United States that the aeroplane was no toy but a serious invention which would one day change the world.

Nevertheless the story of aviation before the First World War is one of remarkable pioneering, of the flights of Farman, Blériot, Santos-Dumont and A V Roe which made the headlines in Europe. By 1914 aeroplanes were being seen as the way forward, and balloons and kites were things of the past.

Man's pursuit of flight probably dates back to prehistoric hunter-gatherers admiring winged creatures that wheeled safely in the skies. For 5,000 years of civilisation, fantasy and fact blended as desire overrode ability and stories of Icarus and Daedalus and Chinese sky chariots represented the sum of aeronautical knowledge. The millennium preceding twentieth-century success was punctuated with the broken bodies of tower jumpers. With no awareness of aerodynamics and the inability of the outstretched human arm to support the body's weight, they tried to emulate the birds by flapping winged arms.

Original ordered thought about flight began when Leonardo da Vinci (1452-1519) designed a helicopter toy and a parachute, and studied bird flight in some detail. In a world of reusable spacecraft and the immediacy of worldwide electronic communication it is difficult to imagine the delay in propagation of information which accompanied the centuries of alchemical aviation. It would be almost 300 years before Leonardo's notes were published. Whilst not causing a great sensation they eventually led to helicopter toys becoming popular in Europe.

Experiments with a similar device fired a young English aristocrat with enthusiasm for aviation. Sir George Cayley spent most of his adult life in experimentation and left designs which contained many sound ideas. Generally accepted as 'The Father of Aerial Navigation', he eventually built gliders which carried first a boy and later a man across a valley on his estate. His major contribution was to lay down the principles of aircraft design with wings to provide lift, joined by a rigid fuselage to tail surfaces which provided directional control in yaw and pitch.

In Victorian England, the Industrial Revolution was generating new riches which allowed able technocrats the resources to match the traditional wealth of the aristocracy. Chard in Somerset was a major lacemaking centre where two skilled engineers, Henson and Stringfellow, worked in the 1840s. Their

Although there is strong cause to believe that the ancients had considered aerodynamics, it was not until Leonardo da Vinci's experiments in the sixteenth century that scientists ordered their thoughts constructively towards the problems of how best to lift man from the Earth. This famous sketch from his notebooks shows a Leonardo dream which later became realised as the helicopter.

design of 1842 is instantly recognisable as an aeroplane in the modern sense, but one must question whether a full-scale machine could actually have carried more than a pilot. The limiting factor mitigating against success was undoubtedly the lack of an efficient power plant – a problem not to be solved until the development of the lightweight petrol engine some forty years later.

The aerial steam carriage

Henson and Stringfellow probably met as teenagers, and became interested in flying whilst both quite young. It is likely that they were influenced by Cayley's pioneering work when they built their own glider models in Chard. In 1842, Henson submitted a provisional patent and specification for 'Locomotive Apparatus for Air, Land and Water'. Generally advertised as the 'Aerial Steamer' or 'Aerial Steam Carriage', the grossly over-optimistic claims of their ill-selected business partners doomed to ridicule what was, in

Sir George Cayley (1774-1857)

Baronet and landowner with estate near Scarborough, in Yorkshire, who was a remarkble inventor in wide-ranging fields. Began interest in aviation about 1796 and was first man to investigate aviation scientifically. Laid foundations of science of aerodynamics, and realised significance of lifting power of plane surface. 1809: Published papers on 'Aerial Navigation', and effectively proved mechanical flight was possible. Established basis of curved aerofoil

section, principles of stability and directional control and the morphology of a practical aeroplane. Was aware of limitations of steam power in aviation and considered that some form of internal combustion engine would be necessary. Called 'Father of Aerial Navigation' by middle of 19th century. 1849: Built full-sized triplane glider which carried a boy a few metres. 1853: Built another which carried a coachman a short distance. His work in aviation influenced all who followed in this field.

The Industrial Revolution reawakened interest in heavier-than-air flight and amongst the most notable of early products of this original thought was the Henson Aerial Steam Carriage, which, though never built, was a portent of the future.

reality, an incredibly far-sighted project.

There were weaknesses in the design. There was no wing dihedral (the wings were not angled upwards, and so provided less lift) nor was there provision for lateral control. In addition, the fuselage was too short for the tail surfaces to be effective, and the power plant

John Stringfellow (1799-1883)
Born in Sheffield, with father who had reputation as mechanical genius. Apprenticed to lace trade in Nottingham and began building own reputation. 1820: Moved to Chard; became well known as outstanding inventive engineer and designer of lightweight steam engines. 1840/41: Began association with Henson and collaborated on Aerial Steam Carriage. Continued after Henson left. 1848: Flew steam-powered monoplane model in disused Chard lace factory, first flight of heavier-than-air powered machine. 1868: Submitted triplane model after style of Cayley and Wenham to Aeronautical Exhibition, and a lightweight steam engine. Triplane reputed to have flown under its own power, and engine won a prize. 1870/71: Built steam engine for airship planned to relieve Siege of Paris. Continued working on his many interests and may have built more models in attempts to improve. One of his outstanding model

John Stringfellow

steam engines was presented to Smithsonian Institution in Washington and prompted Langley's interest in aviation.

was one of Stringfellow's lightweight steam engines. These were adequate for models but not for full-size aeroplanes. But with wings built of ribs on hollow spars, braced by streamline cables from kingposts, and propellers 'handed' to balance out torque reaction the design features were still familiar 70 years later. Four years of unsuccessful work with aircraft models left Henson disillusioned and the partnership broke up. Stringfellow continued with models of his own design, and in June 1848 he succeeded in flying a heavier-than-air powered machine in a disused lace factory in Chard. It is likely, though not known for sure, that two months later he repeated the feat in Cremorne Gardens in London.

Ingenuity in France
In the 1850s, progress moved to France where a number of reasonably practical designs were proposed. Capitaine Le Bris actually built and flew a monoplane glider in 1857, and another seaman, Félix du Temple, built a powered model which took off under its own power and flew freely and stably. His proposal for a full-sized monoplane improved upon Henson and Stringfellow with dihedral wings for lateral stability, a tailplane for longitudinal stability and control, and a large rudder. A retractable tricycle undercarriage was proposed, with a single tractor airscrew, and aluminium was considered as a structural material. In 1852, the Société Aérostatique et Météorologique de France, the first aeronautical society in the world, was founded, later becoming the Société Française de Navigation Aérienne.

The 1860s were an important decade. In 1866, the Aeronautical Society was formed in Great Britain and, in the same year, Francis Wenham published his work which formalised the optimum designs of wing sections and also the concept of superposed planes in a multi-wing configuration, confirming one of Cayley's propositions. The Aeronautical Society organised the first Aeronautical Exhibition at Crystal Palace in 1868. Stringfellow was invited to enter and he built a steam-powered triplane which derived directly from the principles drawn up by Wenham in confirmation of Cayley's ideas. In the same year, Boulton proposed an identifiably modern aileron system for control in roll.

In 1871, Wenham and Browning conducted the first experiments in a wind tunnel, but in the 1870s most of the advanced ideas came from France, with the work of the brilliant if temperamentally flawed Alphonse Pénaud. He first built a number of models powered by rubber bands, which achieved longitudinal stability by a tailplane operating on a sensibly long fuselage, confirming Cayley's 1804 model. His ultimate design was for a full-sized monoplane with dihedral wings, elevators and fin and rudder. It was an amphibian with boat-type fuselage, retractable undercarriage and an enclosed cockpit with a single control column working in the modern idiom.

Another important development came when the Englishman Horatio Phillips registered his patent on cambered aerofoils in 1884. He later built a steam-powered multiplane which ran on a circular track in 1893, lifting itself for significant distances.

Australian ideas

The technically confident late Victorians took up the challenge of combining the three factors of powered, sustained, controlled heavier-than-air flight. Even though the light petrol engine had been invented in 1885 by Carl Benz, it would be some years before the internal combustion engine would supplant steam as the power source favoured by most experimenters. As late as the 1890s, Maxim and Ader built steam-powered contraptions capable of hops rather than sustained flight. But the improving communications of the last decade enabled the efforts of an Australian inventor to influence the mainstream pioneers of Europe and America.

Beginning in the late 1870s, Lawrence Hargrave conducted a large number of experiments. Much of his work was misdirected, but some of it led to the successful rotary engine and, equally important, to the cellular boxkite, which flew extremely stably and which would provide the basis for many early European aeroplanes. Hargrave used his designs to lift himself below a chain of kites in November 1904. His work was reported by Octave Chanute, the great aviation communicator. Although he did build and test some gliders, Chanute's greatest contribution was the dissemination of information worldwide. Knowing of and known to all the pioneers at this time, Chanute rated Hargrave as 'the man most likely to succeed'.

During the 1890s, Otto Lilienthal built and

The balloon came into vogue in the late 1800s, both for pleasure and as a military machine. Sadly to go out of favour after the First World War, it has regained popularity again since the 1960s, both for sport and for commercial use.

flew a series of man-carrying hang-gliders in Germany. Their control was achieved in a very rudimentary way by the pilot swinging his body. He achieved some measure of success in sustained man-carrying, but was more successful in that he rekindled Europe's interest in aviation. A disciple of Lilienthal's, Percy Pilcher, became Britain's most successful glider of the 1890s, but by the end of the decade both men had been killed in gliding accidents. At the time of his death Pilcher had been working on a petrol-engined machine, and may have beaten the Wrights but for his untimely crash.

Otto Lilienthal (1848-96)

Interested in aviation from his teens, Lilienthal became professional engineer, and continued researches into aviation. 1889: Published *Bird Flight as the Basis of Aviation*, and confirmed efficiency of cambered wings and propeller action of birds' wings. 1891: Began building and flying gliders, and became outstanding figure in aviation at end of 19th century. Gliders mainly of monoplane hang-glider form, directed by shifting pilot's body, but with no provision for control in pitch. Began experiments gliding from hills. 1892: Built artificial hill in open country to benefit from stiller air. 1895: Built biplane glider which could climb in flight, but still not fitted with positive control in pitch. 1896: Built and flew glider with movable rear elevator operated by harness around pilot's head. Also glider with flapping wing tips powered by carbon dioxide. Never tested due to his death on 10 August 1896 after gliding accident previous day.

Lilienthal in one of his gliders.

American pioneers

By the turn of the century the pendulum of likely success had swung across the Atlantic. At Washington's Smithsonian Institution, Langley's interest had been stimulated by some Stringfellow models and engines which the Institution had acquired, and with his brilliant assistant, Charles Manly, he had embarked upon a series of model trials himself. Quaintly named 'Aerodromes', his designs seemed well-poised when his quarter-size model of 1901 became the first petrol-powered aeroplane to fly successfully. With official US Government backing, Langley set to work on a full-size machine.

Also in America, but with less distinguished credentials, the Wright brothers began their interest in aviation in the late 1890s. From the outset they favoured an aeroplane which would be positively controlled by the pilot as opposed to the inherently stable machine favoured by most of their contemporaries. Wing warping to give control in roll was incorporated in their first glider of

A remarkable sight on the Potomac River, Washington DC: Langley's optimistic Aerodrome. It was to crash twice, causing an unusually myopic US Government to withdraw from aviation research and give Europe a lead it held for 25 years.

1900. Every aspect of their machines was subject to fundamental research carried out in their own time and at their own expense and to practical evaluation in hundreds of gliding trials.

Having solved the problems associated with aerofoil shapes and propeller design they were faced with finding a suitable motor for their 1903 powered machine. Unable to find one light and powerful enough they designed and built their own, possibly based on the Pope-Toledo automobile engine.

The race to success

In the autumn of 1903, it appeared that the acclaim for the first flight would go to Langley. In October he announced to the Press that an attempt would be made to fly the full-size Aerodrome. The watchers gathered on the banks of the Potomac River to see the machine being catapult-launched from the roof of a houseboat. Fouled by the catapult, the machine fell into the river. After repairs a second attempt was announced early in December. A second failure convinced the Press that flight was impossible, the US Government withdrew its support and Langley's aeronautical work was over.

By this time, the Wrights were at their regular flying ground at Kitty Hawk, North

Samuel Pierpont Langley (1834-1906)

Distinguished scientific career led to appointment as head of Smithsonian Institution in Washington DC. 1880s: Built large number of rubber-powered models based on Pénaud's designs. 1891: Began building steam-engined flying models based on Stringfellow lines. 'Aerodrome' was name given to all his machines. 1896: Succeeded in flying model at 48 kph (30 mph) for three-quarters of a mile – the first really satisfactory heavier-than-air machine. 1898: US Government provided funding for full-sized machine. 1901: Quarter-scale model became first petrol-driven machine to fly successfully. Assistant Manly made the outstanding petrol engine of the era for full-size machine, and was pilot for attempted flights of 8 October and 8 December 1903. Both failures, due to primitive control and Langley's obsession with catapult launching. 1914: Aerodrome was rebuilt by Curtiss with Smithsonian's blessing to belittle Wrights' achievements, and long-running feud began, settled finally in 1948.

Wilbur Wright (1867-1912)

The driving force of the Wright brothers and the one who decided to enter aviation. Provided most of ideas and design work. Investigated every aspect of an aeroplane and carried out hundreds of gliding experiments in company with Orville. Made first accurate measurements of aerofoil section characteristics and developed very efficient propeller designs. Crashed 1903 machine on first attempt at powered flight, but flew 260 m (852 ft) after Orville achieved first flight on 17 December. More introverted and serious than Orville, but impressed all he met. Decided on policy of secrecy when early successes received scant acclaim. 1908: Outstanding pilot who created sensation at first public display in France in August, demonstrating how far ahead the Wrights were of European aviators. 1909: Won prize for outstanding flight up Hudson River in September, during tercentenary celebrations of New York. 1912: Died of typhoid.

Orville Wright (1871-1948)

More extrovert than Wilbur, was very popular and was both technically capable and an excellent pilot. Made the first powered flight of 12 seconds at Kitty Hawk on 17 December 1903. Provided sound discussion partner for Wilbur, helping both in problem-solving and construction of aeroplanes. 1903: Was major worker on aero engine, assisted by Charles Taylor. 1908: Was pilot of machine undertaking US Army trials and was seriously injured in crash which killed Selfridge. 1912: After Wilbur's death sold aeroplane interests and became recluse. 1914: Disgusted by Smithsonian Institution and Curtiss's attempts to exaggerate value of Langley's work and support Curtiss's defence against his infringement of Wright patents. 1928: After years of misrepresentation by Smithsonian lodged original 1903 Flyer in Science Museum, London, as protest. 1942: Smithsonian finally retracted and Orville consented to Flyer returning to USA. Died January 1948 before Flyer installed in Washington DC, in December.

Committed to monoplanes at a time everybody else was interested in biplanes, Blériot succeeded in crossing the English Channel in 1909.

Sitting beneath the mainplane, Alberto Santos-Dumont demonstrates his Demoiselle monoplane of 1907. The aeroplane was controlled by an early idea that used wing warping, rather than ailerons, and a combined rudder/elevator at the tail.

Carolina, and on 14 December judged that conditions were suitable for a flight. Wilbur won the toss of a coin and climbed aboard. The Flyer chugged off its launching rail, stalled and crashed.

On 17 December, Orville tried in the repaired machine. His flight lasted twelve seconds. The first powered, sustained, controlled man-carrying flight in history had taken place. They made three more flights that day. The following year, they progressed further, flying regularly from Huffman's Pasture near Dayton, and by 1905 had flown 38 km (24 mi) in one flight of 38 minutes.

But their fantastic achievement was generally shunned by the Press, who were convinced of its impossibility by Langley's failure. The Wrights offered their invention to the US Government and it was turned down three times; they visited Europe and received no encouragement there either. Four years were to elapse between that historic first flight and any sort of positive response.

European reaction to American success

Reports of the Wrights' success filtered through to an incredulous France, where the aviation fraternity almost refused to believe that their self-convinced lead in aviation had been snatched away from them. Major prizes were offered for the first flights in France. Numerous pioneers struggled to follow the Wrights, most eschewing the brothers' controllable approach in favour of inherently

Many failed, but the pioneering brothers, Orville and Wilbur Wright, succeeded at Kitty Hawk, North Carolina in December 1903, with the first sustained man-flight in a powered machine. The age of the aeroplane had arrived.

stable designs based on Hargrave's boxkites. In this way, Alberto Santos-Dumont succeeded in making the first powered flight in Europe in his 14bis on 23 October 1906 at the Bagatelle in Paris. To enormous patriotic acclaim his odd front-tailed canard staggered an estimated 60 m (197 ft) at 3-5 m (10-16 ft) altitude to win the Archdeacon Prize of 3,000 francs for the first aeroplane in Europe to fly more than 25 metres. On 12 November 1906, after fitting ailerons to the outer cells of his wings, Santos-Dumont managed to win the Aero Club of France prize of 1,500 francs for a flight longer than 100 metres, when he flew for 220 m (722 ft) at about 38 kph (24 mph), the flight lasting 21 seconds. More than a year before, Wilbur Wright had flown 38 km (24 mi) in a 38 minute flight in a fully controllable aeroplane but had received no reward from his countrymen.

Other workers in France at this time were Ferdinand Ferber, who had done much to prompt other pioneers whilst building and

Louis Blériot (1872-1936)

Born Cambrai, trained as engineer. Invented automobile headlamp and developed into successful businessman. 1902: Became interested in aviation and designed unsuccessful man-carrying ornithopter. Lost interest for next few years. 1905/7: Sponsored number of unsuccessful designs, including some by Voisin. 1907: Inspired by Vuia's bizarre machine produced reasonably successful tractor monoplane. Became committed monoplane man in era of boxkite-inspired biplanes. 1908: Made first public

testing his own undistinguished gliders, and Gabriel Voisin, who began by building machines for the patrician Archdeacon and later for Louis Blériot. Most of these were of the boxkite form, but Blériot changed his allegiance to the monoplane, accompanied by Robert Esnault-Pelterie and also by Léon Levavasseur, who would soon produce the successful series of Antoinette engines and monoplanes.

In England, Horatio Phillips experimented with a remarkable multiplane, making a hop of about 150 m (500 ft) in 1907. In the same year S F Cody, whose interest in aviation had begun around 1900 with man-carrying kites based on Hargrave's work, was now working at the Royal Aircraft Factory (RAF) at

demonstrations in current machine and also flight of 28 km (17 mi). 1909: First across English Channel, on 25 July, beating only serious competitor, Hubert Latham. Developed Pénaud's single control column. Set up flying schools. Became major figure in French aviation and sold machines to Italian and British militaries. 1914: After major scandal over activities of Deperdussin was appointed to take over the company. Continued to lead major aircraft manufacturing business until his death in August 1936.

Leading the British development of aeroplanes in the pre-1914 period, Samuel Cody was actually an American. He is pictured at Farnborough, the British aviation *alma mater*, where a Cody tree still stands as a monument to him.

Farnborough on a motorised glider-kite and J W Dunne, also under the RAF's patronage, was experimenting with inherently stable swept-wing gliders in great secrecy.

Farman takes the lead in Europe

But the French were undoubtedly leading the chase to catch up on the Wrights. Towards the end of 1907, Henri Farman became Europe's premier aviator. On 26 October he covered 771 m (2,530 ft) in his modified Voisin, and won the Archdeacon Cup for a flight of 1,030 m (3,379 ft) on 9 November. He further improved on this on 13 January 1908 with the first official kilometre circle in Europe, winning the 50,000 francs Deutsch Archdeacon Prize. But, like all European aviators, he was still unable to bank his machine, and made wide yawing turns using rudder alone.

In the summer of 1908, A V Roe attempted hops in his own biplane, but it was Cody, in his British Army Aeroplane No 1, who made the first accredited flight in Britain, covering 420 m (1,380 ft) in 27 seconds on 16 October. A few days later, on 30 October in France, Henri Farman made the first proper cross-country flight in Europe, covering the 27 km (17 mi) between Bouy and Rheims.

Described as Europe's premier aviator, Henri Farman produced aeroplanes for the rich and the military. But in the 1900s aircraft were still fragile constructions of wood and fabric.

Public triumph for the Wrights

The year of 1908 saw the major activities of another serious American group, the Aerial Experiment Association, founded by Alexander Graham Bell and Glenn Curtiss, a motorcycle engineer, in October 1907. Their aircraft picked details from both the Wrights and European pioneers, and were reasonably successful by the standards of the time. Red Wing was followed by White Wing and then the June Bug of 1908, which won the Scientific American Prize for the first public aeroplane flight of one mile. Whatever public kudos the Wrights might have lost to the AEA by their policy of secrecy would soon be regained. Orville would shortly eclipse June Bug's achievements.

During this period, the Wrights did not fly for two and a half years between October 1905

and May 1908, since they were awaiting a serious purchaser. They did not waste that time, but built up a small stock of improved engines and airframes, one of which was sent to Le Havre in July 1907. This period of reluctance on the part of authority finally came to an end in early 1908 by the signing of contracts with the US Government and also a French company. The brothers spent the spring of 1908 practising at Kitty Hawk where they carried the world's first aeroplane passenger, C W Furnas. Orville went to carry out the US Army trials at Fort Myer, Virginia, whilst Wilbur went to France. Having rebuilt the crated machine, in August 1908 he stunned the disbelieving French crowds by flying in banked turns under perfect control and, by December, in over 100 flights, had carried many passengers, made nearly 30 flights of longer than fifteen minutes' duration, and one of over two hours.

At Fort Myer in early September, Orville finally won national acclaim from his countrymen. On 9 September he made a flight of 62 minutes, the first flight in the world of over one hour. On the same day he took up Lieutenant F Lahm, who became the first officer to fly as an aeroplane passenger. The crowds flocked to the US Army post to see the wonder of the age as Orville flew almost every day.

But their well-deserved international success was to be clouded by tragedy. On 17 September, Orville was flying with Lieutenant Thomas Selfridge as passenger, an aviation-minded officer who was also the member of the AEA responsible for Red Wing. A few minutes into the flight a propeller cracked and damaged the airframe. In the inevitable crash both men were seriously injured. Selfridge died from his injuries later in the day, and became the first aeroplane fatality in history. Orville, though, later made a complete recovery.

Moving from sporting to military uses, several companies saw that their designs, although fragile, could be used in wartime, initially for rapid aerial scouting over enemy lines to observe troop movements. The Morane-Saulnier Scout is illustrated.

The most important year

After millennia of fantasy and one century of scientific progress, the year 1909 was probably the most important for the aviation pioneers. The events of this year meant that all succeeding decades would accept aviation as a reality. The young Englishman Moore Braba-zon had bought a Voisin and learned to fly it in France. He moved it to the new Aero Club flying ground at Eastchurch, Kent, and in April 1909 became the first Englishman to fly properly in his own country.

But in Europe, the epoch-making event stemmed from two groups poised on the Channel coast of France. After Hubert Latham had been forced down in the English Channel in his Antoinette and was preparing for a second attempt, on 25 July 1909, Louis Blériot succeeded in crossing to Dover, in his Type XI. In so doing, he won the Daily Mail £1,000 prize and established the monoplane as a viable type, with great benefits for his business.

In August 1909, the world's first aviation meeting was held at Rheims, France, where Farman set an endurance record of 3 hours 4 minutes, and Glenn Curtiss won two speed prizes at 75 km/h (47 mph). The highest speed was Blériot's single lap at 77 km/h (48 mph). Farman's machine was powered by a Gnome rotary engine, which would assume major importance in the Great War. Latham, in the Antoinette which had been retrieved from the

English Channel after his second unsuccessful attempt, set an altitude record at 155 m (508 ft). Aviation meetings became regular occurrences over the next few years all over Europe and America. The first British meetings were held at Doncaster and Blackpool in October 1909, and the first American one in Los Angeles in January 1910. Also in October 1909, Moore Brabazon won the Daily Mail £1,000 prize for being the first British pilot to fly a circular mile in an all-British machine, a Short-Wright.

Sport and war

In April 1910, the competition for the £10,000 Daily Mail Prize for the first London-Manchester flight became a race between Claude Grahame-White and Louis Paulhan. Despite Grahame-White's first ever night take-off, the prize went to Paulhan. Prestigious competitions grew up for the Schneider Trophy and Gordon Bennett Cup and air racing became almost commonplace.

In the next few years, dozens of aircraft builders would emerge, the best of which would become manufacturers of the thousands of aircraft which would be demanded by the Great War. After their desultory interest at the outset, the major powers would set up military aviation forces with varying degrees of urgency and interest. The first-blooding of the aeroplane in war would be given by the Italians in 1911, with a small unit operating against Turkish forces in North Africa. Military developments in France and Germany would spur Britain to establish its Royal Flying Corps in 1912, but ironically the Aeronautical Division of the US Army Signal Corps would throw away the lead which they could have had in 1903, and certainly had had in 1908. When the United States decided to join the European carnage in 1917, it would have to accept the ignominy of begging combat-worthy aeroplanes from its continental allies. And sadly, little more than ten years after it had achieved man's greatest ambition, the work of the Wrights would provide an additional means by which young men could be sent to their deaths.

That remarkable Frenchman, Louis Blériot (*extreme left*) arrives at Dover on 25 July 1907, earning himself a prize of £1,000 for the first non-stop crossing of the English Channel, landing at a site near Dover Castle.

2

MILITARY AVIATION

FROM THE GREAT WAR TO THE COLD WAR

The First World War saw the development of the sporting aeroplane into a fighting machine, but almost by accident. The stalemate of trench warfare proved to be a catalyst for aerial combat. The first aces were created, and men like Richthofen, Fonck, Mannock, Bishop and Udet became national heroes.

The aeroplane developed slowly in the inter-war period, but the outbreak of the Second World War spurred new technology, including the development of radar, the jet engine and air-launched guided bombs. For the first time, civilians commonly became wartime targets with the mass bombing of cities. After 1945, the Cold War continued the development of military aircraft and led to the jet age.

Aircraft were first used for military purposes in 1911, when the Italians had employed an air flotilla of nine aeroplanes for both small-scale bombing and reconnaissance during their war against Turkey, in Libya. Several new techniques were born there; messages were dropped to report on the accuracy of artillery fire; photographs of Turkish dispositions were taken; and leaflets and small bombs were dropped. Such was the success of this flotilla that a second flight of aircraft was sent to further the work.

At the outbreak of the First World War in August 1914, the major powers had all made provision for the use of heavier-than-air machines. Germany had some 260 ready for service, and a fleet of airships. France had slightly fewer aircraft, but, more importantly, had the world's largest aircraft industry, which supplied aircraft to Britain, Russia, and the United States. In Britain the Royal Flying Corps (RFC) and the Royal Naval Air Service (RNAS) possessed 63 and 91 machines respectively.

Although Germany had already considered plans for the bombing of vital industries and cities in southern England, and French Farman bombers flew on raids behind enemy lines from the commencement of hostilities, the air arms were envisaged by their controllers, the Army commanders, as primarily scouting forces – a sort of aerial cavalry. Therefore, although some aircraft carried observers or small hand-held bombs, none

Below: in service throughout the First World War, the BE 2 series was ubiquitous in the Royal Flying Corps and Royal Air Force squadrons even after it had become obsolete; this is a BE 2c. *Bottom:* another type with a long association with the RFC/RAF was the Bristol F2 Fighter. It entered service in 1917, and in 1918 flew the first RAF sortie, continuing in operational service in the British Empire during the 1920s.

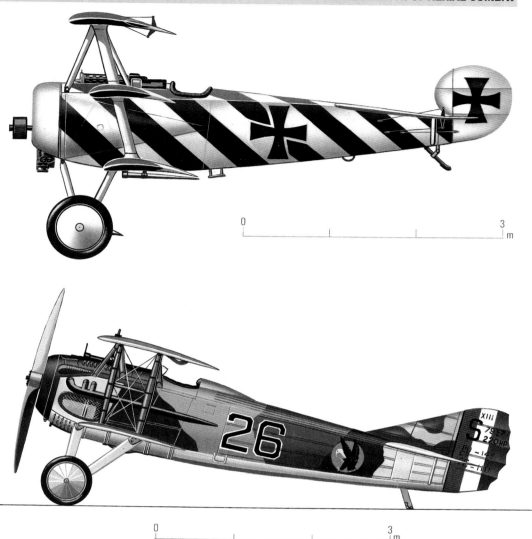

had been given provision for guns.

As the war on the ground began to degenerate into the stalemate of trench warfare, artillery reigned supreme in the military mind. To assess the effects of the enormous tonnage of shells hurled at each other, and to report on concentrations of munitions and men near the railheads that would foretell of an offensive, more and more aircraft were used for reconnaissance.

Two main classes of machine were employed on these duties, the single-seater scout, such as the RFC's Bleriots and French Moranes, and two-seater general-purpose aircraft, such as Germany's Taubes and Albatros and the British BE2c. Typically, the two-seaters were clumsy; powered by 60-100 hp engines they lacked performance and climb, with top speeds of 80-145 km/h (50-90 mph). They were not built to be flown aggressively, the BE2c, for example, being specifically designed to provide the steadiest possible platform for observation, and so possessing little manoeuvrability.

When the inevitable happened, and opposing sides met over the battlefields, the first air-to-air fighting took place, caused by pilots' and observers' attempting to deny the enemy their reconnaissance by firing on them with service carbines, pistols, or even shotguns.

The growth of aerial combat

Whilst the observers of two-seater machines, and 'pusher' types with the propeller at the rear of the engine such as the Vickers Gunbus, could be equipped with machine guns, they were too slow and unwieldy to be fully effective as destroyers of other aircraft. It was to be the fastest, more nimble single-seater scouts that provided the basis for the fighter.

Many designers thought that the ideal position for a machine gun was on the centre-line of the fuselage, firing forward. With the frail structures then in service, this would involve mounting the gun on the engine, as the strongest part of the aircraft, and

Fokker's Synchronised Gun

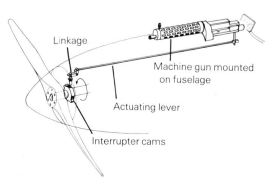

Linkage

Machine gun mounted on fuselage

Actuating lever

Interrupter cams

Top: Fokker Dr 1 of the German air service, *c.*1918. Built to counter the Sopwith Triplane, the Dr 1 was good in dogfights, although Baron Manfred von Richthofen lost his life in the aircraft after an encounter with Captain Roy Brown's Camel.

Above: Spad XIII of the US Army expeditionary force, *c.*1918. One of the dominant types of 1917-18, this French fighter was flown by many nations, including the United States. It later served with Belgian, Polish and Japanese units.

firing through the propeller arc. However, this also meant shooting your own propeller. The Frenchman Roland Garros solved the problem somewhat daringly by having wedges of steel bolted onto the blades of his Morane to deflect the bullets, and in March 1915 started shooting down Germans. When his machine was forced down behind enemy lines, the Germans realised the advantages of this gun position, and had Antony Fokker's factory find a better solution to the propeller problem. They produced a working interruptor gear which timed the shots so that they passed between the blades and fitted it to Fokker's EI monoplane.

What followed became known as the Fokker Scourge. The first Allied aircraft was shot down by an EI on 1 August 1915 and these small aircraft ruled the skies until the spring of 1916. Until they could get their own synchronising gear into production, the RFC used improved pushers such as the FE2b, and the French fitted machine guns to the top wings of their Nieuport 11 scouts. From that point on, fighter ascendancy moved from one side to the other as each improved the

performance of their machines. From the British came the Sopwiths: the Pup, the triplane, and Camel, as well as the powerful in-line-engined SE5a. The French produced more powerful Nieuports, and the excellent SPAD. Germany developed better Albatros designs, the DIII and the Pfalz DIII, and, best known the Fokker Triplane and the remarkable DVII. The two-seaters, likewise, improved in performance and armament, such as the Bristol F2 fighter and Germany's Hannover, LVG, and Halberstadt.

Germany had found the Zeppelin airships too vulnerable to operate over well-defended battlefield areas and replaced them with the Gotha bomber. The British, as well as using very good light, fast bombers such as the 1½ Strutter, the DH4 and DH9A, learnt a sharp lesson from the raids on England by the Zeppelins, Gothas and Staaken bombers, and set about producing machines suitable for long-range attack. The 0/400 from Handley-Page could carry 816 kg (1,800 lb) of bombs, and led to the production of the larger V/1500, with the intention of bombing Berlin. The Italians used the excellent Caproni series of

Captain Albert Ball (1896-1917)

After training as a flying officer and instructing at Gosport, UK, Albert Ball was sent to No 13 Squadron in France in February 1916. Flew two-seaters until transferred to single-seat squadron in May. Achieved most of his victories flying Nieuport scout, often on solo patrols. His careful but aggressive methods had great effect on his contemporaries, and he did much to revive the flagging morale of the RFC after the losses of the 'Fokker Scourge'. 1917: He had been flying with No 56 Squadron for a month when he was shot down and killed on 7 May. The first RFC ace, his total of 43 victories, and his example to the others in the service led to the posthumous award of the Victoria Cross to add to the Distinguished Service Order (two bars), and the Military Cross which, at the age of 20, he had already received.

Sopwith's Camel was the mount of Biggles in fiction and of many of the true Flying Aces; in 1917-18 it destroyed more aircraft than any other type. It was flown by the Royal Naval Air Service, including being launched from ships.

bombers, which proved highly effective.

In the hostile air above the trenches, whilst being able to withstand a large volume of small-arms fire. The outstanding specialised machine produced for this purpose was undoubtedly the Junkers Ju 1, all metal, with semi-cantilever wings, a radio, and thick armour plate to protect its two-man crew.

Throughout this period, the numbers of aircraft used by both sides increased dramatically, and air power became firmly lodged in the public as well as the military mind. The names of the fighter 'aces' were on everybody's lips, as each country showered adulation on men such as Von Richtofen, Boelcke, Udet, Ball, Mannock, McCudden, Guyremer, Fonck, and Rickenbacker.

Military aviation at the end of the war

In the four years of the First World War, military aviation had progressed at a startling rate, once the military commanders had realised how the power could be used.

Technically, airframes had become properly robust, and engine performance had improved dramatically. In 1914 the most common engine in use was the rotary, of around 80 hp. Although this had been developed to its practical limit, it had been eclipsed by the in-line water-cooled types from Hispano Suiza and Mercedes, which were providing more than 200 hp, and by the V12 Rolls-Royce Eagle and US Liberty, of 360 and 400 hp respectively.

The design and engine advances had led to a rise in the maximum speed for single-seaters from 128 to 210 km/h (80 to 130 mph) or more, and there were three-, four-, and even five-engined bombers carrying up to 3,175 kg (7,000 lb) of bombs. 160,000 aircraft had been built, establishing aero-industries in many countries, and flying itself was seen as more than an eccentric sport or hobby.

In terms of the roles played by the land-based aeroplane, every task that they were to perform during the following 25 years had been carried out to some extent: air defence, air superiority, close support, strategic and tactical bombing, reconnaissance, and even primitive night-fighting.

Many new ideas had been tried out, but most of these, such as supercharging, flaps, metal and monocoque construction, variable pitch airscrews, and clear cockpit canopies, were to be largely forgotten by most of the newly created air forces for several years.

Independent air forces had been born, such as the RAF, which was no longer an army

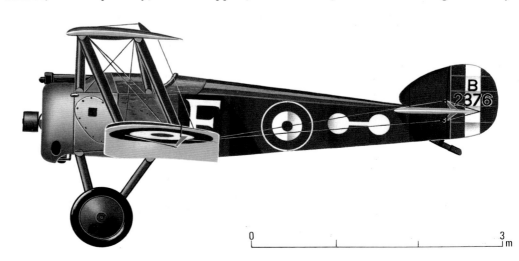

corps, but was as free to pursue its own course as the politicians would allow. The US air arm, although equipped in the front line by designs of other countries, had gained much experience in its short time in France. In particular its commander, William Mitchell, had formed some very firm ideas on air power, and these were to have a profound effect on strategic thinking during the next decade.

Stagnation of military aviation in Europe

Following the Armistice it became clear that for the aeronautical industries of the major powers the only way forward was in the field of civil aviation. The governments of Britain, France, Italy, and the US had been forced to cut expenditure to the bone in an attempt to recover from the terrible financial drain of the war years. Everywhere military forces were demobilised, and air forces drastically reduced

Sopwith Camel 1F1 of Allied Expeditionary Force, *c.*1917, armed with two fixed Vickers machine guns and capable of carrying four 25 lb Cooper bombs underwing. 5,490 Camels were built and it flew in Canada, Russia and Greece after the war.

in size. Huge numbers of engines, airframes and complete aircraft were put into storage or sold off as surplus at knock-down prices.

Germany, with its air force confiscated and prohibited from rebuilding it under the terms of the Treaty of Versailles, was politically weak, but was able to keep a nucleus of military aviation alive by more or less clandestine means. Pilots were trained abroad (in the USSR), and aircraft were manufactured by subsidiary companies in countries where the prescriptions did not apply. By 1925 Junkers, Focke-Wulf, Dornier, Messerschmitt and Heinkel had all opened factories, some in Sweden, others in Italy and Switzerland.

The new Soviet Union was in its infancy, with fighting between the Reds and the Whites in which small numbers of British, Italian, and French designs played a part, concentrating on supplying tactical support at the most important points of the campaign lines. It was not until 1928 that the first Five-Year Plan for Industry included proposals for aircraft of Soviet design and manufacture.

Apart from the well-known examples of pioneering, under military backing, compara-

Another highly successful fighter in France was the SE 5a, built by the Royal Aircraft Factory and pictured here at a Royal Air Force airfield in July 1918. Note the upper wing-mounted Lewis gun; there was an additional Vickers firing forward.

tively little was done in the rest of the world to advance military flying, and the impetus of aircraft development switched to civil aviation as manufacturers tried desperately to create interest for their products in a market glutted by cheap war-surplus equipment.

In Britain no orders for aircraft of new design were placed until 1924, the RAF making do with models in production or ordered at the end of the war, such as the Sopwith Snipe, the ubiquitous DH4, and the Handley-Page bombers. Then, after six years of stagnation, the orders began trickling in, mainly to keep alive the remains of a struggling industry. The designs were not startlingly new, however, and even when the RAF ordered their first all-metal fighters they were little more than the wooden designs translated into steel tube or strip, nor was much advantage taken of the properties of new

Viscount (Hugh) Trenchard (1873-1956)

Having failed the entrance exams to Britannia Royal Naval College, Dartmouth, Trenchard joined the army and served in India and South Africa, where he received serious wounds. 1912: Learned to fly at Sopwith's school at Brooklands and was sent as instructor to RFC Central Flying School. 1914: Given the task of creating a new air force from the elements that had not been posted to France. Joined the RFC in France in December and in August of the following year took command of all active service units. His leadership, fostering an aggressive spirit, and new equipment, helped the RFC to recover the initiative after the Fokker's supremacy. 1918: Became the first Chief of Staff of the new Royal Air Force (RAF) in April, but resigned and returned to France to found the first air arm independent of army control, the Independent Force. Becoming Chief of Air Staff again after the Armistice, and remaining in that position until 1929, he was responsible for assuring the complete autonomy of the RAF within the services. Laying great stress on the importance of proper training and reserves, setting up staff and cadet colleges, the Auxiliary Air Force and University Air Squadrons, he ensured a firm foundation for the expansion after 1934.

Military transport

Aircraft saw little use as military transports during the First World War, and there was no real concept of organised air lift or air supply. Compared to the static war of the Western Front, with dense networks of road and rail, the vast distances and undeveloped terrain of the British Empire offered greater need and opportunity for air transport.

Gradually aircraft were used more and more. British officers on postings flew in 0/400 bombers, or with suitcases strapped to the wings of fighters. Then bombers like the Vimy were converted for other loads, and the first squadron of aircraft designed specifically for transport entered service in India in 1922. This was the Vickers Vernon, developed from the Vimy bomber, which carried a total of 14 crew and passengers. Its successor, the Vicker Victoria, capable of carrying 22 troops, flew the first proper air lift following riots in Kabul, when in 10 weeks 586 civilians and over 10 tonnes of baggage were lifted to safety.

In the 1930s bomber transports such as the Bristol Bombay and Handley-Page Harrow, Italy's Capronis, and the Junkers Ju 52 became popular for mixed duties. They were to prove unsuccessful for the bombing role in all-out war, but laid the foundations for large transport arms in their respective air forces. In their transport duties, particularly within the RAF, they had helped open up many air routes, establishing airfields as stopping posts along the routes, and these were made available for the civil operators to ferry mail and passengers.

The full impact of military air transport was not to be felt until the Second World War, and it was to be the United States which provided the greatest fleets, and many of the aircraft for

America's first twin-engined bomber was the Martin MB-1, but it was its development, the MB-2 (above), which was to achieve fame, when the US Army's Billy Mitchell demonstrated in Chesapeake Bay the vulnerability of ships to air attack.

Below: opening the British Empire to air transport after the First World War were such transport aircraft as the Vickers Victoria III. The open cockpit and biplane tailplane derive from the late wartime Vickers Vimy bomber family.

lighter alloys. Neither was the RAF encouraged to be forward looking, being prohibited by government edict to 'prepare for a war in which close-support aircraft, bombers, and fighters would play a part'.

Things were moving a little faster in the United States. The radial engine was being extensively developed, and it was to power all the US Navy's aircraft, but the US Army Air Corps (USAAC) was still very much a subordinate unit, dedicated to close support of the US Army alone. But the USA was progressing in the design and production of its own aircraft, both military and civil. The competence and size of the industry increased very rapidly, when one considers that American pilots had flown only European-designed combat machines in 1917-18. They also began to develop the new tactics of dive-bombing, and whilst the French provided the yardstick by which major air forces were measured the Americans were soon to supplant them.

its allies. These were not converted bombers, or dual-purpose bomber transports, but civil types, proven on civil routes, and carrying military loads and markings. Fully participating in and launching military operations, they carried in over 60,000 troops and their equipment and supplies during two days of the Allied landings in Normandy.

The best known of these transports must be the Douglas DC-3, in its military guise: C-47, or Dakota to the RAF. Carrying 27 troops, or 4.5 tonnes of load, it served throughout the world from 1942 onwards; over 10,000 were built, and thousands are still flying today.

Just as the British transports had helped open up routes through the Middle East and India, so the vast aerial fleets of the Second World War were to provide the foundations for the tremendous post-war boom in passenger carrying. Not only did they accustom people to being flown, but reached a high standard of reliability in service, and the hundreds of runways built to operate them later became airports and commercial airfields.

Air control

The aeroplane saw considerable use in a policing role between the wars. Britain, France and Italy had large areas, either colonies or territories under League of Nations Mandate, which had to be administered and kept peaceful.

The advantage of the plane was shown in 1922 when, in an attempt to lift the burden from a shrinking army, and find a less costly method of quelling unrest, eight squadrons of RAF aircraft, one brigade of British and Indian troops, four squadrons of armoured cars, and a number of native levies took over

responsibility from two divisions of troops in Iraq. Over the next ten years a simple and successful system was developed there and applied to many other troublesome hot-spots. Formerly, on an outbreak of unrest, an army column would be despatched, at great expense, often over inhospitable terrain. Fighting, casualties on both sides, and widespread ill feeling would ensue. Under the new policy, a warning or summons was given to the miscreant, and if no suitable response was forthcoming, a further two warnings were issued that the village in question would be bombed on a particular date, and that it should be evacuated to avoid bloodshed. Then the bombs would be dropped, followed by an air 'blockade' of the village, disrupting the lives of the villagers until they surrendered. Then teams of police, and often medical staff, would be flown in to disarm the tribesmen and generally 'rehabilitate' the area. It worked well, casualties were few, and little bitterness was felt. As news of the method spread, often the mere threat of aerial bombardment was

The so-called air police role in defence of the British Empire was carried out by a number of types in the 1920s and 1930s. These Westland Wapiti I and IIA aeroplanes are typical of RAF operations conducted in co-operation with the Army.

Italy developed a series of biplane fighter aircraft in the inter-War period of which the less-than-successful Fiat CR32 was one. It entered service in 1933 and was used in action in several countries, including Spain in the Civil War.

enough to subdue unrest, with no loss of face to the tribes concerned.

Similar treatment was tried in Kurdistan, where Turkish forces had instigated a revolt. The RAF airlifted a battalion of Sikh troops in two squadrons of Vernons, and following the destruction by bombing of the Kurdish leader's house and HQ, the trouble died down. Troubles were also put down in Saudi Arabia, Aden and Trans-Jordan, both at a fraction of the cost in lives and money that a standard military action would have incurred.

The RAF used many types of plane: the Sopwith Snipe, Bristol F2 fighter, de Havilland DH9A, later the Westland Wapiti, Hawker Hardy, Fairey IIIF, and Vickers Vincent. All had to stand up to a variety of tasks in often primitive conditions, and it was common to see them loaded down with spare wheels, water skins, and other equipment necessary for repair or the crew's survival if forced down in the rugged areas of Africa and the Middle East.

Of the other European nations, France used her air force in a very similar manner in Syria and the Lebanon, and as part of a greater military campaign in Morocco. Italy, too, was involved in Libya, and later the Regia Aeronautica played a large part in the conquest of Ethiopia, but these were again much larger campaigns, using the air force as an adjunct to military operations.

Generally, air control proved highly effective as a method of policing where the insurrectionists were relatively primitive, and could easily be identified and located. It was very cheap; a few aircraft could achieve much provided that the tribesmen had no proper anti-aircraft weapons and the pilots showed a degree of respect for the often very accurate small-arms fire from the ground.

Of course, it could not be expected to work against what would now be termed 'urban terrorism', or to properly counter mobile forces of infiltration, but similar tactics have been used, under favourable circumstances, by the RAF since the Second World War.

The rise of naval aviation

Prior to the First World War experiments had taken place to see if an aeroplane could fly from a platform mounted on a ship. In the earlier stages of the war seaplanes equipped with floats were used for fleet reconnaissance; launched from platforms on cruisers or converted ferries, they were craned aboard after landing in the sea at the end of the flight. The Royal Naval Air Service (RNAS) had used seven Short seaplanes for a raid on the Zeppelin sheds at Cuxhaven on Christmas Day 1914; the attack was ineffective but did serve to show that the aeroplane could be employed to extend the striking range of the fleet. Two months later, in the Dardanelles, a Short 184 managed to torpedo a Turkish merchantman.

In order to provide a counter to the Zeppelins reporting on the activities of British warships in the North Sea, higher-performance aircraft, capable of shooting them down, had to be carried. That meant landplanes such as the Sopwith Pup, which were initially fitted with flotation bags and recovered after ditching. By March 1918, however, HMS *Furious* had had much longer platforms fitted, first forward of the superstructure, then aft, and Pups had made both take-offs and landings with the ship in motion. Encouraged by this development, two hulls were taken for refit, and the now familiar 'flat-top' aircraft carrier emerged.

Although Britain, with the most powerful navy in the world, had gained an early lead in naval aircraft, it was quickly overtaken by the USA and Japan. The 1921 Washington Treaty

The Blackburn Dart (1923-33) was used for aerial torpedo trials designed to slow an enemy warship down and allow battleships to approach and sink it. Later it was realised that aircraft were capable of sinking a ship alone.

acted as a brake on expansion of conventional naval power, but both these nations saw the aircraft carrier as an effective way round the restrictions on the size and number of cruisers and battleships; they also realised that in the potential battle ground of the Pacific, the carrier-borne strike aircraft represented a far greater striking radius for the relatively slow battlefleets. Building on the hulls of cancelled capital ships, their carriers had an average complement of 70 aircraft, whereas the British, until 1936, had no plans for carriers to take more than 48, with a normal capacity of 30.

It was not until 1939 that the British Admiralty changed its official view that the sole purpose of carrier aircraft was to inflict enough damage on enemy warships to slow them down for the fleet to catch and destroy them by gunnery. Perhaps it was wishful thinking that led the possessors of so many capital ships to refuse to contemplate that the aircraft could sink those same ships in their own right. In addition to this seeming short-sightedness, the British ships tended to be

equipped with a preponderance of general-purpose aircraft, to the extent that on the outbreak of war with Japan they had to rely on imported Grumman F4 Wildcats (Martlets in FAA service), as the only match for the Mitsubishi A6 Zero.

The USA and Japan might not have had any certain method of telling how well their naval air arms would perform, but when war came they at least had suitable equipment, and were able quickly to put into effect the lessons they learnt.

Europe rearms

The economic slump and the legacies of the post-war settlements brought political instability to Europe, and it became apparent that the 'War to end Wars' might not be that continent's last conflict. In 1932, Britain's Air Staff, in view of the clash between Japan and China in Manchuria, and Britain's commitment in the Far East, and also of the unrest in Germany and Spain, declared that the official policy of there being no foreseeable involvement in any major war during the next ten years must be abandoned.

The government's response was to wait and prepare to act upon the outcome of the disarmament conference sponsored by the League of Nations. The results from Geneva were not promising. France, which had not disarmed to the extent of Britain and Germany, was worried by the threat of the latter's obvious discontent at the terms of the Treaty of Versailles and Britain reiterated its need for a strong air force to control the Empire and the Mandated Territories. No practical method could be found to distinguish between aircraft made for aggression and those for defensive purposes. Nobody was willing to relinquish the use of air power altogether.

Before the conference ended in the summer of 1934, it had become clear that to expect any worthwhile results would be a waste of time. The Japanese had walked out of the League of Nations, following criticism of their campaign in Manchuria, Hitler had attained power in Germany, and Britain had been informed that the Germans were already building the aircraft to form a new air force, and were generally

Developed for the shipboard use, the Sopwith Pup was first successfully landed on a deck by Squadron Commander E H Dunning RN in 1917, and after his untimely death further landing trials were carried out with the aircraft.

Japanese imperialist designs were supported by such aircraft as the Kawasaki Type 88, used as a light bomber and reconnaissance aircraft. It is pictured here during operations against Chinese forces in Manchuria, probably in 1932.

embarked upon a programme of rearmament and military training, planning to leave the League of Nations. Faced with the failure of the conference, and with the realisation that the new Germany might be in a position to fight a major war within a few years, Britain too set about rearming. It was not so much that they expected a direct attack from Hitler, more that the web of treaties and pacts of mutual protection could be seen as the way in which Britain could have war forced upon her indirectly.

By the end of the 1920s, Germany's civil aviation network was larger than those of France and Britain combined, and the creators of the clandestine military arm put it to good work. The German Air Transport School took the best of the 50,000 members of the numerous gliding clubs and trained them to fly, and by the time the Nazis came to power, there was no shortage of potential recruits. Officially announced in the spring of 1935, the Luftwaffe had begun by ordering over 4,000 aircraft, mostly trainers, and Hitler then demanded that the home industry produce the largest number of combat aircraft in the shortest possible time. As heavy long-range bombers take much time and money to develop, the Luftwaffe concentrated on single-engined aircraft, and light to medium-weight bombers. Impressed by the USAAC's demonstrations of dive-bombing, it was specified that all their aircraft should be stressed to perform this task, and the idea of the Stuka aerial artillery grew. But before they seized the opportunity of trying out their newly fledged air force in Spanish skies, the proud Luftwaffe commanders had boasted of its size and prowess to the British. This was to prove a mistake.

If the Nazis had not made so much of their air force, and if Hitler had not, at the beginning of 1935, told the British Foreign secretary that the Luftwaffe was already as strong as the RAF, perhaps the British would not have reacted quite so strongly and rearmed in such a determined way. Although Germany's progress was known to be less than the Nazis had planned, the British preferred to believe the unsubstantial reports of their air power and, trying to deter Hitler by showing that any attempt at domination would not go unopposed, set about an accelerated expansion greater than originally planned. This programme called for the completion by 1937 of a larger home-based Metropolitan Air Force, of some 1,500 front-line aircraft, nearly two-thirds of which would be bombers. It was in turn followed by an even more comprehensive and realistic programme, which was begun in 1936. But none of these preparations served to curtail the Germans' plans, or to deflect them from their path. Other schemes were investigated to find a better defence against the fleets of bombers that were expected to devastate the cities in any forthcoming war. From these researches were developed Fighter Command, and radar.

Elsewhere, the Italians had not truly come to terms with Hitler, and themselves rearmed under the expansionist policies of Mussolini. They were hampered by a lack of funds, and despite starting early, they soon became involved in an expensive war in Abyssinia that could teach them few of the skills needed for war against industrialised nations but alerted Britain and France to their general intentions.

The French had begun re-equipping in the 1930s, as advances in design and materials rendered existing types ineffective. They produced a proliferation of new designs, but failed to achieve the results they wanted for a variety of reasons. The French Army was dubious of the value of an air force, seeing it primarily as a first strike weapon, whose importance would be short-lived in any war; the aircraft industry was not geared for mass production, and suffered from sabotage when the government refused to intervene in the Spanish Civil War. Scandals among the

The terror of the Spanish Civil War, Poland and the Low Countries was the Junkers Ju 87 Stuka. It met its match in the British fighters during the Battle of Britain, but it later achieved success on the Russian (Eastern) Front.

In the 1930s the US Army Air Corps was keen to develop a long-range, heavily armed bomber but to take them only in small numbers. Even in 1938, it was planned to operate less than 30 of what became known as the Flying Fortress.

organisers undermined the importance of the programme, and France chose the wrong aircraft to produce. Additionally, even in the face of a clear threat from across the border, it failed to develop any coordinated plan of defence.

Across the Atlantic, a policy of strict isolationism had produced the view that a large air force was unnecessary; throughout the 1920s the strength of the USAAC had not at any time exceeded one fighter group, one attack group, and one bomber group. In 1931 the USAAC had been given responsibility for coastal defence, which allowed it to increase in size, and this arrangement led to independence from direct US Army control in 1935. But the striking power was limited; only 29 of the new B-17 bombers were planned for 1938. The pursuit, or fighter, arm was a little better off, as increasing responsibility in the Pacific region demanded at least a show of defence; but considering the posture of a militarist Japan, it was remarkable that no real expansion took place in the USAAF (as it now was) until 1940. The President then demanded that industry should be prepared to produce 50,000 aircraft per year: in 1939 it had rolled out under 1,000.

Other influences between the wars

The potential for military aviation had been shown in 1914-18, but in the years that followed nobody could be certain of how it was to develop, and what would be needed in the wars of the future. The fighting above the trenches had clearly shown the need for air superiority, and the ability to conduct operations with acceptable losses, and so it was that all air forces continued with construction of fighter aircraft. The future of the bomber was less sure.

The public had seen the first examples of strategic bombing, by the Zeppelins and Gothas over England. Their effect had been psychological rather than material, but the impression that they left was a deep one.

In the 1920s, more and more attention was paid to the theories of Italy's Douhet, and America's Mitchell. Broadly, these tacticians emphasised that the future war would be fought against the civil populace and munitions industries of opponents. Fleets of bombers would strike at industrial plants, and destroy the will of nations to fight by massive bombing of towns and cities.

Between the wars, attempts to prevent an aerial arms race leading to conflict centred on treaties to limit the numbers and size of the bombers. It was felt that the bomber would always get through, and that it was the primary offensive arm of the air forces. Britain's planned ratio of bombers to fighters in 1930 was two to one. The emphasis on the bomber was helped along by new fast bombers that

Against primitive and low-intensity air defences the Italian Air Force was able to use the Breda Ba65 for ground attack in Abyssinia, but the aircraft was less successful in North Africa. Such designs were used throughout the early war years.

could outpace their fighter opposition. In turn, the RAF brought into service the Fairey Fox, Hawker Hart, and Bristol Blenheim; in Germany the swift Dornier Do17 caused a sensation when it was first shown, being faster than the fighters it was likely to face.

The smaller conflicts between the war gave no particular reason to doubt the abiding concern with bombers. Japan in China, Italy in Abyssinia, and the Condor Legion in Spain all bombed at will, the devastation of civilian targets being the subject of sensational propaganda by all parties, and the military fully

World War One Fighter Tactics

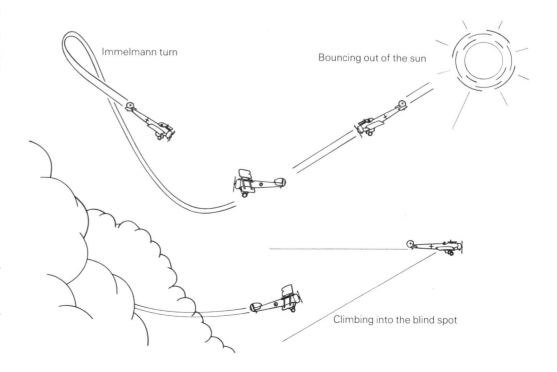

Immelmann turn

Bouncing out of the sun

Climbing into the blind spot

Sir Sydney Camm (1893-1966)

1914: Joined Martinsyde Aeroplanes at Brooklands on the outbreak of war and became a designer. 1925: Transferred to Hawker's as a Senior Draughtsman, and within two years was promoted to Chief Designer. From 1925 onwards he concentrated on military aircraft, and his Hawker Hart (1928) was a major advance on contemporary types. In the early 1930s, more of the Hart's many variants served

with the RAF than all other types added together. 1934: Designed the monoplane Hurricane, with eight machine guns, another leap forward; 1939-45: Designed the Typhoon and Tempest. Post-1945: Designed the Sea Hawk jet and the Hunter. His last design, with Bristol Siddeley and using the revolutionary Pegasus engine, was the P1127/Kestrel, the forerunner of the remarkable Harrier.

An outstanding bomber, reconnaissance, fighter-bomber and night-fighter, the Junkers Ju88 was often fast enough to avoid combat with the slower fighters of many of the Continental nations which the Germans invaded between 1939 and 1941.

appreciating the aid given to their armies by aerial support.

With war unavoidable, and while France was still struggling to produce effective medium-weight and heavy bombers, only Britain decided that it would fight defensively, and that decision was based partly on an entirely new weapon – radar. The clear ground of the English Channel and the North Sea would, it was hoped, give the new system enough time to play its part in an organised, fighter-based defence, that would mean enough time for blockade or assistance from other nations to defeat the potential invader.

This was a last-minute reversal of twenty years' thinking, as previously the RAF had owed its very existence to its firmly held policy of aggressive strategic bombing. The proponents of that theory were to have their chance later, and it would prove much more difficult to achieve than anybody had imagined.

Improvements in design, 1930-39

At the time when many countries were awakening to the fact that another war in Europe was, indeed, possible, military aircraft had changed comparatively little since the Armistice in 1918. The typical fabric-covered European fighter carried two rifle-calibre machine guns mounted on the engine for maximum rigidity, and the pilot sat in an open cockpit, his view obscured by struts and bracing wires. It was designed for maximum manoeuvrability and rate of climb, the two qualities most sought after in a fighting machine. The engine, of four to five hundred horsepower, gave it a top speed of around 300 km/h (195 mph). The bomber, also a biplane, would carry a load of up to 1,400 kg (3,080 lb), powered by two engines at up to 240 km/h (150 mph).

Near the end of the 1930s, when the rearmament plans of the major nations were approaching completion, the picture was very different. The Supermarine Spitfire was a good example of the best of the new fighters developed during the rush to rearm, and incorporated many of the advances made in the civil field. The most obvious difference

HAWKER HURRICANE Mk 1

Country of origin: United Kingdom.
Role: Fighter.
Wing span: 12.19 m (40 ft).
Length: 9.57 m (31.42 ft).
Max loaded weight: 3,000 kg (6,600 lb).
Engine: 1 × 1,030 hp Rolls-Royce Merlin II or III.
Max speed: 521 km/h (324 mph).
Max altitude: 10,425 m (34,200 ft).
Range: 684 km (425 mi).
Military load: 8 × 0.303 Browning machine guns.

One of the classical fighters of all time, the Hawker Hurricane was Britain's first monoplane fighter. It played the major air defence role during the Battle of Britain and destroyed more enemy aircraft than all the other defences put together.

from its forebears was the shape. The all-metal low-wing monoplane, of stressed skin construction, had an enclosed cockpit with a clear canopy, and an undercart that retracted flush into the wings. Those wings, no longer part of a wire-braced structure, were strong enough to provide a firm mounting for the eight machine guns, a number that had been decided upon when the RAF envisaged the speed at which these machines would be fighting. At over 480 km/h (300 mph), it calculated that the pilot would be unlikely to be able to hold his enemy in his sights for more than two seconds; to deliver enough bullets to do significant damage in that short burst needed a battery of eight .303 machine guns. The RAF had not yet found cannon reliable enough to place outboard in the wings at that stage, although

Designed by Reginald Mitchell to bring the Schneider Trophy to Britain, the Supermarine S6B was the forerunner of the Supermarine Spitfire.

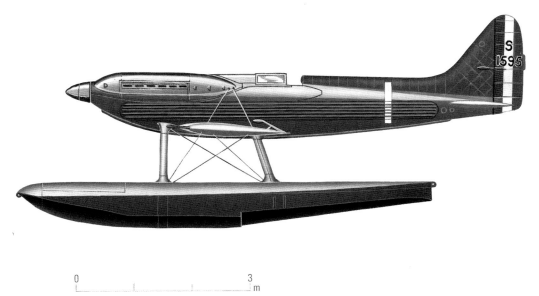

several other nations had them in service. The supercharged, liquid cooled, V12 engine of 1,030 hp was capable of further development, and lacked only the constant-speed propeller that was commonplace elsewhere.

Few four-engined bombers had reached production by the outbreak of war, most that did being conversions of civil airliners. In Europe, the most modern twin-engined medium to heavy type was the Junkers Ju 88. Of similar all-metal construction to the Spitfire, it could carry 2,500 kg (5,500 lb) of bombs at up to 460 km/h (286 mph); immensely strong, it could be used for dive-bombing. Bombers of other nations were now appearing equipped with power-operated gun turrets, in place of the open defensive positions of earlier aircraft, and the Boeing B-17 delivered twice the load of the Junkers.

The race for speed and power, rather than sheer manoeuvrability and ease of handling, had been joined by most of the participating countries. The basic designs of these aircraft of the late 1930s were in many cases to last throughout the six years of war, until jets became practical and useful combat aircraft.

The Spanish Civil War

Following general unrest, the failure of the socialist government to restore order, and discontent among right-wing army officers, civil war broke out in Spain on 18 July 1936. The air force had little equipment and much of that was obsolete. Most of the pilots chose the Nationalist side, but the Republican government was able to retain the bulk of the aircraft. The critical area in the early stages was in the south, as General Francisco Franco's army was waiting in Morocco, and following appeals to the sympathetic rulers of Germany and

Italy, Junkers Ju 52 bomber-transports were sent to participate in their airlift to the mainland.

These were soon followed by 51 biplane fighters, more Ju 52s and from Italy S 81 bombers and CR32 fighters. Volunteer pilots flew with both sides, and during the next six months large numbers of aircraft reinforced each side. Despite an arms embargo, the French quickly sent 50 Potez 54 bombers and 63 fighters and reconnaissance aircraft, but the Republicans received the greater part of their backing from the Soviet Union, which gave the services of its most modern aircraft, as well as technical and political advisers. Hundreds of I-15 biplane and I-16 monoplane fighters helped give the government a measure of control in the air over the Madrid front, the Tupolev SB2 fast monoplane bombers showing they could outrun their biplane fighter opposition.

With the new year came increasing German and Italian support for the Nationalists. The Condor Legion of volunteer pilots received some of the Luftwaffe's most modern machines for evaluation, including Henschel Hs 123 dive bombers, Messerschmitt Bf 109 fighters, Heinkel He IIIb bombers, and Dornier Do17 fast bombers. The Italian Aviazione Ligionaria received the excellent new SM79 bombers, further Fiat CR32s and Breda Ba65 ground-attack fighters.

In the north, where most of the country's industry was based, the Republicans were

Agile and lightweight, the Polikarpov I-16 was the first Russian monoplane fighter aircraft to be built with an enclosed cockpit and retractable undercarriage. It saw service in Spain and on the Russian Front.

FIAT CR32

Country of origin: Italy.
Role: Fighter.
Wing span: 9.49 m (31.17 ft).
Length: 7.46 m (24.44 ft).
Max loaded weight: 1,870 kg (4,112 lb).
Engine: 1 × 600 hp Fiat A30 RA.
Max speed: 375 km/h (233 mph).
Max altitude: 9,000 m (29,525 ft).
Range: 750 km (466 mi).
Military load: 2 × Breda-SAFAT 12.7 mm machine guns.

unable to match the German and Italian units in the air, and had little success. The Nationalists were able to move their air support to the points where it was most needed, and in the spring and summer of 1937 effectively halted government counter attacks.

When the northern area fell to the Nationalists, the writing was on the wall for the government, and Franco's forces were able to bring increasing numbers of aircraft to bear at critical points. The war on the ground did not decrease in ferocity, but up to the fall of Madrid and the end of the war on 28 March 1939, the Nationalists made more and more use of air support, and with the withdrawal of Soviet personnel in the summer of 1938, the Republican government were able to offer less and less effective opposition.

The Spanish Civil War had served as a testing ground for the Messerschmitt Bf 109, Junkers Ju 87 Stuka, and the more modern Italian aircraft. The Luftwaffe, by rotating pilots duty with the Condor Legion had given combat experience to many of their fliers. For the Italians, the combat had been too easy, and they continued to believe that agile, lightly armed biplanes were the best tools for air-to-air combat, despite German experience with fast, cannon-armed monoplanes. But the world had also been shocked and given a foretaste of what was to come with the Nationalists' bombing of Guernica.

The eve of war

As Britain, France, Germany, Italy, and the rest of Europe faced each other in early 1939 it was generally considered that Germany's air force was the finest. Its front-line aircraft had been tested in combat over Spain, and the lessons learnt incorporated into the later models; moreover, many of its aircrews had flown combat with the Condor Legion. Their role was seen as tactical, mainly to support ground forces, and this was reflected in the absence of heavy bombers. Germany was prepared for a short war, and heavily committed to the concept of swift-moving offensives, hence the squadrons of Bf 110 'zerstorer'

Willy Messerschmitt (b. 1898)

One of the most famous names in German aviation. 1921: he assisted with the design of a glider for sporting purposes culminating in monoplane fighter aircraft. 1934: first flight of the Me 108, an all-metal monoplane which led directly to the Bf 109 series of fighters which served with distinction in the Second World War and later in the hands of the Israelis. 1941: first flight of world's first operational jet fighter, Me 262 Swallow. 1944: first flight of Me 163 Komet rocket plane. Other wartime successes included the BF 110 'destroyer', which functioned as a long-range escort fighter and was later converted into a night-fighter aircraft. The Me 323, was also built to carry tanks. Post-1945: Messerschmitt's company combined with Bölkow and Blohm, with government backing, is now one of two airframe manufacturers in Federal Germany, mainly producing helicopters and the Tornado fighter-bomber.

Although late in entering service, the Dewoitine D520 was credited with over 100 enemy aircraft destroyed during the Battle of France (1939-40) but it was often outclassed by the faster and better-flown Bf 109s of the Luftwaffe.

(destroyer) heavy fighters, and Ju 87 Stuka dive bombers were the Luftwaffe's most prestigious units, receiving the lion's share of publicity and public admiration.

As Germany's industry was not on a full war footing, and did not expect to have to produce developing families of aircraft, little priority was accorded to the development of new types. Germany did not expect the war to last long enough to warrant the expenditure of money and effort. As a result, nearly all the major types of aircraft in service in 1939 were to see combat throughout the six years of war, often because there was nothing better to take their place.

In France, despite no lack of new designs on the drawing boards, little really modern equipment had reached operational squadrons even by the time of the German invasion in 1940. The Dewoitine D520 was an excellent fighter, but too few had been produced and the brunt of the fighting was to fall on the older and slower Morane-Saulnier MS406. Impor-

tantly, France also lacked an effective system of control and co-ordination for a defence against determined attack.

In Britain, the hurried but large-scale rearmament programmes had borne greater fruit than in France. Although there were still many obsolescent types such as the Gloster Gladiator, Fairey Battle and Bristol Blenheim in service, the new Spitfire and Hurricane squadrons were becoming operational, albeit still equipped with ancient Watts two-bladed wooden propellers, and the 'phoney war' was to give enough time to build up the industry,

operational systems and reserves to provide an effective home defence.

Along the shores of Britain, new metal radar masts were appearing. The Wellingtons, Hampdens and Whitleys of Bomber Command were to prove of little use in early daylight operations, but larger types had been tested, and as the war changed in character, these became a major participant in Britain's campaign. Generally, given that invasion could be prevented, Britain had greater strength in depth than her rivals, and was in a better position to carry on a prolonged war than they were.

In Italy economic stringencies had prevented the massive expansion of the other European powers, and having met only comparatively primitive opposition in Spain and Ethiopia, the Regia Aeronautica clung to their belief in manoeuvrable light biplane fighters. This was to prove a great mistake, and the Fiat CR32s and CR42s were unable to make any impression on the RAF's more modern types, although the pilots performed well, particularly when pitted against less than the best in the early stages of the war. The Italian industry, hampered partly by the lack of suitable engines of indigenous design, was unable to produce enough of the more modern types such as the Macchi C202, and generally the armament provided, two or four machine guns, was not hard hitting enough. Their bomber squadrons were better equipped, with the SM72 and the four motor Piaggio 108b, but, again, there were not enough, and the latter did not enter service until 1942. The Italian industry, aircraft, and philosophy were better suited to fighting colonial wars and were not able to change rapidly enough to meet the demands that would be placed on them during the next few years.

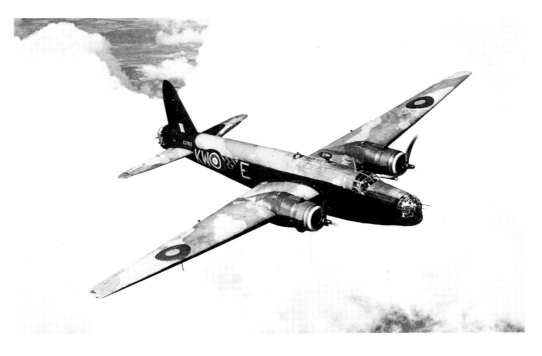

A classic bomber design from Barnes Wallis, the Vickers Wellington was the best heavy bomber available to the Royal Air Force's Bomber Command in 1939. It later served with Coastal Command after replacement by four-engined types.

The Luftwaffe in Western Europe – The Battle of Britain

As Hitler's armies crossed the Polish border on 1 September 1939 the Luftwaffe moved into action to play its part in the Blitzkrieg campaign by which the Germans were to overrun Western Europe and, later, a considerable part of the Soviet Union. The conquest of Poland depended on isolating and encircling the defending armies. Fast armoured columns thrust past their defences, aided by the 'aerial artillery' of Ju 87 dive bombers delivering accurate blows to centres of resistance; tactical strikes from medium-range bombers disrupted forces further behind the lines; and Polish opposition in the air was cleared by Messerschmitt Bf 110 and Bf 109 fighters. Within two days the Polish Air Force was effectively put out of action, as bombing of airfields forced its dispersion and destroyed communications. Though the Polish airmen fought bravely with equipment that was, on the whole, half a generation behind their opponents, they were denied the vital co-ordination, and could not concentrate enough forces to gain superiority at any one point. Outnumbered by three to one, they were mopped up in pockets around the country, a few escaping to fight another day with other air forces. Warsaw surrendered, after heavy bombardment, on 28 September.

Poland was in the hands of Germany and the Soviet Union, which had invaded from the east on 16 September, and the Luftwaffe rested to build up its strength in the West. It had lost around 550 aircraft destroyed or damaged, a third of its force, but the other countries now at war with Germany had seen what the well co-ordinated close-support tactics could do.

In the spring of 1940, the Germans proceeded to do it again. Hitler swept north and occupied Norway, meeting brave but weak opposition in the air, and the 'phoney war' ended in May as he turned to the Low Countries and France. France had 550 fighters, mostly obsolescent MS406 with terribly few of the newer Dewoitine D520s, and 40 Hurricanes and 20 Gladiators sent by Britain. The Polish situation was repeated as dive bombers destroyed strong points and disrupted communications in close co-operation with the ground forces, whilst a strong force of 1,000 modern fighters kept the skies clear for them to operate at will.

The Germans pushed through the demoralised and shocked Allied armies, bombing and straffing them as they tried to re-form. At the beginning of June, the remnants of the British Expeditionary Force (BEF) and detachments of French troops, nearly 340,000 men in all, were taken off from Dunkirk, and by 13 June the Germans were at Paris. Nine days later the French leaders signed an armistice. In weeks, France had fallen – across 35 km (22 mi) of sea lay Britain.

To conquer Britain Hitler needed control of the skies, and if one were to judge by the achievements of the Luftwaffe so far, he would soon have it. But the circumstances were very different, and the contest that followed was to be the Luftwaffe's first defeat. In the Battle of Britain the RAF possessed an advantage the French did not have earlier. Its country was not in chaos; British airfields were not being overrun; when pilots were shot down it was usually over their own territory, so the survivors were able to be returned to their own units rapidly; and the British aircraft had longer endurance over the battle areas. Most importantly, the RAF had radar, the best-controlled system in the world. This permitted the ground controllers to send up the minimum number of aircraft to meet each raid,

Outclassed in the Battle of Britain, the Messerschmitt Bf 110 was designed as a long-range escort fighter, and needed escort itself over Britain. This example was captured intact and used for trials at Farnborough for several years.

husbanding the precious resources of pilots and fighters. It also had the Spitfire, a third of the force, that could equal the performance of the Bf 109, and a well-organised industry that could build and repair aircraft just fast enough, and certainly faster than the German counterpart.

The Luftwaffe commander, Field Marshal Herman Goering, started the offensive in July with attacks on shipping and ports, drawing Fighter Command into combat over the English Channel, and forcing the Admiralty to withdraw naval forces from the area. Building in intensity, the fighting then moved over land with 15 August seeing formations of over 100 aircraft launched on the *Adlerangriff* (Eagle) assault, from Brittany up to Norway.

But these attacks failed to destroy the RAF in the air, and in the following weeks the Luftwaffe could not knock out enough radar installations or airfields. Fighter Command remained fully viable, though stretched to the limit. The Ju 87 Stuka was withdrawn, proving vulnerable in the face of organised and determined defence, and the twin-engine Bf 110, despite speed and heavy armament, lacked the acceleration and manoeuvrability to guard the bombers against the Hurricanes and Spitfires, and had, in turn, to be accompanied by the Bf 109.

The fighting reached it climax around the beginning of September, with hundreds of fighters at a time clashing over south-east England. Air Vice Marshal Dowding at Fighter Command had no reserves, but with the Commonwealth, US, French, Polish and Czech pilots, just enough, but only just, pilots to keep up with the losses. The rate of attrition was too high for the Germans, for whom a far greater proportion of downed pilots had been taken prisoner, and in August and September the total losses for each side were Luftwaffe

Probably the best-known military aircraft of all time, the Supermarine Spitfire played an important role in the Battle of Britain, and over 20,000 were built during the Second World War. It was an outstanding fighter.

The Dornier Do 217 was used for medium-heavy bombing, reconnaissance and later anti-shipping operations. This E model is pictured warming up for a night blitz raid on the British Isles, in the period immediately after the Battle of Britain.

1,606 aircraft destroyed or damaged, 2,691 aircrew killed, missing, prisoner-of-war (POW), or wounded; RAF 927 and 513 respectively.

On 7 September, the first of the night Blitz attacks on London started, and during the next two months daytime activity tailed off, giving Fighter Command a time of relative rest to re-equip and train more pilots. Goering hoped to crush the will of the British to fight with these night raids, but the Luftwaffe was equipped for tactical duties, its bombers were in the medium-weight category, and with no large strategic bomber force to call on, the result was that the United Kingdom survived, and the invasion was postponed indefinitely. Furthermore, the damage sustained by the country's great cities over the next eight months were to be but a small foretaste of what Germany would suffer in return over the next four years.

The Western Desert and the Mediterranean, 1940-44

When Italy declared war on 10 June 1940, Britain's main strength in the Mediterranean lay in the Royal Navy, based at Gibraltar, Malta, Alexandria, and Haifa. The British Army had only just been rescued from Dunkirk and was in urgent need of reorganisation and, above all, equipment. With the threat of German invasion in the wake of France's impending fall, British and Commonwealth forces in North Africa were thinly spread and air power was limited to a mixed

bag of around ten squadrons of Gladiator, Blenheim, Bristol Bombay, and Westland Lysander army co-operation aircraft. Faced with numerically superior air and ground forces, and having to send some of the aircraft to fight alongside the Royal Hellenic Air Force in Greece, where the British were involved in a hard retreating battle against the crushing superiority of the Luftwaffe, the Allies were initially forced back eastwards over the Egyptian border.

After a brief stalemate, General Wavell, reinforced by more Hurricanes, which proved their superiority over the Italian biplanes, rolled the Italians back to the border with Tripoli, taking huge numbers of demoralised Italians prisoner, and destroying many of the Regia Aeronautica's aircraft in attacks on their airfields. In March 1941, however, the Axis forces were reinforced by General Rommel's fresh Afrika Korps troops and the Luftwaffe's Bf 109 and Bf 110 fighters; the Allies, exhausted and at the end of long lines of communications were in their turn driven back across the Egyptian border.

In order to counter the enemy superiority in the air, the newly named (Western) Desert Air Force was given larger numbers of new aircraft, including American land-based Curtiss Tomahawk fighters and Martin Maryland medium bombers, but no decisive advantage could be gained. Heavy fighting in the air failed to break the Germans and Italians as the front line moved back and forth across the desert.

Pictured in the Western Desert after capture by British Commonwealth forces, this Fiat CR42 had been operated by the Italian Air Force in ground-attack and traditional fighter roles. Some CR42s were used in the abortive Italian raid on Britain.

At the beginning of 1942 Japan's entry into the war required the British to divert men and equipment away from North Africa. Another offensive failed, and the Afrika Korps pushed through, past Tobruk, and were finally halted at El Alamein by the British 8th Army, supported by Bristol Beaufighters, Hurricane 'tank-busters' carrying 40mm cannon, and Curtiss Kittyhawk fighter-bombers. The Allies built up strength, reinforcing the air element with American units of fighters and fighter-bombers, and continuing the harassment of Rommel's efforts by striking at supply dumps and airfields. Although the Germans, with the improved Bf 109F aided by the new Macchi C200 and C202, held the upper hand in fighters for a while, they could not adequately protect their own bombers, or prevent considerable damage to their ground forces.

On 24 October 1942, certain that he now had sufficient resources, General Bernard Montgomery launched an offensive. Constantly under attack from the Allied bomber and

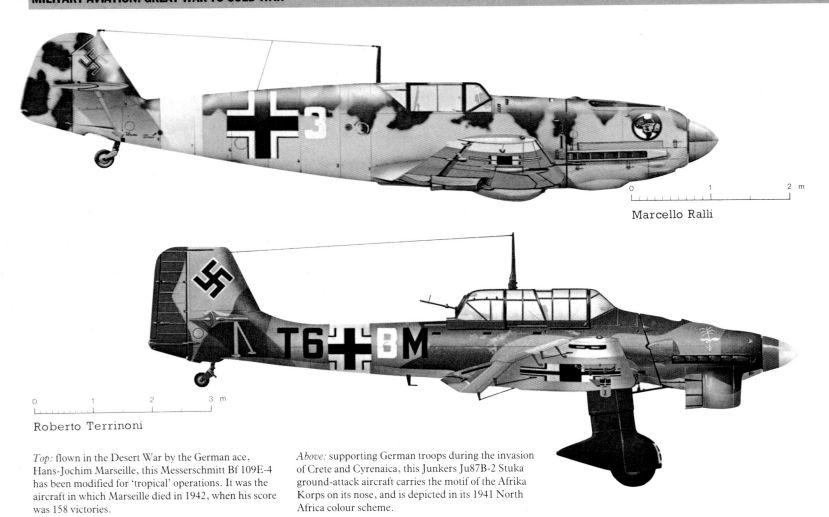

Marcello Ralli

Roberto Terrinoni

Top: flown in the Desert War by the German ace, Hans-Jochim Marseille, this Messerschmitt Bf 109E-4 has been modified for 'tropical' operations. It was the aircraft in which Marseille died in 1942, when his score was 158 victories.

Above: supporting German troops during the invasion of Crete and Cyrenaica, this Junkers Ju87B-2 Stuka ground-attack aircraft carries the motif of the Afrika Korps on its nose, and is depicted in its 1941 North Africa colour scheme.

fighter bomber force, short of fuel and ammunition, the Axis armies were pushed back. Their position was further weakened by the Anglo-American amphibious landings in Algeria and Morocco to their rear. Supported by the USAAF and the RAF, the Allies moved eastwards, linking up with Montgomery's forces south of Rommel's positions in Tunisia in April 1943.

Rommel became unable to supply his army adequately by sea, due to the Allied presence in the Mediterranean, and then was prevented from using air transport by the Allies' massive superiority in the air. The Messerschmitt Gigant and Junkers Ju 52 were shot down in large numbers as they made desperate attempts to fly in the fuel that the Germans so badly needed. In mid-May the fighting ceased,

and over 200,000 German prisoners were taken.

In their years in the desert the British and Commonwealth air forces and, later, the USAAF, which had joined them after America's entry into war in 1941, had learnt many lessons about modern warfare. They succeeded in creating an efficient tactical air force that proved its ability to make a major contribution to the outcome of the war on the ground. The Desert Air Force became the model for Allied tactical air operations for the rest of the war.

The Eastern Front, 1941-45
The Red Air Force was in a poor position to defend the Soviet Union when Hitler launched his offensive in the east on 22 July 1941. Although it had undergone massive expansion, many of its aircraft were obsolescent, and most of the fighters were the same Polikarpov types that had seen action in the Spanish Civil War. The Soviets lacked radar, long-range bombers, and transport aircraft, all of which might have been considered necessary for warfare in such a vast country. The Luftwaffe commanders had a fleet only a third of the size of their opposition, but had to gamble that their superiority in organisation and quality of equipment, and the speed of their attack would tilt the balance in their favour.

In the first few months that gamble paid off spectacularly well. As the three-pronged army

attack swept through eastern Poland, taking 100,000 prisoners on the way to their objectives of Leningrad in the north, Moscow in the centre, and the Caucasus in the south, the Luftwaffe struck at the Red Air Force, whose fighters lay poorly dispersed on their airfields. The Russians admitted to losing 1,200 aircraft in the first nine hours, and by mid-August the total had risen to 40 per cent of their total strength.

As they retreated, Stalin's 'scorched earth' policy ensured that nothing of any possible use to the invader was left behind, and, wherever it could be done, whole industrial plants were moved far eastward, beyond the Urals. These

Operational in the last part of the Second World War against the invading German forces, the Tupolev Tu-2 had a crew of four and carried 2,270 kg (5,000 lb) of bombs. It was to prove itself one of the best Soviet aircraft of the war.

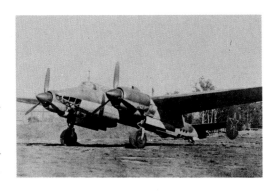

CAPRONI Ca 5

Country of origin: Italy.
Role: Heavy bomber.
Wing span: 23.47 m (77 ft).
Length: 12.6 m (41.33 ft).
Max loaded weight: 5,318 kg (11,700 lb).
Engines: 3 × 300 hp Fiat.
Max speed: 153 km/h (95 mph).
Max altitude: 4,570 m (15,000 ft).
Range: 450 km (280 mi).
Military load: 2 × machine guns; 540 kg (1,188 lb) bomb load.

actions had a profound effect on the Germans, who expected to be to some extent self-supporting, as they had been in Western Europe, capturing fuel, food and transport on their advance. Short of transport aircraft, the Germans started experiencing supply difficulties as their lines of communication lengthened and the worsening weather turned the roads to impassable mud.

As the Germans approached Leningrad, capturing a quarter of a million prisoners in the central region, the first shipments of Allied aid reached the Soviet Union by the perilous sea route to Archangel and Murmansk, having been subjected to determined attacks by the Luftwaffe. In all, Britain, the USA, and Canada were to send over 19,000 aircraft and 9,500 tanks during the next three years.

The bitter winter closed in, and in the north and centre, where Zhukov's sharp counter-attack put Moscow out of immediate danger, the Germans were able to make little progress. In the south less severe conditions enabled them to continue a slow advance towards the huge areas of grain and the oil that was so vital to Hitler's war strategy. Generally, they were badly equipped for the atrocious conditions, and by Christmas 1941 the Luftwaffe was in a sorry condition. Many of the bombers were simply wearing out, having already fought in the west for two years. Their fighters had

suffered from the repeated rough landings on rutted airfields, first baked hard, then frozen solid. Heavy demands were being made on them to support the Italians in the Mediterranean, and now the serviceable strength overall was down to not much above half that with which they had started in 1939. On such a broad front, the Luftwaffe was no longer able to offer the kind of support to which the army was accustomed.

The Russians did not continue the counter-attacks in any great strength the following year, but began to build up their reserves, and the new home-designed aircraft were appearing faster than the German reinforcements. The Il-2 Stormovik, one of the most effective ground-attack aircraft of the war, and the tough La-5 and Yak-7 fighters, showed they could match their opposition in supporting the ground forces.

The Soviet air force grew rapidly in size, strength and experience. The fighters wrested superiority from the Germans during 1943, and the job of destroying the enemy's armies began. Large-calibre cannon, up to 75 mm, were used by both sides to pierce the armour of the enormous numbers of tanks on which the outcome of the battles depended increasingly, but the Germans did not have enough of these tank-busters to prevent their defeat at Kursk.

At the end of the year, Von Paulus's 6th

Army, thwarted in its attempts to take Stalingrad after an epic battle, was cut off and surrounded. Sacrificing hundreds of valuable transports and bombers, the Luftwaffe tried to supply them but failed, and 200,000 troops were captured. It was a turning-point in the war. From then on, the Germans were pushed relentlessly backwards, slowly at first, but then with increasing speed. The Soviets were always able to concentrate overwhelming air power at vital points along the front, and by the end of 1944, their MiG, Lavochkin, and Yak fighters, admirably suited to the combat conditions at low altitude, outnumbered their Luftwaffe counterparts five to one. Despite enormously increased fighter production, Germany was unable to defend the skies on all three fronts at once.

The air war in the Pacific

Japan and China had been at war for some years before the start of the Second World War, and the Western World had been given ample opportunity to build up a clear picture of Japanese air power. But the carrier-based strike against Pearl Harbor, when a large part of the US Pacific fleet was destroyed or damaged, and the simultaneous invasion of Malaya left the American and British forces surprised at the speed and effectiveness of their new enemy. The Japanese Naval Air Force's

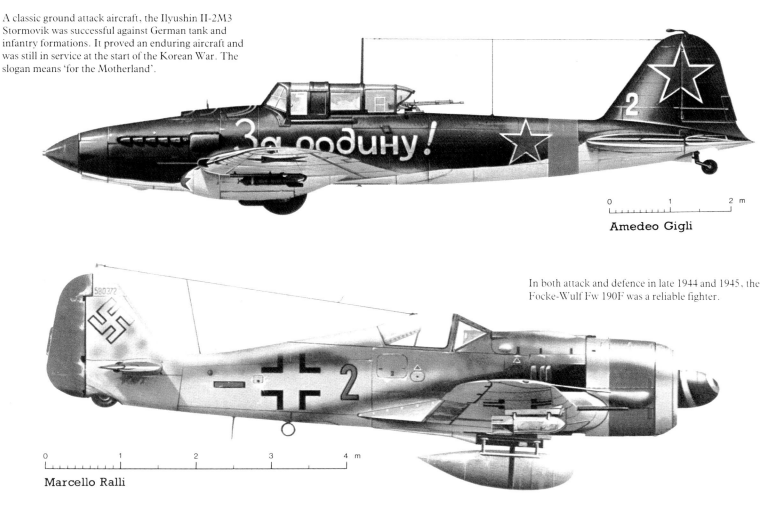

A classic ground attack aircraft, the Ilyushin Il-2M3 Stormovik was successful against German tank and infantry formations. It proved an enduring aircraft and was still in service at the start of the Korean War. The slogan means 'for the Motherland'.

Amedeo Gigli

In both attack and defence in late 1944 and 1945, the Focke-Wulf Fw 190F was a reliable fighter.

Marcello Ralli

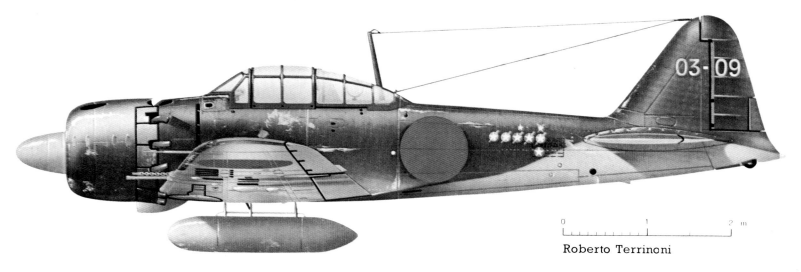

Roberto Terrinoni

Outstanding in the early war years, the Mitsubishi A6M5 Zero fighter had been outclassed by 1944 but was still used for home defence and kamikaze duties in 1945. This one has 'kill' markings.

James Doolittle (b. 1896)

Surely one of the most renowned American pilots, Doolittle first came to the public's notice when he crossed the United States by air in a single day, in 1922, when he flew a DH4B biplane from Florida to southern California, refuelling in Texas; time taken was 22 hours 35 minutes. 1925: entered the Schneider Trophy at Baltimore. 1929: carried out the first instrument landing in adverse weather. 1942: carried out a spectacular raid on Japanese mainland using USAAF B-25 bombers flown from an aircraft carrier.

nimble Mitsubishi A6M Zeroes and the Army's Nakajima Ki-43s swept aside the opposition, in the form of Curtiss P-40s, Hawks, Brewster Buffaloes, .and other obsolescent types that had been deemed good enough for the defence of Britain's Far East possessions.

Within six months, the Japanese were in reach of India and Australia, while still pushing eastwards across the Pacific. During this lightning campaign they had made good use of their large carrier air arm to fill in gaps before the land-based aircraft arrived. They sank the Royal Navy's carrier *Hermes* and delivered a sharp lesson in the importance of

The scene at Pearl Harbor on 7 December 1945 when the US Navy's air arm in the Pacific almost ceased to exist. Prominent in this picture are PBY Catalina long-range flying boats. It only took a few months for the Americans to return to action.

air-cover when land-based torpedo planes and bombers sank the capital ships *Repulse* and *Prince of Wales* within an hour of each other.

However, the USA had also grasped the value of naval aviation, and her carriers *Yorktown* and *Lexington*, which had been at sea and so escaped damage at the bombing of Pearl Harbor, were moved in to halt the Japanese invasion attempt on the south-eastern part of New Guinea, being held at that time by Australian troops.

In the first week of June, the next major encounter, Midway, to stop Japan's expansion along the Aleutian chain in the north, was a decisive American victory. The US Marine Corps established a secure airfield on Guadalcanal after a fierce and prolonged battle, and the tide began to turn against the Japanese.

While the Japanese had over-reached their capabilities against such a mighty enemy, the US aircraft industry was getting into its stride,

producing a flood of improved equipment. Although both sides suffered carrier losses during 1943, the USN was able to receive more and better replacements much faster. Airstrips were built for the USAAF, and new aircraft arrived in large numbers. Vought F4U Corsairs, Grumman Hellcats, Avengers and Helldivers, produced in the light of combat experience, joined the powerful carrier groups.

In 1944, the Americans began to advance in earnest. After taking the Marshall Islands, they captured the island of Saipan to provide a base for the massive B-29 bombers, which otherwise had to fly over the 'Hump' from India in order to attack Japan itself. The battle for this island cost the Japanese upwards of 400 aircraft, while the USA lost only 23, earning the encounter the name the 'Marianas Turkey Shoot'. The following day, strikes from the US carriers sank or damaged nine of the Japanese carriers, and sank a battleship. The advance from island to island continued. USAAF Thunderbolts used napalm for the first time, to winkle out the stubborn defenders. Throughout this 'island hopping', the Japanese air force and navy attacked wherever possible, but they never recovered from the earlier setbacks. Losses on both sides ran high, in part due to the adoption by the Japanese of *Kamikaze* attacks, in which bomb-laden aircraft dived directly on to Allied ships. But even this tactic could not make up for the Japanese lack of aircraft, and the shipping fleet lost to submarines and patrolling aircraft.

In May 1945 the British recaptured Burma, invaded by the Japanese in their initial westward thrust. The battle had been hard fought with slender resources. Following a stand in the north, the British pushed the Japanese southwards, using Spitfires, Hurricanes, Thunderbolts, Beaufighters, and Mosquitoes to break up the Japanese counter-offensives. Considerable use had been made of

Midway and the Pacific Campaign

When the United States of America was drawn into the Second World War, following the Japanese attack on the Pacific Fleet installations at Pearl Harbor on 7 December 1941, the naval aviation element of that Fleet was almost totally destroyed. Within months, however, the vast industrial might of the country had replaced aircraft, and air training programmes had brought crews to readiness.

The Pacific Campaign against the imperial aggression of Japan was to last nearly four years and involve almost 150 air-capable ships and many thousands of aircraft, carrier-borne, sea planes and land-based. There were four major carrier engagements and scores of smaller skirmishes, but the Battle of Midway, 4 to 7 June 1942, is, with hindsight, considered the turning point of the war against Japan. But it was the Battle of the Coral Sea, a month earlier, which saw the power of carrier aviation used for the first time.

By early 1942, the US Navy had managed to break some of the tactical radio codes used by the Imperial Japanese Navy and, believing that the Japanese were planning a major strike against the remaining US units in the Pacific, thus clearing the way to invade New Guinea and Australia, the Americans struck first. This action, which was begun by the launch of 93 naval strike aircraft from the carriers *Yorktown* (fresh from refit after Pearl Harbor) and *Lexington*, turned a potential Japanese victory into a defeat which included the loss of *Shoho*, a light carrier.

The aircraft involved included, on the American side, the Grumman F4F-3 Wildcat, the Douglas SBD Dauntless and TBD-1 Devastator (the latter being the first production-status monoplane designed for carrier air operations). The Japanese, building on their successes in the war with China and Pearl Harbor, were operating the Mitsubishi A6M Type 0 (Zero) or 'Zeke', the Aichi D3A Type 99 or 'Val' dive-bomber and the Nakajima B5N Type 97 or 'Kate' torpedo-bomber. In fact, the odds should have been on the Japanese side, and even though the TBD-1 was badly mauled at Coral Sea and more so at Midway, the USN managed to gain air superiority at sea by mid-1942.

Yorktown had been damaged during the Battle of the Coral Sea and was sent back

The Douglas SBD Dauntless was a successful dive-bomber in the early days of the Pacific Campaign. Unusually for a naval aircraft it did not have folding mainplanes, which made stowage aboard ship difficult and uneconomic.

to Pearl Harbor for repairs. These were completed in a matter of days rather than months, so she was able to rejoin the Pacific Fleet in time for the action to thwart the Japanese plan to invade the Midway Islands. In the Battle of Midway which followed, the US Navy was again able to break the Japanese battle codes and striking first destroyed the carriers *Kaga*, *Akagi* and *Soryu* – ships which had been present for the attack on Pearl Harbor. The Japanese admiral then delayed his counter-blow and lost the initiative, losing the light carrier *Hiryu* as a result.

The sting in the tail for this decisive action was that *Hiryu*'s air group were airborne and succeeded in torpedoing the veteran carrier *Yorktown* – being attacked by 18 torpedo-bombers, covered by six Zeros. Although the anti-aircraft fire put up by the carrier and her escort warships was intense, at least six of the attackers ran the gauntlet and three direct hits were scored by the first wave. In the second Japanese attack wave, two torpedo hits were scored on the carrier's port side by Kates, three of which managed to fly through the 'flak'.

The Battle of Midway is now considered of major importance because it halted the Japanese advance across the South Pacific, but at the time, with two more major carrier battles to be fought, not many people would have suggested that Japan was beaten. In October 1942 came the Battle of Santa Cruz.

In the summer of 1942, Japanese naval submarines were active against US carriers and damaged *Saratoga*, whilst the Battle of the Eastern Solomons, the third carrier-carrier engagement, had left *Enterprise* damaged and she put back to Pearl Harbor. The newly refitted *Hornet* and *Wasp*, fresh from operations to supply Malta with Spitfires in the European War, were the only carriers in the Pacific. But disaster struck the USN on 15 September

Entering combat service in September 1943, the first Grumman F6F Hellcats helped turn the tide against the Japanese in the Pacific Campaign. The aircraft proved highly successful against various targets and was later flown by the Royal Navy.

1942, when *Wasp* was torpedoed and sunk. A little over a month later, the Japanese scored a victory at Santa Cruz when its carrier aircraft disabled *Hornet*, the first time that a suicide aircraft had caused such terrible damage. The *Hornet* was finished off by aerial torpedoes and dive-bombing, but her aircraft managed to damage the Japanese aircraft carrier *Shokaku* severely.

By the summer of 1943, the US Navy was able to counter-attack very successfully with a new generation of naval aircraft, operating from a new generation of aircraft carriers. To counter the Zero menace, Grumman designed the F6F Hellcat, the TBM Avenger torpedo-bomber and the US Marine Corps-manned F4U Corsair fighter. In late 1943, a new dive-bomber, the Douglas SB2C Helldiver, arrived in the Pacific, giving the US Navy, with its 'Essex' Class aircraft carriers and support from British and US escort carriers, the power to win the Pacific War.

Paul Beaver

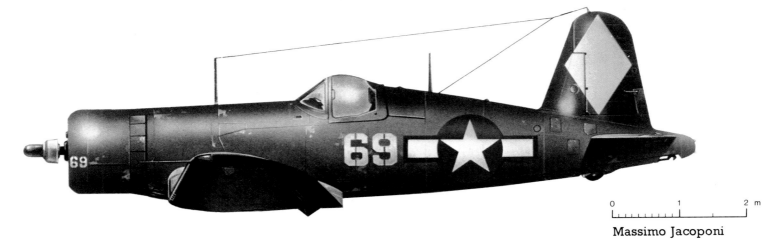

0   1   2 m

Massimo Jacoponi

Flown mainly by the US Marine Corps, the Vought F4U Corsair could be difficult for inexperienced pilots. Shown here in mid-1945 midnight blue, the Corsair was also flown by the US Navy and the Royal Navy's British Pacific Fleet air squadrons.

air transport, to supply cut-off elements of the army and to bring in equipment and supplies from bases in India.

Ever-increasing B-29 Superfortress attacks devastated Japan's industrial capacity, and the night incendiary raids that followed razed vast areas of cities and towns. But a fanatical defence at Iwo Jima and Okinawa, where the Japanese demonstrated quite literally that they were prepared to fight to the last man, persuaded the Allies that a landing on the Japanese mainland would involve immense casualties.

On 6 and 9 August, Superfortresses dropped atomic bombs on the cities of Hiroshima and Nagasaki: the ultimate in strategic bombing. On 14 August Japan surrendered unconditionally.

Carrying the war to the Japanese mainland, the Boeing B-29 Superfortress was an effective bomber for conventional raids, including fire-bombing. It was a B-29, Enola Gay, that dropped the first atomic bomb, on Hiroshima in August 1945.

The Battle of the Atlantic

To counter the blockade by U-boats of Britain in the First World War, a system of sailing in convoys had been evolved. It had also been found that air patrols could restrict the activities of the submarines, which tended to keep away and dived from areas covered by airships, landplanes, floatplanes, and flying boats. The measures taken were crude, but they had shown that submarines could be sunk from the air.

Several valuable lessons had been learnt from this: the attacks had to be immediate to stand a good chance of success; bomb-dropping with any accuracy was extremely difficult; and endurance of patrol was of great importance. With the outbreak of war in 1939, the British again faced the menace of blockade and starvation from the U-boats. The newly formed Coastal Command had patrol aircraft, such as the Lockheed Hudson and Shorts Sunderland, but reliance was placed on the Royal Navy's ability to protect the convoys with the aid of their ship-mounted ASDIC underwater detection, as the aerial umbrella did not stretch far from the coast.

Allied shipping losses mounted as the submarine offensive gathered momentum, and to maintain the flow of vital supplies and equipment, more energy and resources were

Providing air-sea rescue , convoy protection, anti-submarine warfare and reconnaissance, the Consolidated Catalina served RAF Coastal Command in the Battle of the Atlantic. It was also the aircraft which found the battleship *Bismarck* in 1941.

thrown into the fight. In 1941, ASV (air-to-surface vessel) radar came into use, albeit in an undeveloped form, and Wellington bombers fitted with powerful Leigh searchlights entered service at the end of the year. These lights were intended to illuminate the submarine as it cruised on the surface at night, recharging its batteries, enabling the bomber to see its target clearly on the important first run-in with new and much more effective air-dropped depth-charges.

Concurrently with these innovations, the Royal Navy, suffering from the bombing and reconnaissance of the German patrol aircraft such as the FW 200 Condor, had fitted merchant ships each with a catapult, and an obsolescent Hurricane fighter. These catapult-armed merchantmen, or CAMs, launched their fighters to drive off incoming attackers, then recovered the pilots as they ditched alongside, if they were lucky. Secondly, a captured German ship had been converted from carrying fruit to carrying fighters, by refitting it with a flat deck. Commissioned as HMS *Audacity*, it was the first of the small 'escort carriers'. Once these ships had been put to convoy protection, the air power they carried with them helped to close the killing grounds of the gap left by the land-based cover from the Catalinas and Sunderlands from Britain, Nova Scotia, and Iceland. The CAMs

Pino dell'Orco

Also prominent in the Battle of the Atlantic was the Shorts Sunderland III, this version equipped with ASV Mk II radar and flown by a Royal Australian Air Force crew from Plymouth on the English Channel. The bombs are underwing.

and escort carriers also proved their worth when operating under the ghastly conditions from the Arctic convoys to Russia, they were able to offer invaluable protection against the attacks of the Luftwaffe squadrons based in Norway.

In the North Atlantic more destroyers, and the new corvettes arrived in large numbers. Further bolstered by the entry of the US Navy, and Coastal Command's first very long range Liberators, which had an endurance of some 16 hours, the Allies could fight back with greater effect. The turning-point came in the winter and spring of 1943, when, due to heavy losses, Doenitz withdrew his submarines for regrouping. They were never again to gain the upper hand. Faced with over a thousand patrolling aircraft, and denied air cover of their own, they were driven to remain under the surface whenever possible, running their diesels through 'schnorkel' masts, in an attempt to evade the depth-charges and rockets of their radar-equipped hunters. This deprived them of their ability to attack, and better equipment and countermeasures did not reach their bases until too late in the war.

The U-boat arm was defeated at sea, by surface ships and from the air; the strategic bombing of the production facilities and hardened pens in which the submarines sheltered had comparatively little effect. From

1939 to 1945, 892 Axis submarines were sunk, the honours shared between sea and air forces.

Bomber offensive over Germany

Before the fall of France RAF Bomber Command had been employed mainly in harassing the German navy, and supporting the armies on the Continent. For the results gained the losses had been high; the excellent 'flak' protecting the German columns and the fighter cover had found the Wellington and Whitley, even with their powered turrets, easy prey in daylight. The lighter Hampden and Blenheim fared little better, being relatively poorly armed for their low speed.

Night operations were the only answer, and the Whitley was sent over Germany during the winter of 1940, dropping only leaflets for fear of retaliation against British cities. When the immediate danger of invasion had been averted, and feelings for revenge for the Blitz on England were running high, plans for a sustained strategic bombing offensive against Germany were laid. It was for this specific reason that Bomber Command had been created in the 1930s, and the philosophy behind it had persisted since the creation of the Independent Force in 1918. Now it was the only way in which Britain could strike directly at the German homeland, as the army was in no condition to land in Europe.

In 1937 the Bristol Blenheim I was the first British monoplane bomber, but by 1940 it was obsolescent and several were converted to night-fighting, carrying the first airborne interception radar sets.

In the initial defence of Germany against Allied air assaults was the Messerschmitt Bf 109 fighter, both by day and night. Besides cannon and machine guns, some versions were armed with rockets to attack the large American formations.

Initial policy of precision bombing of selected targets of strategic importance produced disappointing results. It was found that accuracy was so poor that these installations were only rarely badly damaged, and so the bombing of industrial towns became standard, known as area bombing. By the end of 1941, when operations were suspended to give the crews a much needed rest, the first of the four-engined heavy bombers, the Short Stirling, had entered service, and Bomber Command had grown greatly in size. The heavily defended industrial area of the Rhur was a prime target, and losses had been high, although the introduction of the 'Gee' navigational aid had helped the accuracy and effectiveness of the attacks.

1942 saw the arrival of Air Marshal 'Bomber' Harris, and the completion of the change to the bombing of towns and cities in an attempt to disrupt the lives and lower the morale of the industrial workers. It also saw the first '1,000 bomber raid', when Bomber Command put all its reserve and trainee crews and machines into the air for an attack on Cologne in such strength that it disrupted the working of the city for months. Area bombing continued, and although still far short of earlier expectations, the accuracy was further improved by the introduction of 'pathfinder' squadrons, specially trained to lead the attacks

RAF bomber offensive against Germany 1939-1945

The seeds of the devastating campaign of aerial attrition conducted over Nazi Germany by the UK Royal Air Force during the Second World War had been sown in the closing days of the First World War. On 5 June 1918 a separate element of the RAF was formed and named the Independent Force. The intention was to create an air arm which could carry the war into the very heart of enemy territory. Mercifully peace came in November of that year, otherwise fleets of new four-engined Handley Page bombers would have attacked Berlin and other German cities from bases in the United Kingdom. Nevertheless the concept of a strategic air offensive in European terms had been established.

During the years that followed and well into the 1930s the technology of the bomber changed little. Various exercises and tactics were undertaken but most of these seemed more fitted to a policing role in the Empire rather than as a serious consideration of another air war over Europe. Successive designs did not vary all that much from the Handley Pages of the old Independent Force. Defensive armament remained light and the techniques of long-range navigation and extended night operations certainly did not receive the emphasis that they deserved.

Airfields were still without paved

The mainstay of RAF Bomber Command and the classic night-bomber of the Second World War was the Avro Lancaster, seen here in the markings of 50 Squadron during a daylight air test. Later versions carried the H2S radar under the fuselage.

runways and in most cases civil airline practices were far ahead of the military. It is easy with the benefit of hindsight to look back and identify the problems. At the time the reason was less obvious and obscured by the growing complexities of political and economic problems.

There were many minds both in the service and in civilian life who were concerned and attempted to address the deficiencies. By the mid-1930s plans were made to expand the RAF and to develop new types of aircraft. These designs later became some of the heavy bombers of the Second World War, yet right up to the onset of hostilities navigation techniques were still more suited to daytime cross-country exercises, with frequent map references, and tactics were based upon daylight formations, which impressed audiences at air shows but showed little regard for a rapidly changing military scenario.

Perhaps many of the difficulties arose from the combination of financial stringency, lack of resolution over the true role of the bomber and even inter-service rivalry, which often made it appear, at times, that the RAF was struggling for its very existence as an independent service.

Whatever the real cause, and even today, there is still room for considerable debate. When war came once more on 3 September 1939 the RAF found itself equipped with five basic types of bomber.

These aircraft were the Bristol Blenheim (MkI & IV); Fairey Battle; Armstrong Whitworth Whitley; Vickers Wellington and Handley Page Hampden. Whatever the qualifications or concerns that were expressed by knowledgeable

observers, the future had to be faced with what was available.

Although in aerodynamic terms the designs were sound, the aircraft were deficient in many of the ergonomic factors that would alleviate the problems of long, and often very cold, flights. Fuel tanks and vulnerable systems lacked protection and the aircraft's self-defensive capability was basically restricted to hand-operated mountings carrying machine guns of limited calibre.

Crews recognised the deficiencies and accepted that their job was to do the best that could be done. War teaches quickly, but the initial tactics were still loosely based upon the daylight formation concepts of the earlier war.

Following the declaration the RAF was quickly into action, the first operational sortie taking place on 3 September, when a Blenheim carried out a photo-reconnaissance mission over Wilhelmshaven. That same night the first leaflet raid was carried out by Whitleys flying over Hamburg, Bremen and the Ruhr. These flights were code-named Nickel and they were to be the established pattern for the next few months so far as targets within the land areas of Germany were concerned. Offensive operations were limited to what were termed naval targets and the first of these took place on 4 September using Blenheims and Wellingtons.

So began what was to become a protracted campaign of attrition that would often test the crews to their limits and instil fear and horror into the very fibre of the many thousands of hapless civilians who would later find themselves under attack in all of the major cities of Germany.

Power-operated turrets became more widely available and by December 1939 it was widely considered that the Wellington

AVRO LANCASTER Mk 1

Country of origin: United Kingdom.
Role: Heavy night bomber.
Wing span: 31.09 m (102 ft).
Length: 21.18 m (69.5 ft).
Max loaded weight: 28,636 kg (63,000 lb).
Engines: 4 × 1,280 hp Rolls-Royce Merlin XX.
Max speed: 450 km/h (280 mph).
Max altitude: 7,163 m (23,500 ft).
Range: 4,345 km (2,700 mi).
Military load: Up to 10 machine guns in three powered turrets; 3,635 kg (14,000 lb) typical bomb load carried internally.

was the best of the bomber force – in a formation it was capable of a considerable degree of self-protection. The theory seemed to be practicable, but the test proved how fallacious it was. The reckoning came on 18 December, when a force of 24 Wellingtons set course for the Heligoland Bight. Two were forced to return for technical reasons and the others continued to the target area, which was clear of cloud, providing ideal visibility for the waiting German fighters. Only 10 of the British aircraft returned to base.

The lesson was brutal and clear: the only measure of survival that the bombers could reasonably expect would be through the element of surprise. For continuing daylight attacks this meant taking advantage of cloud cover, an idea that was not practicable on a day-to-day basis. Any sustained campaign over the European landmass could only be realistically conducted by using the 'cover of darkness'.

In May 1940 the air of realism was further fortified by lifting the ban on targets involving the risk of civilian casualties. On the night of the 15th of that month a force of aircraft attacked targets in the Ruhr and the pattern was established for the future.

By July the first 907 kg (2,000 lb) bomb

had been dropped by Bomber Command on Kiel. Other new equipment was coming along, including the new breed of heavy, four-engined bomber, the Shorts Stirling, which entered service in August. It was closely followed by the Avro Manchester in November. This latter aeroplane was a large twin-engined type and the rather complex engines were less than successful. It rapidly gained an unenviable reputation among the crews for unreliability and in time the airframe evolved into the highly successful Lancaster.

The new aircraft led to the recruitment of specialist crew members so that in addition to the established pilot and observer trades the RAF introduced Navigator, Signaller, Flight Engineer and Air Gunner brevets. This was later joined by other specialists such as Bomb Aimer.

The Handley Page Halifax began offensive operation early in 1941 coincident with the first use of a 1,814 kg (4,000 lb) bomb more generally known as the Cookie. The American-built Boeing B-17 also made an appearance with the RAF, and in view of the massive contribution made by this aircraft later in the war its initial use by the RAF was disappointing.

Meanwhile the onslaught was building

Companion to the Lancaster was the Handley Page Halifax, which was also used for glider-towing and special operations later in the war. The wartime censor has removed the squadron codes so it is not possible to identify the aircraft's unit.

up and at a time when the Allies were facing considerable reverses on all fronts the almost nightly excursions of Bomber Command were a considerable morale booster for the British population. The operations offered the only 'good news' available to the Allied media and this was used to advantage in newspapers, magazines, books and films. These sanitised accounts offered a confident view of the growing campaign and offered an impression that the crews knew was far from reality.

Navigation techniques were still based upon a large degree of supposition and haphazard checking of ground features. With the best will in the world it was almost impossible to pinpoint targets with the required delicacy and the problems of cold, fatigue and mechanical irritation were as persistent as ever.

The problems were certainly recognised and the rapidly evolving science of electronics was beginning to offer some answers. By August 1941 a navigational aid known as GEE was tried out during a raid on München Gladbach and the scene was set for the cycle ploys that characterise electronic warfare to this day.

Meanwhile the German defences were becoming more organised and the loss rate in Bomber Command was steadily escalating. Paradoxically the Command was under a form of attack by its own side. Factions within the services wanted greater participation by the bomber force in the Battle of the Atlantic, where shipping losses through the action of German U-Boats was causing increasing and gravè concern.

It was at this point that the controversial Air Vice Marshal Sir Arthur Harris took over as Air Officer Commanding Bomber Command. This dynamic leader was convinced of the importance and necessity of carrying the war on a nightly basis into the heart of the enemy homeland. He certainly did not want his Command segmented and he realised that at this time in early 1942 that the strength of Bomber Command was only just in excess of 600 aircraft at operational readiness and fewer than 100 of this force were four-engined aircraft.

Typically he moved towards the dramatic gesture that would marshall that elusive element known as public opinion. In a manner which is called propaganda in wartime and PR in peacetime he organised three very intense attacks on major city targets. To gain maximum publicity impact these were billed as 1,000-bomber raids and the first of these attacks took place against Cologne on 30 May 1942. Whatever the actual effect upon the target area the operation was something of a masterpiece of organisation. To achieve this total aircraft were obtained from all sorts of places and training establishments. It was a coup for the new Commander and from that time onwards Bomber Command was allowed to remain an integrated force and was steadily built up in numbers and technical superiority.

A specialist group called Path Finder Force (PFF) was set up in the summer of 1942 to be the vanguard in major attacks and to accurately locate and mark the target area.

The Avro Lancaster had made its first operational sortie earlier in the year and it became an increasingly common sight as Bomber Command expanded in both numbers and expertise during the following year. In May 1943 the famous attack on the Möhne, Eder and Sorpe dams took place, an attack that had all the ingredients for maximum public

For more than three years there was an afternoon ritual of 'bombing up' the RAF's heavy bombers in their East Anglian and Yorkshire bases for the nightly sorties into Germany and occupied Europe. Various bombs were developed for the Lancaster.

consumption.

Other forces were also at work during this frenetic time. Early in the year de Havilland Mosquitoes attacked Berlin in daylight, this nippy and highly manoeuvrable aircraft depending upon its speed to escape from trouble. Nevertheless the main thrust of the heavy bomber forces would continue to be under the protection of night.

New electronic systems were appearing with rapidity and were being used for navigation, target location, warning of attack and self-protection. This game of electronic chess was now obsessing both sides and various attempts were made to jam or distract each of the opponents' equipment. For a while both sides had been aware of the possibility of using small strips of tin foil which would be released in showers of sufficient intensity to jam out radar signals and reduce the radar displays to chaos.

It was a simple electronic trick, but one that was double edged. For many months debate about its use by Bomber Command continued and although the technique had been used in a limited trial in North Africa it was not until July 1943 that permission was finally given for its use over Germany.

The raid was against Hamburg and was to be the first of a series taking place under the grimly prophetic code name Gomorrah. The tin-foil technique was code-named Window and it came up to expectations, resulting in a very successful attack. The second raid of the series took place on the night of July 27/28 and created another horror. A combination of the intensity of the raid coupled with the 'right' kind of meteorological situation caused a massive fire storm. This is created by a number of smaller fires combining and rapidly increasing the circulation of air rather in the manner of a blast furnace. As the inrush of air increases so the increased flow of oxygen can create temperatures in excess of 600°C (1,112°F).

Despite the ever-increasing range of new equipment the attrition among both aircraft and crews continued, many of the losses being caused by meteorological problems on the return flight. Typically a tired crew and possibly damaged aircraft running low on fuel would be caught out by fog at the destination airfield. Once again the 'boffins' created a new idea that eventually turned into a remarkable system known as Fog Investigation and Dispersal Operations – FIDO.

In many ways it was a crude, though effective, device. Petrol-carrying pipes were laid alongside the runway and a trench was flooded with the liquid and then ignited. The resultant heating of the

The de Havilland Mosquito did pathfinding for the bomber stream and took photographs of the damage after the drop. It was also used for specialised bombing missions and escort roles, especially in raids on Berlin.

Defending the Reich was the Messerschmitt Bf 110 night-fighter. It was successful in destroying Allied night-bombers, but not in preventing the destruction of commerce, industry and people's homes by the bomber formations.

surrounding air caused the fog to disperse. It is rather astonishing in retrospect to recall that the system was seriously considered for use at London's new Heathrow Airport during the late 1940s.

From about April 1943 to April 1944 Bomber Command operations were at their height, the Battle of Berlin starting in November 1943. Losses were high and now mostly caused by increasingly sophisticated night-fighting techniques by German aircraft.

At the end of this period the main thrust of the Command's activity came in the form of attacks on transportation systems and infrastructure in France to prepare the way for the D-Day operation. Following the successful landings in Europe, Bomber Command provided tactical support for the advancing armies. As the land forces consolidated their positions the aerial attack turned against Germany once more. By now the ability of Germany to defend itself against aerial bombardment was seriously weakened and raids could be carried out equally effectively by day or night.

The final assault took place on the night of 3 May, when a strong force of Mosquitoes took the battle to the German homeland for the last time. The long campaign was finally over.

Since that time considerable discussion and controversy have been created by the role of Bomber Command. Strong moral voices have been raised and particular operations have been mulled over at length. In particular Dresden seemed to create concern among crews taking part that was not expressed about other targets. Economists prove that German industrial

capacity remained buoyant and Sir Arthur Harris received considerable criticism.

To many of the German civilians who lived through those times the young RAF aircrew were 'terror fliers', while people in Hamburg still speak of the 'catastrophe' in the summer of 1943.

With the passage of time surviving crew members recall the cold, the discomfort and the fear. The majority returned the civilian life and resumed disrupted careers, many joining the airlines. Others bore scars in both mind and body.

Now as memories fade it is still worth

The Boeing B-17 Fortress, operated in daylight from the United Kingdom to targets across Europe. Its US 8th Air Force crews thus prevented wholesale reconstruction after the night raids by Lancaster, Halifax and Stirling bombers.

recalling that at the darkest period of the war for the Allies it was Bomber Command that stiffened morale and took the war to the enemy heartland. The scale and duration of the campaign forced the German war machine to devote massive resources to home defence.

The crews represented not only those of British birth but the many Commonwealth countries as well. The final toll in Bomber Command was more than 47,000 killed on operations and another 8,300 killed on non-operational duties.

Don Parry

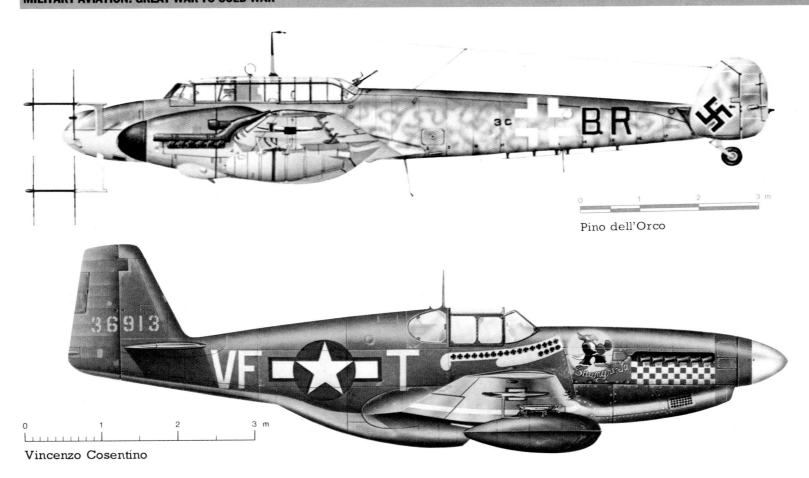

0   1   2   3 m

Pino dell'Orco

36913 VF ★ T

0   1   2   3 m

Vincenzo Cosentino

and mark the targets for the stream that followed.

By the end of that year, the US Eighth Air Force was arriving in strength and preparing to open its own offensive. The plan was for large formations of Boeing B-17 Flying For-

tresses and B-24 Liberators to carry out precision attacks in daylight. At this stage in the campaign, however, the Germans had significantly improved the defence of their country; fighter production, one of the bom-bers' main targets, had grown rapidly. When

Willy Messerschmitt's Zerstorer *(top)*, equipped with 'stag horn' radar antennas and engine-limiting flame dampers. No such problem troubled the North American P-51B Mustang *(above)*, used to escort the American bomber streams across Europe. Later versions of this widely used fighter were fitted with bubble canopies for better pilot vision, especially during air-to-air combat with defending fighters. This aircraft was flown by Don Gentile of the 336th Fighter Squadron, US 8th Air Force.

Shows a typical night for one group of RAF Bomber Command 1944

★ major targets
■ RAF Bomber Command HQ
◉ Bomber Group HQ (Swinderby)

● Bomber Command airfields
➡ Outbound
⬅ Homebound
➡ Diversionary raids

✳ Window dropped to confuse radar
✳✳ Large quantities of window dropped
□ Capital cities

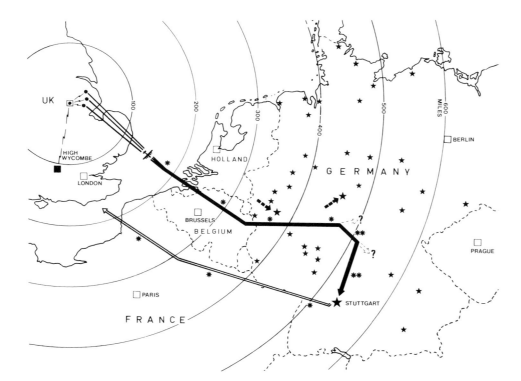

the USAAF flew beyond the range of its Spitfire or Thunderbolt fighter escort it had only the massed 'fifty calibre' guns of its own aircraft flying in box formations to protect them. The FW190s and Bf109s inflicted terrible losses, as the RAF had predicted, but the Americans persisted, their force growing in numbers and skill.

The RAF, now equipped with the big four-engined Avro Lancaster and Handley Page Halifax, plus the fast DH Mosquito twin-engined bomber, started to use new navigational aids, 'Oboe' and 'H2S', but the German system of sector radar location, and ground control of the Messerschmitt Bf110 and Junkers Ju88 night fighters, still made for a formidable opponent. The pace of technolo-gical development increased as the Germans produced equipment to home in on the 'H2S' and 'Monica' tail-warning radar transmissions on the bombers. RAF countermeasures in-cluded electronic jamming of radar, dropping of 'window' to confuse the ground controllers, and broadcasting bogus instructions on the fighters' radio frequencies.

Second World War Aircraft Weaponry

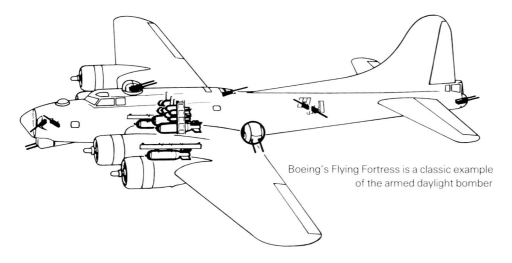

Boeing's Flying Fortress is a classic example of the armed daylight bomber

Defensive armament: up to 13 × 0.50 calibre machine guns
Offensive armament: up to 5,897kg (13,000lb) bomb load
bomb load illustrated: 8 × 454kg (1,000lb) + 2 × 907kg (2,000lb)
2 × 907kg (2,000lb) on wing rack

Marauder, Boston, and Baltimore. Despite the support offered by its bombing and strafing, the German defence lines proved hard to crack, and it was not until the rail and road communications behind the enemy had been destroyed that progress was made in stages past the Gustav Line at Cassino, and the more northerly Gothic Line. German troops from this front did not finally surrender until only six days before the fall of Germany itself.

In preparation for the invasion planned for June 1944 huge tonnages of bombs were dropped on railways and marshalling yards in France and Belgium in the months before in order to aid the ground armies by crippling the enemy's ability to deploy his forces. Coastal batteries also received their share of attention, but the other prime targets were the launching sites for the new V-1 and V-2 weapons. By the end of August over 80,000 tonnes of bombs had been aimed at these installations, but they were not eliminated as a threat until the ground forces overran them on the advance into Germany.

With over 10,000 aircraft available for operation Overlord, the Allies were able to claim complete air superiority; as they came ashore, the troops were at least free from enemy strafing. British and American fighters provided a protective umbrella for the convoys and beach-heads, and less initial resistance was encountered than might have been feared. This was mainly due to the deception tactics employed, preventing the Germans from knowing, until the last moment, the site of the landings. Part of these measures to deny the effective use of reinforcements were the dropping of 'window' by patrols of bombers to

Nevertheless, massive damage was being done by the round-the-clock attacks: in four nights over 40,000 inhabitants of Hamburg were killed, and industry severely disrupted. On the whole, however, civilian morale showed no signs of breakdown, and the Germans reorganised their industries. In the year to the spring of 1944, when the Allied bombers made huge raids on targets like Schweinfurt, Berlin, and the Rhur, defence was stiffening and German fighter production continued to increase, despite the attention of the bombers. Then large numbers of P-51 Mustang escort fighters started to arrive from the USA, and these 'little friends', as the USAAF crews referred to them, could carry enough fuel to fight all the way to Berlin and back. Bomber losses over the heavily defended targets began to fall as the escorts fought hundreds of Luftwaffe fighters at a time in great aerial battles. The destruction of the transportation system, oil refineries, and synthetic oil plants finally began to be felt, and the Luftwaffe lost large numbers of pilots. In the last year of the war, those who had earlier forecast that war would mean widespread civilian bombing were to some extent vindicated; the bomber had got through, but it had been a far longer and more bloody struggle than had been imagined.

European operations to 1945

When Hitler's attention turned to the east after the Blitz on British cities, German air activity in Western Europe tailed off. RAF night fighters, with their new airborne radar sets, chased nuisance raiders over southern England, and by day Fighter Command made sweeps over France to bring the Luftwaffe out to fight; but the offensive initiative had passed to Bomber Command. With the major action on the Eastern front, the Western Desert, and in the Far East, it was not until the Allied

invasions of Sicily and Italy that Western Europe saw the tactical air forces in large-scale combat again.

When they opened the front in Italy, the Allies found well-organised resistance, in the form of a German army that was determined not to be driven back. In the air it was a different story, as much of the Axis air power had been wiped out in the preliminary invasion of Sicily, and in a campaign dogged by foul weather Allied aircraft were able to harass the enemy as they were slowly forced up the peninsula, using the tactics learnt in North Africa. At the end of 1943 the Germans had only 29 bombers available in the Mediterranean theatre, whilst the MAAF (Mediterranean Allied Air Force) could call upon B-17 and B-24 'heavies, as well as the B-26

Second World War fighter tactics

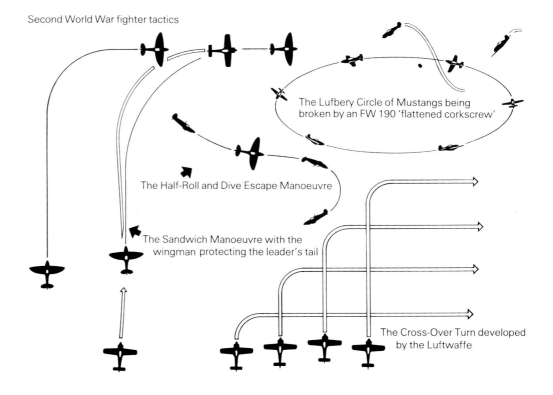

The Lufbery Circle of Mustangs being broken by an FW 190 'flattened corkscrew'

The Half-Roll and Dive Escape Manoeuvre

The Sandwich Manoeuvre with the wingman protecting the leader's tail

The Cross-Over Turn developed by the Luftwaffe

NORTH AMERICAN P-51D MUSTANG

Country of origin: USA.
Role: Long-range fighter.
Wing span: 11.28 m (37.04 ft).
Length: 9.83 m (32.25 ft).
Max loaded weight: 5,500 kg (12,100 lb)
Engine: 1 × 1,790 hp Packard-built Merlin V1650-7.
Max speed: 703 km/h (437 mph).
Max altitude: 12,770 m (41,900 ft).
Range: 2,655 km (1,650 mi) with drop-tanks.
Military load: 6 × 0.50 in machine guns, underwing points for up to 900 kg (2,000 lb) bombs or drop-tanks, or 6 × 12.7 cm (5 in) rockets.

produce bogus 'convoys' on German radar, and fighter sweeps to the north, as well as the dropping of dummy parachutists and small diversionary parties in other areas of France. Behind and around the beaches, many thousands of glider-borne and parachute troops landed to carry out specific tasks and take the pressure off the main areas. The transports also brought fuel and all other sorts of supplies, augmenting the efforts of the ships until artificial harbours could be established.

The tremendous demands being made on the Luftwaffe on the Eastern Front and over Germany itself gave the Allied tactical air forces a great chance. Fighter-bombers, such as the Hawker Typhoon, armed with powerful rockets, flew 'cab rank' patrols which were called down by the men on the ground to strike at points of resistance. Breaking out of the Normandy area after a hard struggle, the Thunderbolts, Typhoons, and medium bombers flew whenever possible to prevent an orderly retreat, but despite the havoc wrought by machine guns, cannon, and rocket, the Germans showed they were not yet beaten. Airborne landings in the north were marred by the British 1st Airborne's failure in their struggle to the bridge across the lower Rhine at Arnhem in Holland, which was part of a set of coordinated attacks involving over 1,500 transports and 2,800 gliders.

By the end of 1944, the Luftwaffe was at a very low level of capability. Though Germany had built fighters in huge numbers, and were introducing potent jet aircraft, the tremendous losses it had suffered meant that the standard of the pilots' training was very poor.

In support of the Ardennes offensive, and in a desperate attempt to redress the balance of air superiority, 800 German fighters, led by the most experienced of their pilots, struck at the Allied tactical airfields on New Year's day, 1945. Records show that 130 aircraft were destroyed on the ground, but the German losses amounted to 200 valuable machines. With the improvement in the weather, the Allied advance started again.

In the final months of the war the destruction of Hitler's oil supplies, and the disruption of means of transportation by Allied bombing, led to the bulk of the Luftwaffe's aircraft being grounded by lack of fuel. The flying bombs ceased landing on Britain and Antwerp, as there was not enough fuel for the launching aircraft, and the permanent launching sites were in Allied hands or destroyed. The supply of the A4 rockets (V-2s), which could not be intercepted in flight, dried up. The bombers continued, and in a final demonstration of their power over Germany, Dresden and Leipzig were burnt to the ground. The Russians entered Berlin, and by the end of April the war in Europe was all but over.

The most advanced fighter of the Second World War was the Messerschmitt Me 262. Unfortunately for the Germans, however, it could not be used to its full advantage because of engine faults.

Wartime advances in technology and equipment

During the war the maximum speed of a front-line fighter aircraft had risen from around 560 km/h (350 mph) to 730 km/h (455 mph). Armament had also changed; in Britain, Germany, Italy, and Japan the cannon had become standard, either on their own, or supplemented by machine guns. Only in the United States had a battery of 0.50 calibre machine guns remained the primary fire-power in aircraft. Fighters had seen much action working as fast, light bombers, for close support. A Blenheim bomber of 1939 carried a payload of 450 kg (1,000 lb) at 460 km/h (285 mph); by 1944, an F4U Corsair or Thunderbolt was capable of delivering twice that load, and could fight at over 160 km/h (100 mph) faster. The mighty B-29 Superfortress represented a tremendous advance over its B-17 predecessor in pressurised flight.

The most obvious step forward in aircraft design was the arrival of the jet. The only Allied jet to see active service was the Gloster Meteor, whose speed was little better than the fastest of the piston-engined aircraft, but the new Vampire was also nearing production, and the Bell XP-59A had flown over England and Italy. It was from Germany, however, that most of the new jet types had come. One of the best, and with the best service record, was the Me 262. More than 1,300 of these 870 km/h (540 mph) twin-engined fighters had been produced in little more than a year before the end of the war. The Arado 234 reconnaissance bomber was first used in the autumn of 1944. Another Messerschmitt, the Me 163 rocket-powered interceptor, had severely limited endurance, but could fly at over 900 km/h (560 mph), and had been used to break up American bomber formations. The other jet fighter to see service was the Heinkel Salamander, a light-weight with a single engine mounted on top of the fuselage, which had gone from design to first flight in nine weeks. There was a host of other designs, in various stages of development from completed prototype to sketches and feasibility studies.

De Havilland's Wooden Wonder, the Mosquito, served in many roles, the Mk IV as a light bomber carrying four 227 kg (500 lb) bombs; this aircraft served with 105 Squadron in 1941-42. Unarmed, it was fast enough to outrun fighters.

Vincenzo Cosentino

0   1   2   3 m

Pino dell'Orco

By late in the war heavy fighters, with ground attack ability, such as the Republic P-47 Thunderbolt *(above)*, were entering service. This aircraft, called 'Spirit of Atlantic City NJ', was flown from a British base by the ace Bud Mahurin in 1943. As a counter to the new Allied fighters and bombers, the Luftwaffe was equipped with a series of unusual designs, like the Heinkel He 162 Salamander *(left)*. Built of non-strategic materials and flown by untrained pilots, it was not successful.

Although futuristic in concept at the time, a vast proliferation of designs does not shoot down bombers; the jets were delivered too few and too late to affect the outcome of the war.

The engines of the new jet types, on both sides, stemmed from work done before the war: centrifugal and axial flow turbojets, designed by Britain's Sir Frank Whittle, and the German Von Ohain, respectively. Both types were to be used in their original forms for many years, but the axial flow has proved capable of development to greater power, and has since become the predominant type in military aviation.

In addition to these aircraft, the Germans had brought other entirely new weapons into service: the FZG76, or flying bomb, the V-2 ballistic missile, and the FX1400 and other guided bombs used for anti-shipping strikes in the Mediterranean. The flying bomb was a small pilotless machine, with a one-tonne warhead; it was guided on a set course by gyroscopes, dropping after a predetermined distance, and was powered by a relatively crude pulse-jet engine. It could be said to be the ancestor of the cruise missile.

One pre-war invention that truly came of age was radar. In 1939 it had been used by both sides, but only in a rudimentary form. From the Chain Home, and Chain Home Low stations around the coast of Britain, and the 'Seetakt' gun-laying radar of the German

Built as a medium bomber to support the German Army, the Heinkel He III was used for other tasks later in the war, including anti-shipping strikes and to air-launch the V-I, as was the case with this version, based in the Netherlands.

navy, came whole families of radar, for ground-mapping of bomber targets, night fighters, navigation and marking of dropping zones, and ranging and setting of anti-aircraft shell fuses. All the time, the sets became smaller and more reliable, enabling even fighter aircraft to carry them. The development of radar has continued since then, but even at the end of the war it was possible to see that the true value of an aircraft might come to depend as much on its electronics as much as its airframe.

The unguided air-to-ground rocket had been much used, and had largely supplanted the airborne torpedo for anti-shipping strikes, particularly in the Far East. Other innovations seen were the extensive use of parachute troops, and Rocket Assisted Take-Off Gear, to be known as RATOG, for aiding heavily laden machines under difficult conditions. But, of course, the weapon that had the greatest consequences of all was the atomic bomb. The total destruction of the two Japanese cities, each by a single bomber, raised the importance of military aviation to such a level that the power it wielded ensured that warfare and politics would never be the same again.

0   1   2   3 m

Amedeo Gigli

Deterrence and US supremacy

The two atomic bombs that speeded Japan's surrender, and the power that they represented, brought about the concept of nuclear deterrence. In the immediate post-war period it could be said that a 'Pax Americana' existed; no country was about to start a war against another that had the capacity to obliterate any nation on earth. As the medium by which the destruction would be delivered, the USAAF was in a unique position, but it was not to last. It became obvious that the USSR would soon possess their own nuclear bombs, and with that prospect conditions changed. The two powers were opposed in dramatic fashion; each had definite and immutable ideas on how the world should be shaped, but the threat of the catastrophic consequences meant that neither would risk an all-out war against the other. This was the beginning of the Cold War, when the two great power blocs of the world sought to gain advantage without stepping over the brink to the ultimate military engagement.

The US Air Force had been drastically cut at the end of the Second World War, but America had then shouldered the burden of responsibility for the protection of its allies. These undertakings, the need to possess bases all over the globe from which to overfly the huge territory of the USSR, and the requirement for wide distribution of a more flexible response to conflicts that did not merit the drastic action of the nuclear strike force, meant that the procurement of aircraft grew rapidly for the second time within ten years.

To prevent the other side from pulling ahead in this new race, and to ensure that the nuclear threats remained feasible and deliverable, the USA and USSR embarked on programmes of rapid development. More powerful bombs were built, smaller for ease of deployment, and work was stepped up on improving the ballistic missile as a delivery

In the Cold War period which followed the Second World War, Berlin was blockaded by the Soviets. Aircraft used by the Allies in its support included the Avro York transport, developed from the Lancaster bomber.

B-47 STRATOJET

Country of origin: USA.
Role: Strategic bomber.
Wing span: 35.36 m (116 ft).
Length: 32.64 m (107.08 ft).
Max loaded weight: 79,545 kg (175,000 lb).
Engines: 6 × GE J47-GE-25 each 2,577 kg (5,670 lb) thrust.
Max speed: 1,045 km/h (650 mph).
Max altitude: 14,325 m (47,000 ft).
Range: 6,437 km (4,000 mi) with 4,500 kg (10,000 lb) load.
Military load: 2 × 20 mm cannon, 8,500 kg (18,700 lb) bombs.

platform, both sides having started with scientists and information from the German A4 rocket centres. In more conventional areas, the bombers that would drop the nuclear weapons, and the fighters to keep out the enemy aircraft, all had vast amounts of money poured into their improvement.

While most of Europe tried to recover from the enormous financial drain of over five years of war, the US aircraft industry moved rapidly ahead, introducing plant and equipment that their former allies could not possibly afford. But if the former colonial powers were still dazed from their combat and struggling to resolve the problems of their disintegrating empires, the USA were about to feel the first effects of the position of defender of the free world, and of the many treaties, promises, and arrangements that they had made to turn an isolationist foreign policy into one of expansion of global influence.

The Korean war

On 25 June 1950, North Korea's People's Army swept across the 38th Parallel and moved towards Seoul, the capital of South Korea. The US 5th Air Force, based in Japan, responded rapidly to the plight of its ill-prepared ally, and on 27 June, three La-7 and Yak-9 fighters were shot down by F-82 Twin Mustangs. The next day Seoul fell, and following a UN Security Council resolution, member countries were requested to aid South Korea in repelling the invaders. Of the 16 nations that responded to this call, the United

Supporting United Nations forces in Korea after the Communist invasion was the USAF's Lockheed F-80 Shooting Star. The aircraft was outclassed by the swept-wing MiG-15s, but a training version, the T-33, is still in service in the 1980s.

States, deciding to commit ground forces within a week, were to bear the brunt of the fighting, particularly in the air.

This conflict was the first combat test of the new US jets, but much of the work of ground attack throughout was performed by piston-engined aircraft, mainly of Second World War vintage: the P-51 Mustang, F4U Corsair, B-26 Invader, Hawker Sea Fury and Fairey Firefly.

While UN aircraft struck at the North Korean bases, and succeeded in destroying most of their aircraft on the ground, ground-attack sorties strove to slow down the communist army's steady advance. But the UN forces, under General MacArthur, found that their enemy was still able to fight most effectively without air support, and by mid-August they had been squeezed back into the south-eastern tip of the peninsula, holding a perimeter around the port of Pusan. All available air power was directed to the support, and within four weeks MacArthur's forces were able to break out. After a well-executed out-flanking landing at Inchon, they drove north to the 38th Parallel, reaching it in September and inflicting massive losses.

Continuing his advance, MacArthur captured the Northern capital of Pyongyang on 19 October, and paying no heed to warnings of direct Chinese intervention should he enter North Korea, he began to encounter Chinese troops. The Chinese were backed by the Soviet Union, who helped with equipment and training. At the end of November, the Chinese commander, Lin Piao, launched eighteen divisions on an offensive which drove the UN forces back in disarray. Again ground-attack aircraft were used in an attempt to slow down

MiG 15

Country of origin: USSR.
Role: Interceptor.
Wing span: 10.1 m (33.13 ft).
Length: 11.1 m (36.42 ft).
Max loaded weight: 6,465 kg (14,350 lb).
Engine: 1 × RD45 centrifugal flow turbojet, 2,740 kg (6,040 lb) thrust.
Max speed: 1,070 km/h (665 mph).
Max altitude: 15,550 m (51,000 ft).
Military load: 2 × 23 mm and 1 × 37 mm cannon in pack in lower nose. Attachment points for rockets, bombs, or 2 × 600 1 (159 US gal) drop tanks.

the enemy advance and gain time for an effective defence to be organised. The air power of the UN, demonstrated by sorties of B-29 Superfortress and B-26 bombers, forced the Chinese to adopt the tactic of moving at night, and this was countered by B-26 raids illuminated by flares. Chinese losses were heavy, and by the end of February 1952, their march down the peninsula had been halted 64 km (40 mi) south of Seoul.

In the air, the Chinese presence was most significantly felt by the appearance of the new Russian-built MiG-15, their first truly modern jet fighter. In the first all-jet combats in history, American F-80 Shooting Stars and Royal Australian Air Force (RAAF) Meteors were outclassed by the excellent performance of this nimble swept-wing fighter. The Americans were obliged to deploy their best new aircraft to counter the threat to their operations, and the North American F-86 Sabre soon showed that the MiGs could be beaten.

The superior performance at altitude of the MiG was overcome by the better training and greater aggression of the US pilots, and the

South Africa also took part in Korean War. Its pilots flew the North American F-86 Sabre with great success. These fighters, completed with extended range tanks, were used by No 2 Squadron, the Cheetahs.

Sabre's excellent gun sight system. However, as they were based in Manchuria, across the Yalu river, the MiGs' airfields were safe from attack, as the US pilots were forbidden to cross into China even in hot pursuit, and the MiGs could break off the fight at will by diving to cover across the river.

Nevertheless, throughout the war, the Sabre pilots achieved a kill ratio of better than ten to one in their favour.

The range of both these aircraft was limited: the MiGs were unable to give direct support to their troops, and the Sabres were unable to operate effectively in 'MiG Alley', as the southern airfields were in Chinese hands. To prevent new fields being built in the north, there were bombing raids by the B-29, but they suffered unacceptably heavy losses when the escorting F-80s and F-84 Thunderjets could not offer sufficient protection. The bombing was only resumed when new fields in the south allowed Sabres to fly top-cover. Hundreds of aircraft at a time would then meet at high altitude in fierce dogfights, and this pattern continued, flaring up in spring 1952 when the USAF firmly stamped their authority on the enemy when 83 MiGs were downed for the loss of only nine American aircraft.

In close-support actions, light bombers and fighter-bombers encountered increasingly effective anti-aircraft fire; over 250 pilots had reason to be thankful for the rescue helicopters which snatched them back to safety when forced down behind enemy lines. These helicopters had started on what was to become a massive involvement in US tactical operations. They were used increasingly for rescue, casualty evacuation, observation, and troop movement. This new resource was the biggest 'find' of the Korean war.

On the ground, for the last eighteen months of the war the UN forces were limited to containing the enemy north of the 38th Parallel, and inflicting damage on them in an attempt to bring about fruitful peace negotiations. Extensive destruction of the North Korean infrastructure of roads, dams, rail-

ways, and industry from bombing appeared to make little impression on their will to fight, or on the army in the field. The US in particular should have been learning that against an unsophisticated enemy air power may be better directed at the support of one's own ground forces than on strategic attempts to bomb such a nation into surrender.

Although MacArthur had been replaced in 1951, following disagreements over his desire to attack facilities and plant inside China itself, and so escalating the war, it was not until it had been made known to the Chinese that the United States was prepared to use weapons 'never yet tried in Korea' (thermonuclear weapons), and that the war would be extended into China itself, that the North Koreans signed an armistice on 27 July 1953.

Break-up of empires

Apart from the large-scale fighting in Korea, there were several other places over the globe where military aviation was called upon in the 1950s. In Malaya communist guerrillas, originally from an anti-Japanese force started by the British, began operating among the Chinese farmers of the area. Their aim was to take power from the British, who administered the region; they were well armed, and struck at public and military buildings and rubber plantations from bases in the jungle, as well as killing 'pro-British' civilians.

As part of a highly co-ordinated campaign with the ground forces, the RAF and RAAF began a security operation in 1948. In the 12 years that followed, many methods of guerrilla suppression were tried. Aerial reconnaissance above the dense jungle proved difficult, but where concentrations of guerrillas, or their camps, were located they were bombed and strafed by a force consisting of the de Havilland Hornet, Brigand, Avro Lincoln, DH Vampire and Mosquito, among several other types. Many thousands of tonnes of bombs were dropped, but in the difficult terrain it was almost impossible to assess the results. The most effective action against the guerrillas proved to be the use of small long-range patrols of ground troops who pursued and ambushed the enemy. This was done with the close support of strike aircraft and the use of light aircraft, such as Austers and Pioneers as well as helicopters for air supply and casualty evacuation.

Meanwhile, a 'hearts and minds' campaign among the local Chinese populace starved the guerrillas of support, food and shelter. Self-government brought about the end of the conflict, but by then the guerrillas had lost the initiative anyway.

Similar tactics also proved successful against Indonesian-backed insurgents during later confrontations in Borneo, and in Kenya

First flown in 1941, the Gloster Meteor *(left)* was the standard British fighter from 1944 until the mid-1950s, operating around the world, including fighter-bomber operations against Communist bandits in Malaya and later as a night-fighter. France developed the twin-engined Sud Vautour *(right)* for ground-attack and fighter bomber tasks, but the aircraft was used with great effect in the Six Day War by the Israeli Air Force. This is the Vautor IIa version.

Providing support to the ground troops in Malaya, the Royal Navy's helicopter forces, including this Westland Wessex from HMS Albion, developed jungle flying skills which were later to be adopted by the American forces in Vietnam.

during the Mau-Mau emergency. The Mau-Mau terrorists differed in that they lacked the armaments and training of the Malayan guerrillas, but their methods were equally brutal. RAF Harvards from Southern Rhodesia helped the ground forces, with bombs and machine guns, and at times assistance from regular service units was also available. During the four years it existed, the Mau-Mau lost 14,000 members.

After the surrender of Japan, the French were not welcomed back to Indo-China. Leading the opposition to any sort of union with the former colonial power was the Vietminh (VM). It was in a very strong position in North Vietnam, drawing material and ideological support from the communist

Chinese, and having considerable experience of struggle against the Japanese. Fighting between the VM and the French started in earnest in 1949, with the accession to power in China of Mao Tse-tung. Until the summer of 1951 it was conventional warfare, with the Vietminh driving the French back until halted in the lower coastal region. The Vietminh were defeated there, and withdrew to the Highlands to rethink their methods. Reverting to now classical guerrilla tactics they slowly ate away at the ground the French had gained.

France's air support consisted mainly of aircraft dating from the Second World War and, as in Malaya, there were attempts to bomb the guerrillas out of their home bases. But unlike as in Malaya, the ground forces were unsuccessful in wearing down the enemy. Large areas remained wholly in Vietminh hands, and no amount of modern air weaponry alone could dislodge them. Building up again to regular army status, the communists surrounded the French garrison and

airfield at Dien Bien Phu late in 1953. A large air-supply effort was needed to keep the position safe, but the French lacked the transport aircraft, and the Vietminh were well supported with anti-aircraft guns and artillery in the surrounding hills. US aid was not forthcoming and in the following May the Vietminh triumphed, signalling the end of French possessions in Indo-China.

In Algeria, matters went better for the French, with both ground and air forces put to effect after the hard-learnt lessons of the earlier defeat. The lighter, more flexible Harvard and Trojan aircraft, as well as helicopters were used against the Algerian Army of Liberation. With closer co-operation, more surgical, dedicated operations in a country that offered fewer natural advantages to a guerrilla army, prevented the enemy from reaching the stage where open warfare was viable. By the end of the decade, after six years struggle, the French had virtually won the military struggle, but the financial cost was too high and internal dissent in France, particularly in the person of General Charles de Gaulle, made Algerian independence inevitable.

The Suez operation could not be called an anti-terrorist campaign, a colonial war, or a 'police action'; it is interesting to note as an example of a combined operation amphibious assault. President Nasser's decision to nationalise the Suez Canal brought forth the intervention of Britain and France. In concert with the Israeli forces, who moved across into Sinai, British and French aircraft simply knocked out the Egyptian Air Force in the space of 48 hours. Using carrier-launched Sea Hawk, Sea Venom and Wyvern aircraft, supported by land-based Meteor, Venom, Mystère, Hunter and Thunderstreak strike aircraft annihilated the Egyptian aircraft on the ground, while Canberra and the new Vickers Valiant bombed from altitude. Egypt lost nearly all of its 170 Russian-built aircraft and was powerless to stop the seaborne invasion or the landing of troops by parachute and helicopter. The episode ended in humiliation for the invaders, when political pressure from the United States forced them to withdraw before the intended goals had been achieved, but it proved carrier air power and helicopter assaults as 'limited war' techniques.

Marcel Dassault (1900-1986)

A Frenchman who started life with the name Bloch and under which name his company was responsible for a number of light bomber designs for the French Air Force (1930s and early 1940s). 1940/45: the new family name came from

Resistance activities, under title d'Assault, and was adopted in 1946. From 1950 the company was reformed to build performance jets for the French military and export buyers, including the Mystère and Mirage series, many of which continue in international service.

Right: One of the more unusual post-war strategic bomber designs to come out of America was the Convair B-36J, powered by four turbojets and six 'pusher' piston engines in the wing trailing edges. The aircraft was replaced by the Boeing B-47.

The Acceleration of technology

From 1920 to the mid-1930s, the greatest innovations in aviation tended to be initiated by commercial pressure. It was not until later that, with war seeming likely, the emphasis switched to the military field. From 1945 to the present day, war has been a continuous threat, and the arms race's need to gain advantage over the potential enemy has spurred and funded the rapid technological advancement of military air power.

With the experience gained from the British and American jets designed during the war, and the mass of information from the German jet programme, a new generation of machines emerged, vastly improved in performance and fitness for service. Engine power was doubled, and interceptors such as the North American F-86 Sabre and Saab J29 had swept wings to delay the onset of control difficulties encountered near the speed of sound. Parallel development was taking place in the USSR with the MiG 15. From the USA came the remarkable B-47 bomber, a swept-wing design which incorporated many new features such as a brake parachute, fuel tanks

in the fuselage and tandem landing gear later seen on the massive B-52.

The progress continued rapidly into the 1950s. Ejector seats became standard, radar gunsights, thrust reversers, afterburners, delta wings and tailless designs, and air-to-air guided missiles were all introduced to service. After the subsonic aircraft came the first of the supersonic types, the F-100, successor to the Sabre. Within a few years the increase in available engine power to over 5,000 kg (11,000 lb) thrust, and the better understanding of high-speed aerodynamics led to fighters such as the F-101 Voodoo, Chance-Vought Crusader, and Saab Draken, all capable of over 1,600 km/h (1,000 mph).

By the end of the 1950s, many of the designs still in front-line service in the mid-1980s had already flown, and work on the advanced electronic systems that would keep them there was beginning to show results. It

Used as a fighter by the USAF, Air National Guard, Nationalist China and the Canadian Armed Forces, the McDonnell F-101 Voodoo first flew in 1954. It had a top speed of Mach 1.85.

was at this time that the policy-makers in Britain and the United States took stock of the new guided missiles and radar defence systems. They foresaw that the rocket- and ramjet-powered missiles offered a combination of cheapness and efficiency that would render the manned interceptor unnecessary in the future.

Similarly, the powerful and high-flying bombers such as the B-52, Vulcan, Victor, and even the astonishing B-58 Hustler that was soon to enter service, would be unable to live in the skies in the presence of such weapons, and their payloads could be delivered with less risk of interception by ballistic missiles, launched from the safety of blast-proof silos or from deep under the ocean.

But electronic countermeasures find it more difficult to jam the combination of eyeballs and cannon than guided missiles, and bombers can fly underneath a defensive radar net. The designs of the 1950s were not to lead to a dead end.

First of the Century Series fighters, the F-100 Super Sabre was flown by the US Air Force and NATO forces until the late 1970s. This is the F-100F two-seat operational and continuation trainer.

The classic British post-war fighter was the Hawker Hunter which first entered service in 1954 and is still used today by several air forces, having fought in the Six Day War.

3

MILITARY AVIATION

THE JET AGE

From the development of the early jets, military aviation has been on the leading edge of technology. Today's Mach 3 aircraft, armed with guided missiles and stand-off weapons, are deadly, highly effective, and tremendously expensive. However, military aviation is not all fast jets. In recent years, famine relief operations and sea rescues have become an increasingly important secondary role for the military. There has still been a place, too, for piston-engined aeroplanes and, increasingly in the last two decades, for the helicopter. In some countries, military aircraft in peace time are seen as an extension of the civilian power. There is no doubt that the military aircraft is here to stay – it cannot be replaced by the missile as was predicted just three decades ago.

The past quarter of a century has been one of mixed fortunes for military aviation. The post-war years promised so much, began to deliver the goods, were beset by governmental prevarication over strategy and cost, yet still presented an amazing cross section of military aircraft. The key, perhaps, has been the continual input from the only sure arbiter in the effectiveness of military aviation – operational aerial combat.

The period has seen the rise and fall of the Mach 2+ bomber, then its partial re-emergence; the fallacy of missile-armed fighters which had to have guns designed back in; the durability of aged designs, which are still in production and selling; the relearning of lessons of air combat that should never have been forgotten.

Design and technology have been at the forefront of military aviation at all times, yet at the whim of politicians, strategies have changed the operational requirements. This has led to cancellations and the adaption of existing aircraft into roles never initially conceived for them. The success rate has been higher than perhaps the probabilities would have led one to expect. Necessity, after all, is the mother of invention.

The basic driving force behind military aviation has been the threat and counter-threat of the Soviet Union and the United States. Between 1960 and 1980, the Soviet Union's technological inferiority was made up by quantitative superiority to the United States. Now its technology is on a par, yet quantitative superiority is still with the USSR. This will be a thread which continues through the development of the modern military aircraft.

The classic fighter of the post-1960 era is the McDonnell Douglas F-4 Phantom. Among the export customers have been the UK RAF and RN, receiving the F-4K (illustrated) and F-4M variants. This Phantom is seen in the colours of 43 Squadron, RAF.

The rules of air combat are as true today as they were in the early days of 1915, and this must be borne in mind when tracing the development of modern combat aircraft.

The fighter and friends

In basic concept, the fighter is a simple aircraft: its job is to intercept and destroy enemy aircraft. Like many simple things, however, it has been refined and developed. By adding weapons under the wing, the fighter-bomber was created, and by using guns and air-to-ground rockets against troops and installations on the ground, the ground-attack fighter evolved. Later the fighter-bomber went on to become a strike or attack aircraft.

The pure fighter has had, over the past 25 years, two major tasks – the traditional air defence role (as interceptor fighter) and initially for offensive operations; in the combined land-air incursion, the concept of air superiority arose. Air defence, as it is known today, is a multi-layered military science of which interceptors are but a part. Integrated within a complex network are ground radars, anti-aircraft artillery (AAA), surface-to-air missiles (SAMs), operations centres, communication systems and the fighter itself. In recent years, the concept of airborne early warning has taken an increasing slice of importance as the perceived threat has changed.

In 1960, Western Europe was four minutes away from nuclear attack from Soviet missiles. Fighters tended towards the interceptor type, such as the Lightning, which could be deployed against any Soviet bomber attempting to complete the nuclear destruction. Today, while still possible, the threat of nuclear missiles being indiscriminately exchanged has faded. Air attack by fighter-bombers as well as strategic bombers delivering lower-yield nuclear, biological or chemical

Above: the Soviet MiG-23 Flogger single-seat all-weather interceptor fighter. It can carry air-to-air missiles as well as being armed with guns. It first appeared in 1967.

(NBC) weapons, as well as conventional high explosive weapons, is a much more realistic threat. While still having a place, quick reaction alert interceptors are no longer the ultimate in air defence.

Air defence has evolved into a more sophisticated art. Along one path, the layered concept is retained, although the fighter element tends to be deployed at longer ranges out from the air defence zone and directed by either ground or airborne radar, with the objective of destroying the enemy as far away from the target as possible. Lower down the scale, the UK for example has adopted a war-role plan for its BAe Hawk trainers, armed with a 30 mm cannon and a pair of AIM-9L Sidewinder AAMs, to defend airfields and certain vital fixed points against any aircraft which penetrate or 'leak' the other elements of the air defence screen.

Air superiority now is more akin to the dog fights over the trenches during the First World War. It means eliminating enemy aircraft in the air over the battlefield, whether over land or sea, thus denying the support of air power in

such conflicts. The new European Fighter Aircraft, now in the early stages of its development by Federal Germany, Italy, Spain and the UK, is such a fighter. Designed primarily for air-to-air combat, it should be able to match anything the Soviet Union can field in the mid-1990s, as well as having a secondary air-to-ground role.

So far as NATO tactical thought is concerned, the change of emphasis of fighters from interception/air defence to air superiority/air defence came about in the early 1970s,

Above: the latest development in fixed-wing aircraft-carrier operation is fighters like the A/F-18 Hornet designed to attack and to achieve air superiority according to current 'flexible response' strategies.

when the 'tripwire' policy of reacting with nuclear weapons to the first wave of Soviet troops who cross into the Federal Republic of Germany was replaced by one of 'flexible response', which matched the level of reaction to the level of threat. Use of nuclear weapons, whether strategic or tactical, is now further down the options open to SACEUR (Supreme Allied Commander Europe).

Whichever bias the fighter takes, four points prevail: it must be fast; it must have the means to detect its target, it must have the means to destroy the target and it must be capable of manoeuvring into a position to fire the first shot. Translated into technology, these points equate to propulsion, avionics, weapons and aerodynamics.

First entering service in 1960, the English Electric Lightning was still in use 25 years later. The Lightning was the first radar-equipped, guided-missile-armed, supersonic fighter to reach squadron service with the RAF's Fighter Command.

Propulsion

In 1960 the afterburning turbojet engine was the ultimate in fighter powerplants. Early gas turbine engines, as the turbojet is more correctly described, had a centrifugal compressor (a fan at the front and rear of the engine to suck in and compress the air on entering the engine and then to compress the burnt gases at the rear), which pushed the air outwards from the centre of the engine. This was later refined to push the air parallel to the axis of the engine, and was thus known as an axial flow compressor. Depending on the degree of air compression required, the number of sets of fans – known as stages – can be varied. This gives a better compression ratio, in the region of 7:1, but is not efficient at low speeds.

A typical example of the axial-flow gas turbine is the General Electric J79, which powers the F-104 Starfighter and F-4 Phantom. The J79-GE-19, which powers the F-4E version of the Phantom, is rated at 8,083 kg (17,820 lb) static thrust (st) with afterburning. Afterburning, or reheat as it is also known, is the process of injecting fuel into the already burnt gases in the exhaust to produce further thrust on ignition by the hot gases. This enables the fighter to reach its highest speeds, but it is very heavy on fuel consumption, and as such is used sparingly. During the Falklands campaign, if the defending Sea Harriers could induce their Mirage opponents to engage reheat, then it was fairly certain that because of fuel capacity, the earlier would not have enough fuel to return to base.

During the early-1970s, the augmented turbofan engine was developed. The turbofan is a gas turbine engine which generates most of

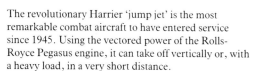

An afterburning turbofan engine, the Rolls-Royce/Turbo Union RB 199 is used to power the Panavia Tornado air defence variant (UK and Saudi air forces) and the interdictor-strike version (UK RAF, Italian and Federal German naval and air forces).

The revolutionary Harrier 'jump jet' is the most remarkable combat aircraft to have entered service since 1945. Using the vectored power of the Rolls-Royce Pegasus engine, it can take off vertically or, with a heavy load, in a very short distance.

its thrust by the use of a large diameter cowled fan (compressor) with only a small part of the thrust coming from the jet core. The term augmented means that fuel is burnt in the air from the fan in a chamber around the core, as well as in the hot gases from the core in the same manner of afterburning in the core gases of a turbojet.

A typical example of an augmented turbofan engine is the Turbo-Union RB 199, which powers the Panavia Tornado multi-role combat aircraft. The RB199 Mk101 has a dry thrust (without afterburning) of 3,570 kg (7,870 lb) st, and 7,137 kg (15,735 lb) st with reheat.

The advantages of the turbofan were twofold: smaller size (and thus weight) with a lower specific fuel consumption (sfc). Taking the J79-GE-17: it has a length of 530 cm (208.7 in), a weight of 1,745 kg (3,847 lb) and an sfc of 1.97. The RB199 Mk101 has a length of 322.5 cm (127 in), a weight of 898 kg (1,980 lb) and an sfc of 1.5. Thus, today, the augmented turbofan has become the major powerplant of fighter aircraft.

However, the exception to this rule is the vectored-thrust engine, exemplified by the Rolls-Royce Pegasus, which powers the Harrier STOVL fighter. The engine gases are deflected through two pairs of nozzles on either side of the fuselage, to produce a four-poster concept of thrust. The four nozzles are capable of being rotated in unison to provide downwards thrust for lift or lateral thrust in the traditional way, or a combination

Called 'the manned missile', the Lockheed F-104 Starfighter entered USAF service in 1958 as a fighter. In Europe, the need for a multi-mission strike and reconnaissance aircraft was fulfilled by the F-104G Starfighter; these are Dutch.

of both, depending on the angle of rotation.

The Pegasus is a turbofan engine – one of the first – introduced in 1959. During the past 26 years it has been improved steadily. The Pegasus 3, which powered the Hawker P1127 V/STOL prototype, was rated at 6,123 kg (13,500 lb) st, while the Pegasus 11 Mk104 of the Sea Harrier FRS1 is rated at 9,742 kg (21,500 lb) st. These engines are 'dry', although a form of afterburning – to increase thrust – known as plenum chamber burning (PCB) is being redeveloped. PCB was originally conceived for the cancelled Hawker P1154 project in 1965. The concept of PCB has, however, been resurrected in recent years as part of a low-key development programme for a supersonic STOVL aircraft. A further development of vectored thrust, using the three-poster or 'lobsterback' concept, is discussed in the section on the future of military aviation technology.

Avionics

In spite of the relatively low weight, the avionics (a contraction of aircraft electronics) can account for over half the unit cost of many types of aircraft. For the fighter, the principal element of the avionics suite is its fire control system (the radar and its associated processors and interfaces) followed closely by its electronic warfare system.

Airborne interception (AI) radars of 25 years ago tended to divide into two areas: small units able to be installed and operated by single-seat interceptors, and larger units, with more capability and range, used in larger two-seat aall-weather fighters and operated by a second crew member. In the RAF, the second crew member is a navigator, in the Fleet Air Arm, the observer and in the USAF, the weapons system operator.

In 1960 a typical radar was the Ferranti AI23 Airpass (AI radar, pilot attack sight system) installed in the English Electric Lightning. Operating in the X-band, it sent

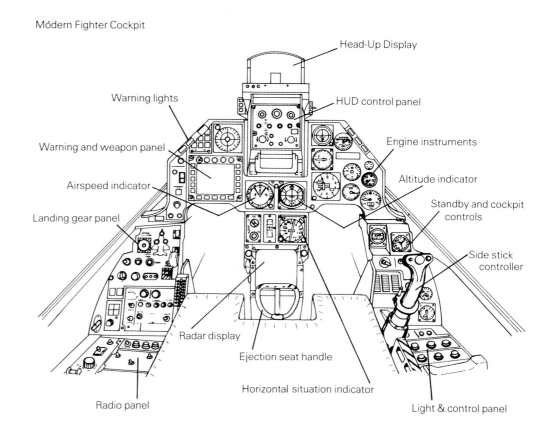

Modern Fighter Cockpit

Head-Up Display
HUD control panel
Warning lights
Engine instruments
Warning and weapon panel
Altitude indicator
Airspeed indicator
Standby and cockpit controls
Landing gear panel
Side stick controller
Radar display
Ejection seat handle
Horizontal situation indicator
Radio panel
Light & control panel

McDONNELL DOUGLAS F-4E PHANTOM II

Country of origin: USA.
Role: Two-seat multi-role fighter.
Wing span: 11.68 m (38.33 ft).
Length: 19.2 m (63.0 ft).
Max T/O weight: 28,055 kg (61,795 lb).
Engines: Two 8,127 kg (17,900 lb) General Electric J79-GE-17 afterburning turbojets.
Max speed: Mach 2.17 at 11,000 m (36,000 ft).
Max altitude: 17,907 m (58,750 ft).
Range (ferry): 2,593 km (1,610 mi).
Military load: 1 × 20 mm M61A1 rotary cannon, four AIM-7E Sparrow AAMs or up to 1,371 kg (3,020 lb) on centreline pylon; plus up to 5,888 kg (12,980 lb) of stores on four underwing pylons.

out four overlapping beams, two in azimuth and two in elevation, and when the target was central between the four beams, the output voltage of the set was zero. Known as a monopulse radar, it had the advantage that it did not require a CRT radar scope in the cockpit. A head-up sight enabled the pilot to observe a computer-driven aiming mark, and when this covered the target, or visual acquisition was made by the pilot, the perfect interception was possible.

The F-4B Phantom (1962) must rank as the typical two-seat fighter of the early period; armed with up to six AIM-7 Sparrow medium-range AAMs, it was fitted with a Westinghouse APQ-72 radar with a 91.4 cm (36 in) antenna dish. As the Phantom developed, so did the radar, the ultimate version being the APQ-120 of the F-4E. The F-4B also featured an infra-red detector system in a fairing just below and behind the radome. More discriminating than radar, it is capable, for example, of identifying the number of aircraft in a formation some 110 km (60 nm) distant.

During the 1960s, radar benefited from the advancing miniaturisation of components, which played a part in reducing weight and volume of radar systems, or in offering

Designed originally as a naval fighter, the Phantom has seen service all over the world and it has a reputation as the best fighter design of the post-war era. Besides the US production line, a large number have been built in Japan.

increased capability. Two other techniques were also instrumental in improving the radars' efficiency – the use of pulse-Doppler and software control. Pulse-Doppler radars use the apparent shift of wave motion frequency if the source is moving with respect to an observer, aided by filters and mixers to show targets of interest standing out from background clutter. The need to discriminate such targets has increased vastly over the past 25 years with the dramatic switch from high-level to low-level penetration techniques used by bomber and strike aircraft. Defending fighters need to 'look-down' for their targets.

Software control of the radar, using a small, fast and reliable digital computer, can alter many characteristics of the radar, such as the pulse repetition frequency, wavelength of signature (how the radar emissions look to an enemy). It does much to enhance the flexibility of radar systems, and can be retro-fitted to earlier systems.

Another development in radar technology is the introduction of planar array antennas, replacing the parabolic or Cassegrain dish reflectors of the 1960s. The planar array is flat as as such, for a given radome size, the scanner can be some 10-12 per cent bigger.

To offer comparisons with the AI23 the APQ-72/120 radars cited as examples of 1960s state-of-the-art, the Hornet's Hughes APG-65 radar and the Tornado F2/3s' AI24 Foxhunter can be considered. The former system offers both air-to-air and air-to-ground capability as befits a multi-role aircraft, with a range of about 185 km (100 nm) and for air-to-air use it offers range-while-scan and track-while-scan facilities, enabling 10 targets to be tracked simultaneously, while displaying eight of them to the pilot. Doppler-beam sharpening in the air-to-ground mapping mode gives enhanced resolution, of great value in launching conventional and precision-guided ground attack munitions.

The GEC Avionics Foxhunter system has a range of about 180 km (95 nm). It also has track-while-scan with multiple target capability to control its armament systems of two AIM-9L Sidewinder and four Sky Flash AAMs, plus an internal 27 mm cannon.

Weapons

Captain Louis Strange, a pioneer First World War fighter pilot, wrote '. . . every man who goes into the air in a fighting machine is a gun layer – first and last . . .' His words have stood the test of time and air combat. However, during the late 1950s and early 1960s, this truism was almost lost as a result of an almost blind faith in the abilities of both air-to-air and surface-to-air missiles. American, British and Soviet designers all produced interceptor fighters in this period without an internal gun armament. The F-102, F-4B/C/D Phantom, the Lightning F3/6, the Tu-28P and Yak-28P serve as examples.

Internal guns of 20 mm and 30 mm calibre were standard fighter armament for Western aircraft up to the late 1950s; 23 mm was preferred by the Soviet Union, and remains so to this day. Then, for a short time, some fighters were actually designed without an internal gun; the air-to-air missile (AAM) was

For 25 years, the skies of North America were defended by the Convair F-102A Delta Dagger which carried its missiles in an internal weapons bay. After leaving USAF Air Defense Command service in 1973, the aircraft served in Greece and Turkey.

DASSAULT-BREGUET MIRAGE IIIE

Country of origin: France.
Role: Single-seat fighter-bomber.
Wing span: 8.22 m (26.96 ft).
Length: 15.03 m (49.29 ft).
Max T/O weight: 13,500 kg (29,760 lb).
Engines: One 6,200 kg (13,670 lb) SNECMA Atar-9C afterburning turbojet, with provision for one 1,500 kg (3,307 lb) SEPR 844 rocket motor.
Max speed: Mach 2.2 at 12,000 m (39,375 ft).
Max altitude: 17,000 m (55,775 ft).
Range: 2,400 km (1,490 mi).
Military load: 2 × 30 mm DEFA 552 A cannon; 1 × Matra R530/Super R530 AAM or 2 × 454 kg (1,000 lb) bombs (centreline); plus four pylons (underwing), each capable of carrying 1 × 454 kg (1,000 lb) bomb, JL-100 rocket pod, Matra R550 Magic I/II or AIM-9 Sidewinder AAM – 4,000 kg (8,818 lb) of stores.

thought all-powerful.

It was, perhaps, the Indo-Pakistani War (1965) which precipitated the return of the internal gun. This was followed by the Israeli experience in 1967, 1970 and 1973. America learnt the hard way – in air combat over Vietnam. There the main escort fighters for strike bombers were the F-8 Crusader (four 20 mm) colt Mk12 cannon plus two AIM-9 Sidewinder AAMs) and the all-missile F-4 Phantoms. When the latter were met by North Vietnamese MiG fighters, all armed with guns and missiles, the missing American ingredient in a dog-fight situation was to prove nearly disastrous. Ironically, one of the principal strike aircraft, the F-105 Thunderchief, did have an internal 20 mm cannon and at one stage the aircraft was almost put back into production. McDonnell Douglas rapidly developed the Phantom F-4E, armed with a 20 mm Vulcan cannon. The RAF had done away with two internal 30 mm cannon after the Lightning F2, but fitted the guns in the F6 version, relocated in the forward part of an enlarged belly fuel tank.

Of course, not every fighter manufacturer was lured by the apparent 'invincibility' of the air-to-air missile, and many kept the gun as standard. This was particularly so with the Dassault Mirage series of fighters. From the Mirage IIIC (1960) to the Mirage 2000 (1978),

The latest generation of European combat aircraft includes the Tornado, built jointly by Germany, Italy and the UK. This is the RAF Air Defence variant, equipped with Foxhunter radar and armed with Skyflash air-to-air missiles.

Leon 'Lee' F. Begin Junior (1924-1984)

Educated at Pasadena City College (California). 1942: joined the Northrop Corporation, where he remained, except for US Navy Service (1944-46). Worked on

P-61 Black Widow, B-49 Flying Wing and F-89 Scorpion. Key member of design team for T-38/F-5/F-20 Freedom Fighter and Tigershark. 1983: became Vice-President for Advanced Programmes and Planning.

Above: designed around the GAU-8/A 30mm cannon in the forward fuselage, the Fairchild A-10A Thunderbolt II is especially adapted for a close air-support role, particularly in Europe where the aircraft work closely with helicopters.

twin 30mm DEFA cannon were *de rigueur*. Perhaps one reason was the need to offer dual-role capability, particularly with a view to the export market.

In Europe, the tri-national Tornado is armed with a new 27mm IWKA-Mauser cannon, which has a dual air-air and air-ground role. The UK RAF's F2 and F3 air defence versions of the Tornado have one such gun, while the GR1 interdictor strike version has two. However, for the foreseeable future, 20mm and 30mm remain the preferred calibres for air defence fighters.

Air-to-air missiles

The air-to-air missile (AAM) began life during the Second World War, and by 1960 technology had developed to a state whereby the AAM could guide itself on to its target. Two principal methods of guidance have found favour with manufacturers and users alike – infra-red (IR) homing and semi-active radar homing (SARH). Despite the impressions given previously, the missile is a potent weapon, especially when used in conjunction with a fire control radar. Due to the two different methods of homing employed, the AAM is either a short-range, dogfight weapon with ranges up to 18.5km (10nm) using

Ultramodern Fighter Cockpit (based on Saab JAS 39)

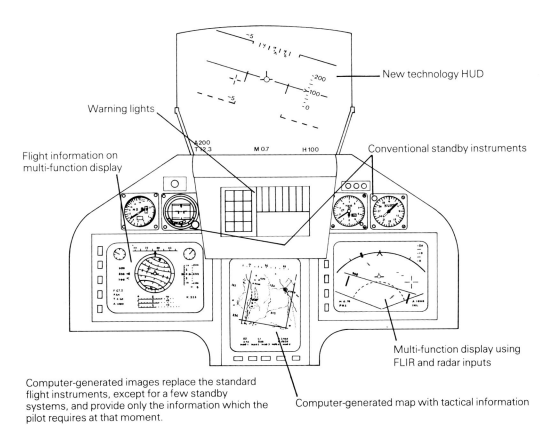

New technology HUD

Warning lights

Conventional standby instruments

Flight information on multi-function display

Multi-function display using FLIR and radar inputs

Computer-generated images replace the standard flight instruments, except for a few standby systems, and provide only the information which the pilot requires at that moment.

Computer-generated map with tactical information

IR-homing; or medium to long range, typically 44km (24nm) to 200km (108nm) using SARH. The latter bracket of medium-range missiles now tend to be known as 'Beyond Visual Range' types.

The advantage of the IR-homing missile is that it can be used without the need for a fire control radar and, being small and relatively inexpensive, it can be carried by most types of aircraft. Indeed, during the Falklands conflict, RAF Nimrods were hurriedly converted to carry four AIM-9L Sidewinders for self-defence.

The IR-homer seeks the heat sources generated by the target aircraft – mainly the engine and its exhaust. Early missiles of this type tended to be misled, on occasions, by the Sun or its reflections from glass or water on the ground. The traditional fighter attack came from astern of the target aircraft, with kill probabilities of up to 70 per cent, but as the technology has advanced, the sensitivity of the IR-seeker is such that attacks from all aspects (including head-on) of the target aircraft have become possible.

Typical of this genre is the AIM-9 Sidewinder series of AAMs, which was conceived as a simple and inexpensive fire-and-forget missile by the US Naval Weapons Center. In production since 1951, it is in service with over 30 nations. Some 12 major variants of Sidewinder have been produced with varying degrees of improvement as seeker, electronics, warhead and propulsion technology has developed. The most recent versions, from the AIM-9H model, have all-aspect capability. During the Falklands conflict, the AIM-9L was officially credited with 16 kills, although other sources suggest 18 kills. The current

Developed for the Soviet Air Force and also exported, in the case of the one illustrated here to India, the infra-red AA-2 Atoll air-to-air missile is very similar to the Sidewinder. It is seen here on a MiG-21PFM interceptor.

Top: in an effort to provide cost-effective air defence for British airfields and to develop the idea of a single-seat variant, many of the Hawk T1 trainers of the Royal Air Force have been made compatible with the AIM-9 Sidewinder missile.

Above: aiming for an export market, the Northrop F-20 Tigershark is a development of the earlier F-5 Freedom Fighter and F-5E Tiger, used in various conflicts, including Vietnam. This Tigershark is shown launching a Sparrow medium-range AAM.

version is the AIM-9M, which is built in Europe for NATO forces, and the end of Sidewinder production is not yet in sight.

America does not possess the monopoly on this type of weapon, as the Soviet AA-2 (K-13) Atoll has demonstrated; it is possibly based on reverse technology which has been progressively developed. The Advanced Atoll is known to exist in both infra-red and semi-active radar homing versions.

France, too, has met with success in this market with the Matra R550 Magic AAM. Designed to a higher specification than the

original Sidewinder, including a launch bracket of up to 18,000m (59,000ft) within a forward 140° hemisphere, it has met with acceptance in the world market with exports to 14 countries.

Although basically similar to the AIM-9 in appearance, and indeed interchangeable with it, Magic is easily identified by the double set of forward fins: the first (leading) set being fixed winglets, the second set, movable control surfaces. While not evolving in as many variants as Sidewinder, the Magic 2 is now available, with all-aspect capability and advanced electronics, to be compatible with modern dogfight aircraft.

For the future, a European consortium, led by Bödenseewerk Geratetechnik of Germany, is developing an Advanced Short Range AAM (ASRAAM) with the designation AIM-132. While development progresses, whether it will ever see operational service is open to question due, simply, to the longevity of Sidewinder.

Beyond the dogfight missile comes the medium-range type requiring its launch aircraft to have a radar able to lock on to the hostile target, thus illuminating it for the missile. A sensitive radar receiver in the missile's nose then locks on to the radiation reflected back or scattered from the target. This enables the missile to 'home in' on its target until a hit is made or a proximity fuse is

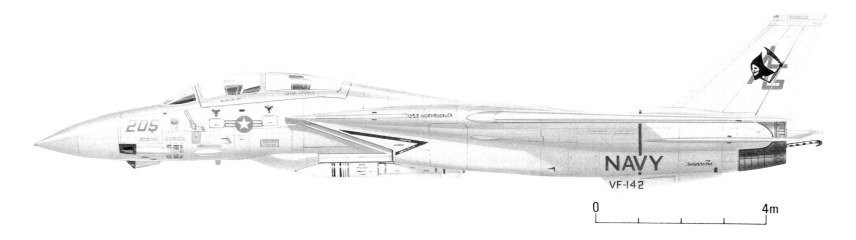

triggered as the missile passes its target.

A typical medium-range AAM is the AIM-7 Sparrow series, which began life in 1946 as Project Hot Shot and has been developed in three distinct versions – the Sparrow I (AIM-7A), a beam-riding missile, the AIM-7B Sparrow II with active radar homing for the F5D-1 Skylancer and Avro Canada Arrow (both applications of which were cancelled along with the missile), and the Sparrow III, from which came the AIM-7C/D/E/F/M versions, all with SARH guidance. The AIM-7E2 also provided the start point for a British development known as Sky Flash with a new homing head, and the Italian Aspide. The AIM-7F/M versions, Aspide and Sky Flash remain in production today, but will eventually give way to the AIM-120 AMRAAM.

Other medium-range missiles in this bracket include the French Matra 530 and its development the Super 530, the Soviet AA-5 Ash and AA-6 Acrid and the American AIM-54 Phoenix, probably the longest-range AAM in service in the West. Its only application to date is on the US Navy's F-14 Tomcat, coupled with the AWG-9 fire control radar. Both the Phoenix and AWG-9 have capability out to 185 km (100 nm), which makes them the ultimate in AAMs in service today.

Aerodynamics

In the 1950s 'design for speed' was the maxim and this accounts for the Mach 2 designs that entered service in the early 1960s. By the late 1960s, experience over the Middle East and Vietnam had suggested that although Mach 2 was sometimes called for by way of locating and positioning, air combat then was carried out in the speed range 280-450 kt. The higher the speed, the greater the radius of turn is the rule, so the need for a slower speed to give a better rate of turn was appreciated and designers began to look for a better balance.

High sweep in the wings versus delta configuration was the battle of the 1960s, with each type giving superior performance in differing areas. Next came the variable-geometry (or swing) wing, which enables the

Armed with the new British Aerospace Sky Flash medium-range air-to-air missile, the Tornado ADV *(above)* is a powerful interceptor with good loiter time. In Sweden, the Saab Viggen *(below)* has been developed for a similar role, as well as armed maritime reconnaissance, ground-attack and strike missions. The tactical camouflage for the aircraft helps it operate from small air-strips or the sides of major roads. The Viggen has good short take-off and landing (STOL) performance.

Part of the US Fleet's air defence is the Grumman F-14A Tomcat *(above)*, armed with the amazing Phoenix air-to-air missile which can engage targets at ranges up to 314 km (170 nm). This particular aircraft served with VF-142 in 1979.

pros and cons to be evened out and maximises advantages in many areas. Today, cambered fixed wings are in vogue. These alter the aerofoil section by means of leading and trailing edge flaps, controlled by a computer, allowing the wing profile to be altered for maximum performance at take-off, cruise, supersonic flight, high-g manoeuvres and landing. For the future, the mission-adaptive wing is being developed, which can change profile and planform by means of various hinges and pivots, again controlled by computer.

SAAB (Sweden) pioneered the use of canard foreplanes on the AJ37 Viggen. These negate many of the control disadvantages of the tailless delta format, principally the small effective moment-arm provided by the control surfaces (elevons in this case) for take-off. On an aircraft such as the Mirage III series, these have to be large, which adds weight to the airframe. By using a flapped canard foreplane, as on the Viggen, SAAB Gripen and EFA (European Fighter Aircraft), lift is derived not only from the main wing and its elevons but also from the flapped canard foreplane. They also have the added advantage of causing

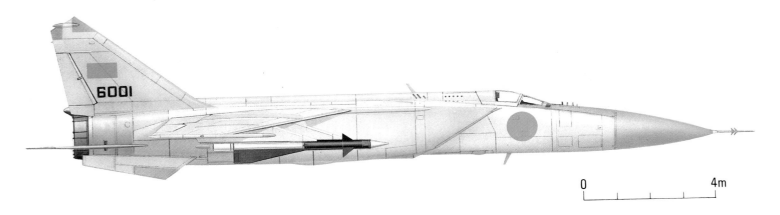

0   4m

MIKOYAN GUREVICH MiG-25 FOXBAT-A

Country of origin: USSR.
Role: Single-seat interceptor fighter.
Wing span: 13.95 m (45.75 ft).
Length: 23.82 m (78.75 ft).
Max T/O weight: 36,200 kg (79,800 lb).
Engines: Two 12,251 kg (27,010 lb) Tumansky
R-31 (R-266) afterburning turbojets.
Max speed: Mach 2.83 at 11,000 m (36,000 ft).
Max altitude: 24,400 m (80,000 ft).
Range: 2,900 km (1,800 mi).
Military load: 4 × AAMs (underwing): typically
4 × AA-6 Acrids, or 2 × AA-7 Apex and 2 × AA-
8 Aphids.

Called Foxbat-A by NATO, this MiG-25 interceptor
fighter is shown in the markings of the Libyan Arab Air
Force and carries the AA-6 Acrid infra-red homing
missile. The aircraft is in widespread service with
Warsaw Pact and client states.

It may be fairly stated that, in 1960, the
fighter was in its final phase, but the missile
did not prove to be the panacea for all
problems, and its costs kept on rising. The
operational experience gained first, second
and third hand in the late 1960s and early
1970s caused second thoughts which have
been confirmed in the late 1970s and early
1980s. Realism has returned and the fighter's
place in the air force over the next 25 years is
no longer in doubt.

The fall and rise of the bomber

In 1960 the strategic bomber was approaching
its zenith. Post-war developments had taken it
to the position of the major component part in
the offensive armoury of the nuclear powers,
being the means of delivering 'The Bomb',
and thus the linchpin of nuclear deterrence. Its
decline in overall importance is matched by
the rise of the intercontinental ballistic missile
(ICBM) with a nuclear warhead. Today, for
the nations who can afford nuclear delivery –
the Super Powers – they form part of a
combined-arms formation for delivering nuclear warheads. Apart from the manned
strategic bomber, they comprise the ICBM,
the submarine-launched ballistic missile
(SLBM) and the cruise missile. This latter
weapons system is a recent addition and can be
one of three types – the air-launched cruise
missile (ALCM), the ground-launched cruise
missile (GLCM) and the naval variety, which
can be ship-launched or submarine-launched
(SLCM).

The USSR, United States, China, France
and the United Kingdom had strategic bombers in service by 1960. Of those five
countries, only two can now be said to be
continuing to produce such bombers – the
Soviet Union with the Tu-22M (Tu-26)
Backfire and Tupolev Blackjack, and the
United States with the Rockwell B-1B. France
is giving up its force of Mirage IVs in favour of
a smaller aircraft, the Mirage 2000N, in a more

The return of the manned bomber is marked with the
continued development, under President Reagan, of
the Rockwell B-1. The aircraft will feature high-
technology equipment, including electronic warfare
systems of a 'stealth' type.

limited role; while the UK's V-force has gone
and only Tornado remains, but not in the
strategic context. Exactly what China is doing
is uncertain, although their version of the
TU-16 Badger, the Xian H-6, a product of the
late 1950s, remains in slow production and
service.

For the main part of this quarter-century
under consideration, strategic bombers in
service were designed during the late 1940s
and early 1950s, and although their role may

ROCKWELL INTERNATIONAL B-1B

Country of origin: USA.
Role: Four-man strategic bomber.
Wing span: 41.67 m (136.71 ft) 15° sweep; 25.6 m
(84.0 ft) at max 60° sweep.
Length: 45.78 m (150.21 ft) incl probe.
Max T/O weight: 216,800 kg (477,954 lb).
Engines: Four 13,600 kg (30,000 lb) General
Electric F101-GE-102 afterburning turbofans.
Max speed: Mach 1.2 above 7,620 m (25,000 ft).
Max altitude: 14,600 m (47,900 ft).
Range: 12,000 km (7,460 mi).
Military load: Up to 84 Mk82 227 kg (500 lb)
bombs internally, plus 44 × Mk82 bombs on
multiple-ejector racks underwing; or 24 × B-61/B-
83 nuclear weapons internally, plus 14 externally;
or 8 × AGM-86B air-launched cruise missiles
internally, plus 14 externally.

vortices, which add energy to the boundary
layer air on the outer wings, giving more
effective control.

Such vortices can also be provided by
wing-mounted vortex generators, the jagged
discontinuation of leading edge plan (known
as dogtooth leading edges), wing-body strakes
or leading edge extensions to the wings.
Examples of these have been progressively
applied in fighter designs over the past 25
years.

In 1960 tail control surfaces were usually in
the form of fins with hinged rudders and
tailplanes with hinged elevators. Since then
the fixed and movable surfaces have tended to
become all-moving (or slab) control surfaces.
For high-angle-of-attack régimes, a twin-fin
layout has proved to be preferable, as in-
stanced by the F-14 Tomcat, F/A-18 Hornet,
F-15 Eagle and the MiG-25/31 Foxbat/Fox-
hound, and is found to be just as effective in
other areas of the flight envelope.

For the future, fighters will forgo natural
stability, being designed unstable and hence
more manoeuvrable. While the pilot will
signal his intentions by means of conventional
controls, these will be interpreted by a
computer which controls a fly-by-wire flight
control system. The whole system operates at
speeds impossible for a normal pilot to react to,
as well as offering other advantages. The
whole concept of the control configured
vehicle (CCV), as this type of aircraft is called,
is to achieve instant reaction to control input.

BOEING B-52H STRATOFORTRESS

Country of origin: USA.
Role: Six-man strategic bomber.
Wing span: 56.39 m (185.0 ft).
Length: 49.05 m (160.92 ft).
Max T/O weight: 221,000 kg (488,000 lb).
Engines: Eight 7,720 kg (17,000 lb) Pratt & Whitney TF33-P-3 turbofans.
Max speed: 957 km/h (595 mph).
Max altitude: 16,700 m (55,000 ft).
Range: 9,200 km (5,717 mi).
Military load: Up to 84 Mk82 227 kg (500 lb) bombs internally, plus 24 Mk82 bombs on multiple-ejector racks underwing. Tail-mounted 20 mm ASG-21 rotary cannon, directed by ALQ-153 pulse-Doppler tail-warning radar.

Mainstay of the USAF's Strategic Air Command since the late 1950s, the Boeing B-52 Stratofortress is a remarkable eight-engined bomber. It can carry nuclear, conventional or air-launched cruise missile weapons over 20,000 km (10,800 nm)

have changed, especially as far as the *modus operandi* is concerned, and their equipment and weapons updated, they remained in service far longer than anyone would have expected. Indeed, pilots of the USAF converting onto the venerable B-52 Stratofortress are wont to talk about flying the bomber their fathers flew. With current programmes for updating avionics and weapons on the B-52, there is every likelihood that it will achieve 50 years in service . . . and that is akin to having First World War biplanes leading the air war in Vietnam in 1965.

The first wing of USAF B-58 Hustler Mach-2 bombers entered service in 1960, and, together with early models of the B-52, began replacing the B-47 Stratojet. Production of both types ceased in 1962, and by 1970, the B-58s and early model B-52s had been replaced by the FB-111A, to partner the B-52D/G/H versions, of which the D model has only recently been withdrawn from ser-

vice. The B-52G/Hs will continue to serve in both nuclear and non-nuclear roles for the foreseeable future, with the B-1B in prospect for the nuclear role from 1986. Beyond then, the Northrop Advanced Technology bomber – incorporating every stealth technique which can be applied to an aircraft – should finally see the B-52 retire, but well into the 1990s.

Within our quarter century, America has had four major bombers, plus two attempts to replace the B-52. These were the XB-70 Valkyrie, cancelled in 1964, and the B-1A, cancelled in 1977. The latter reappeared in 1982 in a modified form as the B-1B, and at the same time the Northrop Stealth bomber was authorised.

Examination of USAF bomber development will show the trends, which are applicable worldwide. In 1960 the strategic bomber was the instrument of deterrence and the B-47E was in worldwide service with the

USAF. Of swept-wing configuration, its six J47-GE-25 turbojets, mounted in underwing pods, gave it a speed of 1013 km/h (547 kt) and a range of 5,150 km (3,200 mi). It represented a great leap forward in bomber development aerodynamically, structurally and operationally. It was the first high-altitude (12,344 m/40,500 ft) jet bomber to enter large-scale service.

Its big brother was the Boeing B-52 Stratofortress. Again of swept-wing configuration, its engines were again pod-mounted TF33-P-3 turbofans and gave a speed of 1,039 km/h (561 kt) and a range of 20,120 km (12,500 mi), with a crew which had doubled from the three of the B-47 to six. Most important of all was its ability to be refuelled in flight, which had been pioneered with the B-47, and enabled it to be based in the continental United States, yet able to strike almost anywhere in the world. Its longevity is due to a comprehensive series of updates and improvements over eight production versions. It offered (and achieved) great flexibility and was used in action during the Vietnam War as a conventional 'iron' bomber.

The Convair B-58 Hustler was a delta-winged supersonic aircraft with dash speeds in excess of Mach 2. Powered by four podded J79-GE-5B turbojets, the three-man Hustler could fly 3,862 km (2,400 mi). Unlike the B-47 and B-52, its bomb load was carried in a specialised 'mission pod' under the fuselage. Its place with Strategic Air Command was taken by the smaller, two-man General Dynamics FB-111A.

During the 1960s, the efficiency of long-range surface-to-air missiles led tactical thought to bring the bombers down to low-level penetration. For this, the FB-111A, with its variable-geometry wings and terrain-following radar, was ideal. This change of tactics was one of the reasons for the cancellation of the XB-70 Valkyrie in 1964. Although

The General Dynamics F-111 swing-wing bomber was not too successful in the standard bomber role but has proved very capable in attack. Only operated by the USAF and the Royal Australian Air Force, the F-111 is mainly based in Europe and Australia.

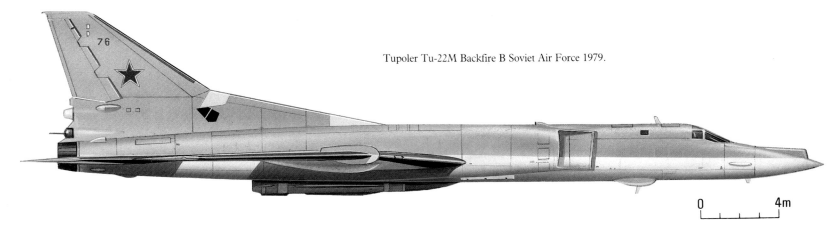

Tupoler Tu-22M Backfire B Soviet Air Force 1979.

0   4m

an advanced aircraft of delta-canard configuration and Mach 3 capability, it became just too expensive – even for America. So the FB-111 was the only boost the on-going B-52 force had during the 1970s. Powered by two TF30-P-7 turbofan engines, it can fly at Mach 1.1 at sea level or Mach 2.2 at 12,190 m (40,000 ft), with a similar range to the B-58. Its weapons are carried under the wings.

A second attempt to produce a B-52 replacement began in the early 1970s with the Rockwell B-1A bomber, the first of four prototypes flying in 1974. With variable-geometry wings and four F101-GE-100 turbofan engines, underslung on the fuselage centre-section in two pods of two either side of the fuselage, it was supersonic at sea level and at 12,190 m (40,000 ft) capable of Mach 2.2.

Cancelled by President Carter in 1977 in favour of cruise missiles, it was revived in modified form in 1982 by President Reagan. The first production B-1B aircraft flew in May 1985, and in 1986, the first SAC squadron achieved initial operating capability (IOC). Only 100 of these bombers, optimised for low-level penetration of enemy territory, are to be built, pending the arrival in the 1990s of the Northrop Stealth bomber.

Of the four types mentioned, all but the FB-111A had a rear-facing 20 mm cannon to provide some measure of self-defence, but the day of the tail gunner in US aircraft is long gone: his place has been taken by a rear-mounted radar system. Radar, avionics and electronic warfare have increasingly made their impact on the bomber, making possible the long life of the B-52.

Having been designed for high-altitude work, it has not altogether been surprising to find that some have been turned out as reconnaissance aircraft, although this later became a field for specialist design. The B-52 has had other applications added to its role, such as maritime reconnaissance and, when equipped with AGM-84 Harpoon anti-ship missiles, maritime strike as well. Delivery of conventional bombs, which took place over Vietnam, has returned to its repertoire and can now be called up in support of ground forces dropping conventional high-explosive bombs. These non-strategic roles will continue after the B-1B has taken over the strategic role.

Soviet trends

Just as the B-58 was entering service in 1960, so was the Soviet equivalent, the Tupolev Tu-22 Blinder. This highly swept-wing, conventional-tailed bomber with engines mounted either side of the fin is capable of Mach 1.4 at 12,190 m (40,000 ft).

Again, like the B-58, it supplemented and eventually replaced earlier types, such as the TU-16 Badger and Myasanchev M-4 Bison in the strategic role. These types have now been relegated to duties as tankers, photographic or maritime reconnaissance aircraft, where their numbers still make them a potent force. The TU-95/142, a swept-wing turboprop-powered aircraft, has also been seen in these secondary roles, but is no longer a strategic system. The Americans now believe that an entirely new version of the Bear, the Bear-H, is now operational with the AS-15 long-range cruise missile. This is the first new production of a strike version of the Bear for 15 years.

The bomber to enter service after the Tu-22 was the Tu-22M Backfire. First reported in 1971, this variable-geometry bomber is larger than the FB-111A and thought now to be 'Euro-strategic' in capability, rather than intercontinental. Capable of speeds up to Mach 1.8/2.0, its range (hi-lo-hi) is thought to be about 3,200 km (1,988 mi). It has a maritime strike role (armed with AS-4 Kitchen anti-ship missiles) as well as conventionally-carried HE or nuclear weapons.

The Soviet answer to the B-1 is the Tupolev Blackjack and is significantly larger. In flight test now, it will probably replace the remaining Bison and Bear-A bombers in service in this role. Of variable-geometry configuration, its unrefuelled combat radius of action is thought to be 7,300 km (4,536 mi) and its maximum speed 2,224 km/h (1,200 kt) or Mach 2.09.

Marshall of the Royal Air Force Lord (Neil) Cameron (b. 1920)

Educated at Perth Academy (Scotland). 1940: joined the Royal Air Force serving in fighter and fighter-bomber squadrons during the Second World War. 1945: directing staff, School of Land/Air Warfare. 1955-56: CO University of

The European solution

In the strategic stakes, only France and the UK were to develop the medium- to long-range bomber. In the UK, by 1960, all three V-bombers – Vickers Valiant, Handley Page Victor and Avro Vulcan – were in service, with the nuclear deterrent in the hands of the latter two aircraft. The Valiant, having been developed as an 'insurance policy' and somewhat faster than the other two types, had been pushed out of the strategic role into conventional medium bomber, photo-recce and airborne tanker duties. This high-wing bomber, powered by four Rolls-Royce Avon 201 turbojets, could reach a speed of 902 km/h (487 kt) at 9,144 m (30,000 ft) and carried no defensive guns. It was in the process of being re-roled in 1964 when discovery of metal fatigue led to the whole fleet being withdrawn from service. Fortunately, by then the Mk2 versions of both the Victor and Vulcan were in service, and Mk1 versions of the former were converted to take over the Valiant's tanker role.

The Victor is a mid-wing monoplane with a tee-tail configuration, powered by four Armstrong Siddeley Sapphire 200 turbojets (B1) and four Rolls-Royce Conway 201 turbofans (B2). The Victor B2 could reach Mach 0.92 at 12,192 m (40,000 ft) and had a combat radius of some 2,776 km (1,725 mi). It could be armed with conventional or nuclear bombs, or carry a Blue Steel stand-off nuclear bomb semi-recessed into the lower fuselage. It entered service in 1961.

Avro's Vulcan was a pure delta-wing bomber powered by four Rolls-Royce Olympus 301 turbojets (B2) capable of Mach 0.94 at 16,764 m (55,000 ft). Like the Victor it was in service in the late 1950s in B1 form, with the B2 coming in by 1961, again armed with Blue Steel as part of the UK's nuclear deterrent.

London Air Squadron. 1964: Staff Officer to Deputy Supreme Commander SHAPE (Paris). 1968-70: Assistant Chief of Defence Staff (Policy). 1972-73: Deputy Commander of RAF Germany. 1976-77: Chief of the Air Staff. 1977-79: Chief of the Defence Staff.

Entering service in the late 1950s, the Vulcan remained in RAF service long enough to complete the longest bombing mission in history, to the Falkland Islands in 1982. First used for nuclear strike, it later carried conventional bombs.

Both Victor and Vulcan adopted low-level penetration techniques from 1965-66, still in the nuclear deterrent role until June 1969, when responsibility was handed over to the Royal Navy's Polaris-armed submarines. Their usefulness did not end here, however, as the Vulcans continued in the tactical and conventional bomber role until their final withdrawal in 1982, after having made their operational debut on the 'Black Buck' raids of the Falklands campaign. Six Vulcans, hastily converted to tankers, remained in service for a further 15 months. The Victors were re-deployed in the strategic reconnaissance role until 1974 and also in their present role as tankers. Their days in this role, however, are

now numbered due to the increased airframe fatigue hours incurred by the demands put on them during the operations of the Falklands campaign of 1982.

France was the only other nation to develop a modern strategic bomber after the Second World War. The Dassault Mirage IV was a scaled-up version of the Mirage III fighter. Between 1964 and 1968, some 62 aircraft were built and continue to serve as part of the French strategic nuclear forces. Powered by a pair of SNECMA Atar 9K turbojets, it is a Mach 2.2 aircraft at 18,288 m (60,000 ft). Although 36 remain in strategic service, 12 serve in the reconnaissance role. Current plans call for 18 Mirage IVP aircraft, equipped with the new nuclear ASMP missile, to remain in service for some years to come.

The smaller successor

The Canberra was Britain's first jet-powered bomber, powered by two Rolls-Royce Avon turbojets. It replaced the Mosquitoes, Lincolns and B-29 Washingtons in the early 1950s. It is an uncomplicated aircraft, of conventional construction. The wings, mid-set in the fuselage, housed each engine in a

nacelle about one-third the way from wing root to tip. This leaves the fuselage free for an internal bomb bay, able to carry 2,722 kg (6,000 lb) of weapons and fly at 917 km/h (495 kt) at 12,192 m (40,000 ft). First flown in 1949, it remains in front-line service with several foreign air forces.

The success of the Canberra was legion, and it was built in the United States as the B-57, modified with tandem seating for the crew, and later adapted for reconnaissance purposes with longer-span wings. It was also built under licence in Australia.

During the early 1960s, however, the bomber as such was phasing out in its tactical role, being replaced by smaller aircraft able to carry similar payloads, including tactical

The Canberra was designed as a bomber but has undertaken many other tasks, including photo-reconnaissance and target facilities as shown here with a Royal Navy version taking off to support ship training.

Left: The sole French bomber aircraft of the post-war era is the Mirage IVA, part of the nation's nuclear deterrent force. The design was produced in 1964-8 and should soon be replaced as nuclear-powered submarines take over the role.

nuclear weapons, and more like fighters in appearance. Known as strike or attack aircraft, the weapons were principally carried on external weapons pylons.

The 'classic' of these types, another development of the early 1950s, which remained in production until February 1979, was the Douglas (later McDonnell Douglas) A-4 Skyhawk – known affectionately as the 'Bantam Bomber'. It was designed to be faster and lighter than the US Navy's attack aircraft of the time. During the late 1960s and early 1970s, this delta-winged but conventionally tailed aircraft secured several export successes, as well as serving widely with the US Navy and Marine Corps.

Almost ahead of its time, it was progressively improved and re-engined, fitted with more effective and comprehensive avionics and weapons, and has been used to effect in action more than once in the Middle East. Refurbished examples of early models in Argentine service carried out many attacks on the British Task Force during the Falklands campaign. In its ultimate production variant, the A-4M Skyhawk II, it was powered by a Pratt & Whitney J52-PW-408 turbojet capable of 1,040 km/h (548 kt) with a 1,814 kg (4,000 lb) bomb load.

The Skyhawk was to give way in the mid-1960s to the Vought A-7 Corsair II, another single-seat attack aircraft, capable of carrying twice the disposable load of an A-4 twice the distance. Derived from the F-8 Crusader fighter it was not dissimilar in appearance, although shorter and fatter. This gave rise to one of the several nicknames applied to the A-7 – the 'short, little, ugly feller' or SLUF for short. Initial production variants were powered by a non-afterburning version of the Pratt & Whitney TF30 turbofan used in the F-111. A-7A Corsairs were in action over the Gulf of Tonkin in 1967.

Like the Skyhawk, the Corsair was re-engined, initially for the USAF (making it the second major US Navy type adopted by that service), with an Allison-built version of the Rolls-Royce Spey turbofan, designated the TF-41. It also introduced an internal M61 20 mm cannon for strafing and self-defence. The more powerful TF-41 engine enabled performance figures of 1,112 km/h (600 kt) at 1,524 m (5,000 ft) to be attained. During operations over South-East Asia, A-7s of both services flew in excess of 100,000 sorties, proving the effectiveness of the concept. The US Navy adopted the TF-41-powered version of the Corsair as the A-7E, as did Greece (as the A-7H). Some early A model Corsairs were refurbished in the late 1970s and early 1980s for use by Portugal as the A-7P.

At the heavier end of the USN's carrier-borne attack force, in service four years ahead

VOUGHT A-7E CORSAIR II

Country of origin: USA.
Role: Single-seat shipboard strike fighter.
Wing span: 11.81 m (38.75 ft).
Length: 14.06 m (46.17 ft).
Max T/O weight: 19,051 kg (42,000 lb).
Engine: One 6,804 kg (15,000 lb) Allison TF41-A-2 (Rolls-Royce RB168-68 Spey) turbofan.
Max speed: Mach 0.92, 1,115 km/h (693 mph).
Max altitude: 13,100 m (42,979 ft).
Range: 740-1,790 km (460-1,112 mi).
Military load: 1 × 20 mm M61A1 Vulcan cannon plus 6,804 kg (15,000 lb) on eight hardpoints; typically 2 × AIM-9 Sidewinder AAMs and 6 × 907 kg (2,000 lb) bombs.

of the A-7 and still in production today, is the Grumman A-6 Intruder. A two-seat low-level attack aircraft, design was begun in the late 1950s, with a first flight in April 1960. The A-6 was powered by two Pratt & Whitney J52-P-8A turbojets, and in its current A-6E version it can reach a maximum speed of 1,043 km/h

For more than 30 years the Douglas A-4 Skyhawk *(top)* has been in front-line military service, first with the US Navy (as illustrated) and later with other states. Today, countries like Malaysia and Singapore depend on the aircraft for national defence. Another excellent naval strike aircraft is the Vought A-7 Corsair II *(above)* which entered service ten years after the A-4. In the Vietnam War the A-7 was supplied to the USAF and afterwards was also sold to Pakistan, Portugal and Greece.

(563 kt) at sea level. Although slower than the A-7, the Intruder's advantage lay in its ability to carry a maximum weapon load of 6,804 kg (15,000 lb) of external stores. The first A-6A Intruders began their combat career over Vietnam in 1965.

The ability to carry a heavy payload led to a specialised electronic counter-measures (ECM) version being developed as well as an air-to-air tanker conversion. Currently, plans are in hand for an advanced version, designated A-6F, which will take advantage of developments in materials for construction (lighter weight composites) and advanced

Grumman have a fine reputation for building naval aircraft in the United States, designs such as the A-6 Intruder. It is a Vietnam veteran, and has seen combat service in Lebanon and Grenada. It is also flown by the US Marine Corps.

avionics. This should see the Intruder remain in service well into the 1990s.

The British response

The British involvement in smaller tactical strike aircraft has centred on the naval strike requirements. In 1960, the standard Fleet Air Arm strike fighter was the Supermarine Scimitar – the first naval aircraft capable of exceeding the speed of sound in a shallow dive. Although able to operate as a fighter with four 20mm cannon and early-mark Sidewinder AAMs, the role was better performed by the de Havilland Sea Vixen, and as radar and surface-to-air missile (SAM) threats increased in effectiveness, the only credible role for strike aircraft was low-level weapons delivery. To do this effectively and give the aircrew a sensible chance of survival called for a terrain-following radar. The Scimitar had no such radar, and thus its days were numbered.

Its successor was already flying in 1960. The Blackburn Buccaneer made its maiden flight in 1958, and was designed to penetrate at

low level under the enemy defences. It was not designed to be supersonic, but rather high subsonic, its two de Havilland Gyron Junior turbojets giving it a speed of some 1,158km/h (625kt) at sea level. The high subsonic performance was achieved by applying the principles of area rule to the design of the Buccaneer.

The first Buccaneer S1s were dogged by a lack of thrust. The two Gyron Juniors only producing 1.596kN (7,100lb) almost jeopardised its operational career, and so before the S1 even entered service in July 1962 a re-engined version powered by two Rolls-Royce Spey turbojets, each rated at 2.495kN (11,100lb) st had flown; the substitution of engines was carried out with the minimum of structural changes. As well as increasing

thrust, the weapon load of the S2 version was increased 100 per cent to 7,257kg (16,000lb); up to 1,814kg (4,000lb) in the internal rotary bomb-bay and up to 5,443kg (12,000lb) on four underwing pylons. Its range was some 3,704km (2,000nm), although this could be extended by air-to-air refuelling. Indeed, the range problem with the S1 was such that on aircraft carriers it had to be supported by a flight of Scimitars, modified for in-flight refuelling.

Although developed for the UK's Fleet Air Arm, the Buccaneer was offered to the RAF in 1965 following cancellation of the TSR2 – the long-awaited and politically controversial replacement for the Canberras and V-force. It was turned down in favour of the F-111K. This was, in turn, cancelled in 1968, and coincidentally with the run-down of the Royal Navy's carrier force, the RAF found itself taking over a number of ex-RN Buccaneers as well as ordering new production aircraft. The aircraft remains in service and currently 60 are being upgraded and fitted with the new British Aerospace Sea Eagle anti-ship missile (ASM) for service well into the 1990s.

Developed as a low-level nuclear strike aircraft for the Royal Navy (and only exported to the South African Air Force), the Blackburn Buccaneer entered service in 1962, and later was taken by the RAF after the cancellation of TSR-2. It still serves in Germany and will continue past 1990.

Hero of the Falklands conflict, the British Aerospace Sea Harrier FRS 1 is a unique naval fighter aircraft. It can use a small deck and yet still provide air superiority to the fleet. Being very agile, it can out-fly almost any known fighter.

0   4m

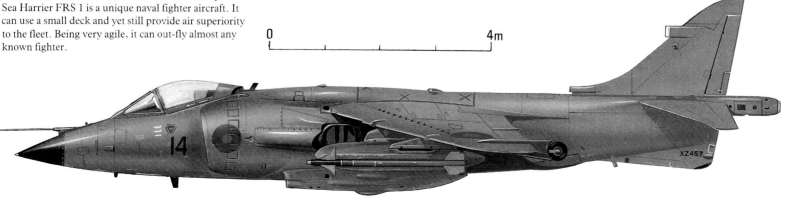

Enter the Tornado

The acquisition of the Buccaneers by the RAF was seen as being a replacement for the Canberra B(I)8 interdictor bombers only, and as providing the RAF with a means of providing the Royal Navy with the anti-ship capability it lost with its carriers. The Vulcans soldiered on and in 1969 the UK, together with Germany and Italy, began work on the aircraft which would replace the Vulcans in RAF service. Initially known by the acronym MRCA (Multi-Role Combat Aircraft) it became the Tornado in 1976, on the signing of the Memorandum of Understanding between the three governments involved for series production of the type.

The roles which MRCA was to fulfil show the enormity of the task undertaken by the three major contractors – BAC (British Aerospace), Fiat (Aeritalia) and Messerschmitt-Bölkow-Blohm (MBB), who formed the Panavia consortium to built the aircraft. The roles were close air support and battlefield air interdiction; interdiction/counter-air strike; naval strike (from land bases); reconnaissance; air superiority and interception/air defence. The latter two tasks were assigned to the air defence variant – ADV – being developed specifically for the RAF.

Normally aircraft are designed for one or perhaps two primary roles; to encompass six tasks, four of which are devolved to one basic type, the interdictor-strike version (IDS), is no mean feat. History records the three partners had only one unilateral agreement in configuration at the start – the need for two engines!

An air combat fighter needs a large wing area to achieve a minimum radius of turn, yet in order for the crew and aircraft not to critically vibrate, a small wing area is needed for ground attack. Short take-off and landing (STOL) requires a large wing span with slats and flaps almost impossible to fit in the classic subsonic wing, yet for supersonic speeds a short span, minimal frontal area and a relatively thin wing is the order of the day. In the early days, it was only the RAF and German Navy who required a two-man crew, and two versions – the Panavia 200 and the Panavia 100 respectively for the two-man and single-seat versions – were proposed. All these interactive variables were eventually translated into one variable-geometry two-man aircraft. The electronics and avionics fit, apart from the radar of the IDS version and a new air intercept radar for the ADV, were left to the individual service, as was the weapons fit.

The wings are all-metal construction with a fixed inboard portion of 60 degrees sweep on the leading edge, while the outer variable-sweep panel has a fully-forward leading edge sweep of 25 degrees and a fully-swept angle of 68 degrees. There are no conventional ailerons in the design. In their place are two spoilers on the upper wing, forward of the trailing edge flaps, which augment roll control at the full forward and intermediate sweep positions. At touchdown, these spoilers act as lift dumpers.

High-lift devices are fitted to the wings. The fixed wing roots each have a Kruger flap on the leading edge, while the moving portion of the wing has a three-section leading edge slat across the full span. The entire trailing edge of the moving wing (apart from the wingtip) has full span, fixed-vane, double-slotted Fowler flaps in four sections.

The wing root houses the wing glove box, an integral part of the centre section of the Tornado. It contains the single Teflon-plated bearing which is the pivot point for the outer wing panel, together with the ballscrew actuator driven by a hydraulic motor. A surface seal between the moving and fixed portions of the wing uses fibre-reinforced plastics and an elastic seal. This design has been so successful that it has been adopted on the American B-1B design.

The large fin and rudder is of conventional metal construction and design, while the

---

PANAVIA TORNADO GR1

Countries of origin: Italy/Germany/Great Britain.
Role: Two-seat interdictor strike aircraft.
Wing span: 13.9 m (45.6 ft) 25° sweep, 8.6 m (28.21 ft) 68° sweep.
Length: 16.7 m (54.79 ft).
Max T/O weight: 27,215 kg (60,000 lb).
Engines: Two 7,257 kg (16,000 lb) Turbo Union RB199-34R Mk101 afterburning turbofans.
Max speed: Mach 1.4 at 11,000 m (36,000 ft).
Max altitude: Exact figure not available; operates low level.
Range: 2,780 km (1,726 mi).
Military load: 2 × 27 mm IWKA-Mauser cannon, plus 9,072 kg (20,000 lb): typically (airfield denial) 2 × JP233 weapons, 1 × Sky Shadow ECM pod, 1 × BOZ-107 chaff/flare dispenser pod and 2 × 1,500 l (330 Imp gal) drop tanks, with 2 × AIM-9L Sidewinder AAM.

Photographed here in Italian Air Force service, the Panavia Tornado is the remarkable result of European co-operation to provide a strike aircraft and a good interceptor using basically the same airframe and powerplant. It has been exported to Saudi Arabia.

Serving with the Federal German Luftwaffe and Marineflieger, the Tornado has replaced the Starfighter in service for interdiction and strike roles, including anti-shipping in the Baltic approaches and offensive raids against airfields.

swept horizontal tail surfaces are of the all-moving taileron type. These can be operated collectively as elevators for pitch control, or differentially as ailerons for roll control. The wing spoilers operate in conjunction with the tailerons when the wings are not fully swept.

In an aircraft with many sets of systems, for example electrical, hydraulic and fuel, one of which is worth mention here is the flight control system. In operational service, the Tornado will spend most of its time at low level and high speeds, so the pilot needs some assistance in actually controlling the aircraft. This is provided by a triplex command stability augmentation system (CSAS), incorporating active control technology (fly-by-wire) and auto-stabilisation, controlled by two computers – one for lateral use and one for pitch. Instead of the pilot moving his controls around continuously for a smooth ride – which is practically impossible – the CSAS does this for him.

The engines, too, were a tri-national effort: Rolls-Royce (UK), MTU (Germany) and Fiat (Italy) forming the Turbo-Union consortium to develop the RB199 turbofan engines essential for the small size of the Tornado relative to some of the aircraft it was to replace. The RB199 has good fuel economy in the maximum non-afterburning (dry) mode for long-range interdiction and a very high augmented (afterburning) thrust for combat manoeuvres, short take-off and Mach 2+ speeds.

The RB199 is a compact engine. Among its many features are the dual afterburner, with an inner manifold and gutter similar to a ramjet, and an outer burner, fuelled separately, in the fan airflow. The engine intakes are also worthy of note, being of the two-dimensional, horizontal, double-wedge type. Their hydraulically actuated variable intake ramps are maintained at the optimum setting for the best engine performance in all flight conditions by a fully automatic digital control system. The engine itself has a three-stage fan, and for reverse thrust, two clamshell doors deploy over the exhaust nozzles to redirect the thrust

forward. Current technology, including electron beam welding, is used in its construction.

The Tornado programme has been something of a political football during its life, but it is well up to expectations and not too far above projected costings. Its only debit has been the long time scale of development and production, which can only be attributed to an optimistic original schedule and a longer-than-expected learning curve for collaboration. The success of the RAF competing in the USAF SAC 'Prairie Vortex' bombing competition in 1984 and 1985, where the Tornados came first in two of their three categories, and second in the third, did much to enhance its reputation.

Sukhoi's Fencer

The Soviet's answer to both the F-111 and Tornado is the Sukhoi Su-24 Fencer, a variable-geometry attack aircraft described in 1974 by US sources as 'the first modern Soviet fighter to be developed specifically for the ground attack mission'. Smaller and lighter

The Soviet equivalent to the Tornado or F-111 is the Sukhoi Su-24 Fencer, widely used by Soviet and Warsaw Pact attack squadrons. It is capable of carrying AA-2 Atoll or AA-8 Aphid self-defence missiles, and 1,000 kg (2,205 lb) of external bombs.

Below: Sukhoi's Su-24 was a completely new design, built to variable geometry (swing wing) configuration, and it can be looked upon as the modern version of a wartime heavy bomber in terms of role, range and payload.

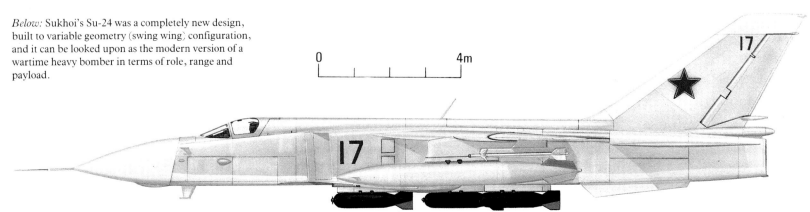

Sir Frederick William Page (b. 1917)

Educated at Rutlish and St Catherine's College (Cambridge, UK). 1938: joined Hawker Aircraft. 1945: joined English Electric and became Chief Engineer. 1959: appointed Director and Chief executive (Aircraft), English Electric Aviation. 1965:

appointed Managing Director, Military Aircraft Division, British Aircraft Corporation, and elected Divisional Chairman (1967-77), becoming Chairman BAC in 1977, then Chief Executive and Chairman of the British Aerospace Aircraft Group before retiring in 1983.

than the F-111, the Fencer is credited with having 'the capability to deliver ordnance in all weather within 55 m (118 ft) of its target'.

In configuration it follows the F-111, and to a certain extent the Tornado. Unlike previous Soviet swing-wing aircraft, with up to half of its wing span fixed (only the outer panels having variable sweep), only the wing roots/ glove boxes are fixed. The sweep is pilot-selectable at 16, 45 or 68 degrees, and the wing itself is equipped with the slattery and flappery akin to that described for the Tornado. Similarly, tailerons are used. The engines are afterburning turbofans, believed to be R-29Bs, rated at 122.3 kN (27,500 lb) with reheat.

The weapons pylons configuration is a mix of F-111 and Tornado: four fixed stations under the fuselage, plus two fixed stations under the wing root and two variable-sweep pylons under the wings. For a hi-lo-hi mission with 2,000 kg (4,400 lb) of weapons and external fuel tanks, the mission radius of action is 1,800 km (970 nm). The fuselage width is dictated by the need for a large multi-mode

pulse-Doppler radar and allows side-by-side seating for the crew, like the F-111, rather than Tornado's tandem arrangement.

The Su-24 Fencer is a copy of Western designs in concept, but probably derived from original thought within the USSR in detail. Although classed as a fighter, its place within the bomber section of this book is justified by its operational role.

From ground attack to CAS

During the Second World War, many fighters were modified to carry air-to-ground rockets or small bombs and so the fighter-bomber was born. By 1960, this particular type of aircraft had become known as a fighter ground-attack (FGA). By 1985, the terminology had changed yet again, and close air support (CAS) became the approved description of the role. Indeed, alongside CAS comes the next level of operations, known as battlefield air interdiction (BAI).

In 1960, ground attack was just that – aircraft unloading weaponry on to targets designated by ground forces. Today it has

taken on a new sophistication. NATO's Tactical Air Doctrine Manual defines CAS as 'air action against hostile targets which are in close proximity to friendly forces and which require detailed integration of each air mission with the fire and movement of those forces'. Such actions are tasked against enemy formations between the forward line of own troops (FLOT) and the fire support co-ordination line (FSCL), a distance which can vary (depending on the range of any ground artillery in use) but for all practical purposes may be said to be between 15 km (9.3 mi) and 25 km (15.5 mi). Action beyond this line to, say, 80-100 km (50-62 mi) is described as battlefield air interdiction, which although demanding co-ordination with the ground commander's fire support plan, does not require integration.

This sophistication is tailored for the current worst case scenario of an air war over NATO's central front, and today SACEUR does not possess the aircraft resources available in 1945 or even 1960. He has a pressing need, therefore, to make every sortie cost effective against a potential enemy every bit as sophisticated as his own forces, but with a vastly greater number of aircraft at his disposal.

In less sophisticated scenarios, usually in remote parts of the globe, the threat and its counter may not be of the same number and

Using unguided rockets to saturate a ground target, the Harrier, like its US Marine Corps sister-aircraft, the AV-8A, is typical of the close air-support aircraft of the 1970s and 1980s.

quality as that expected in the NATO first league. Indeed, the situation is usually such that the air commander fights with what resources are available. Thus history shows that a number of aircraft usually associated with other tasks have found themselves dropping 'iron bombs' on targets in support of ground forces.

A second role for fighters

The fighter was always the starting point for ground-attack aircraft. If provision for air-to-ground weaponry was not available initially, it was soon made. Thus in 1960, the majority of the world's ground-attack aircraft were made up of 1950s-vintage day fighters able to carry the appropriate armaments, with the balance made up from armed training aircraft. Perhaps the classic types in this bracket are, from fighters, the Hawker Hunter FGA series and the Douglas A-1 Skyraider, which was conceived as a US Navy attack bomber, but widely used over Vietnam in the CAS role by the USAF, and from trainers, the North American AT-6 Harvard/Texan.

During the 1960s, intense brush fire wars began to erupt around the world. In South-East Asia in the early 1960s, the United States began its involvement in the Vietnam War. Air power was one of the first levels of assistance, and the Skyraider began its second lease of life, being supplied to the Republic of Vietnam in increasing numbers for counter-insurgency (COIN) operations.

By mid-1964, the US Navy was looking for a new aircraft specifically for this new, low-intensity type of operation. Six designs were submitted before the North American division of what is now Rockwell International was selected to build the OV-10 Bronco. In the event, the US Marine Corps and the USAF were to adopt the two-seat, twin-boom, twin

A classic counter insurgency (COIN) aircraft, the Rockwell OV-10A Bronco is the only US-designed one, and it has served with the US Marine Corps and the US Air Force. It has also been exported to Asia, South America and Europe.

Although it was originally designed as an air superiority fighter, the missile-carrying McDonnell Douglas F-15 Eagle has also been developed for a ground-attack role, as the Strike Eagle.

turboprop aircraft, employing it for forward air control (FAC) duties rather than as a ground attack aircraft. In recent years, the USMC aircraft have been updated with new avionics and night vision systems to offer night attack capability. The Bronco was the only specific COIN aircraft developed in the West, the role being taken on by armed trainers in the main.

Fighters assigned the ground attack role were given increasingly sophisticated systems, including air-to-ground U/VHF communications equipment. Weapon sights became more specialised and weapons themselves were adapted for use against softer ground targets. The escalation of the Vietnam War saw the increased use of air-to-ground rocket projectiles, retarded bombs, napalm and, later on, the first generation of 'smart' bombs. In American terms, the demarcation between what the UK called a fighter ground attack (FGA) and the US designated as an attack aircraft was very wavy. The USAF flew F-100 Super Sabres and F-105 Thunderchiefs, while the US Navy and Marine Corps relied on the A-4 Skyhawk. Later in the conflict, the F-4 Phantom served all three services in this role, supplemented in the US Navy by the A-7 Corsair. Range and payload of weapons became major design criteria as a result of operational requirements. This, at least in

part, was one of the reasons for the renaissance in the use of the Skyraider, a propeller-powered survivor in the jet age.

Within Europe during the late 1950s, Fiat of Italy produced the G91, which it had been hoped would have been adapted by most, if not all, NATO air forces as a light tactical strike aircraft. In the event, this single-engined aircraft was adopted only by Italy, Portugal and Federal Germany. Although well armed with a pair of 30mm cannon, it could only carry two 227kg (500lb) bombs or equivalent air-to-ground rockets to a range of 628km (339nm).

In the late 1960s, Anglo-French collaboration led to the development of the Jaguar attack/trainer aircraft. In the single-seat configuration, it was used as a replacement for the venerable Hunter in the RAF, as well as some Canberra tactical bomber roles. Again the dividing line between ground attack and strike was indistinct. As with the G91 it was armed with two 30mm cannon, but could carry a maximum weapon load of 4,763kg (10,500lb), with a tactical radius of action of 852km (460nm) in a hi-lo-hi profile using internal fuel. With external fuel tanks, reducing the weapons payload by 2,268kg (5,000lb) its radius is 1,408km (760nm). In addition, its avionics and systems were more sophisticated: a digital computer-driven navigations and weapons-aiming sub-system, a moving map display, head-up display and a laser ranger and marked target seeker, plus a radar warning receiver and radio equipment compatible with its air-ground role.

South-East Asian Air Operations

The Air War in South-East Asia – a defeat for technology

Although the outcome of battles can be influenced by air power, ultimately it is the determination of the foot soldier that leads to victory or defeat. The 30-year war fought to determine the future of Vietnam witnessed an unprecedented use of advanced technology – stopping short only in the use of nuclear weapons. However, despite countless examples of valour and the use of innovative techniques, air power could not provide a victory for corrupt and ineffective leaders.

When France tried to reassert herself as a colonial power in Indo China, various 'freedom fighters' were determined otherwise. However, backed by an increasing use of air power – including the employment of napalm for the first time – the French forces managed to inflict a number of defeats on the Viet Minh insurgents.

Piston-engined combat aircraft such as the B-26 Invader and F8F Bearcat were capable of delivering punishing blows against enemy forces but were unable to lift the 55-day siege of Dien Bien Phu. Some 62 aircraft were lost during this battle, which proved to be a turning point in French involvement in South-East Asia.

With Vietnam partitioned in 1954, America came to regard the South as a bastion against communist expansion, providing increasing quantities of military aid. By 1956, the last of the French troops had left South Vietnam and slowly United States forces began to take their place. Initially, this took the form of advisors but gradually it became clear that a direct involvement would be necessary.

Practical American aid started with the supply of vintage AT-6 Texan and T-28 trainers adapted for a ground-attack role.

The doughty C-47 provided an invaluable transport lifeline to ground units, which were given the task of fighting the Viet Cong guerrilla forces but as North Vietnam committed her regular army to the 'liberation' of the South, the American war machine fed increasingly sophisticated weaponry into the fray.

To avoid a direct confrontation with ground forces near the 17th Parallel, North Vietnam established the Ho Chi Minh Trail through neighbouring Laos and Cambodia to feed men and supplies to the South. The United States responded by fielding B-52 bombers from Guam,

pounding the trail and inflicting considerable damage but failing to stem the flow of material. Although thousands of tons of bombs were dropped on the routes taken by the North Vietnamese, the thick jungle provided a cloak for their activities, making accurate attacks difficult. One measure taken by the Americans to remove this natural camouflage was to spray defoliants from aircraft flying over the area to be cleared. Often C-123 Providers would fly in formation to ensure that a wide swathe of jungle would no longer provide shelter.

Such action proved helpful in clearing designated areas, but the Viet Cong were adept at overcoming such problems and simply redoubled their supply activity by night. The Americans had an answer for this too. Sensors dropped into the jungle could monitor the movement of men and vehicles, sending signals to specially equipped aircraft which automatically relayed the information to Infiltration Surveillance centres. The Air Delivered Seismic Intruder Device (ADSID) could operate on its batteries for up to 45 days, and the information provided by these devices was finally processed by a computer to produce target guidance for attack aircraft. Types as varied as the Beech Bonanza – designated QU-22 – and the Lockheed EC-121 acted as airborne switching stations for the data picked up by the ADSIDs.

Many of the American air operations were initially limited to the support of units fighting in South Vietnam. Army units in particular made extensive use of helicopters to support combat bases and to insert troops into areas which were to be

cleared of the enemy but these units in turn were supported by attack aircraft operated by the USAF, US Marine Corps and US Navy, as well as South Vietnamese Air Force units.

Various versions of the McDonnell Phantom bore the brunt of the ground-attack duties, although the Republic F-105 Thunderchief also carried out many support missions too. Not all the ground-attack aircraft were capable of supersonic speeds, however; the Cessna A-37 Dragonfly, originally designed as a trainer, proved to be an effective and manoeuvrable support fighter.

The Douglas A-1E Skyraider, although a piston-engined design, gave good service throughout the conflict, having been used by the French Air Force before it withdrew and continuing in service with the USAF. Capable of absorbing a lot of punishment, the Skyraider could out-manoeuvre MiG fighters, shooting one down during an aircrew rescue support mission.

The American forces would go to considerable lengths to rescue pilots shot down in the course of a mission. Helicopters such as the Sikorsky HH-53C (which can be refuelled in flight) would often be tasked with lifting a downed pilot to safety, flying to the site escorted by Skyraiders and directed by a Forward Air Controller.

The concept of using a light aircraft to fly close to the battlefront in order to

Ground-attack support for the Allies in South Vietnam included the A-1 Skyraider, which had also served with distinction in Korea. It could carry 3,630 kg (8,000 lb) of underwing ordnance, including various types of bomb and napalm.

In the early part of the Vietnam War various types were used for ground-attack missions, including the F-100 Super Sabre. As the threat of infra-red shoulder-launched missiles grew, so the aircraft had to be less vulnerable and more advanced.

provide direction to ground artillery or attack aircraft had been developed before the war in Vietnam, but the technique was perfected during this conflict.

Often an unarmed Cessna O-1E Bird Dog would fly at tree-top height to locate an enemy position unseen by ground forces. It was a hazardous job which required the pilot/observer to use manoeuvring skills to avoid enemy fire. The Rockwell OV-10 Bronco proved to be ideal for the Forward Air Control role, carrying many special devices to enable it to be operated at night as well as by day. The twin-boom design fitted the Bronco to carry a variety of stores, including weapons with which to defend itself or protect a downed pilot until a rescue helicopter arrived.

In retaliation for a North Vietnamese attack on two USN destroyers in August 1964, America decided to launch attacks on military targets in North Vietnam. Carrier-borne aircraft and fighter-bombers from bases in South Vietnam bombed important targets but did not operate unopposed. It did not take long for SA-2 Guideline missiles to be brought to bear on the attacking aircraft, shooting some of them down.

Various methods were employed to defeat missile and fighter defences, including the use of electronic counter-measures to jam enemy radar. To provide a warning of enemy fighter activity long-range ground and ship-borne radars were supplemented by airborne radar carried on board EC-121s. These aircraft later took on the role of airborne fighter direction posts, ensuring American fighters gained the best position to counter enemy MiGs.

Although used for some time to provide tactical support in Vietnam, the Americans did not commit B-52s to bomb North Vietnam until 1966, but although considerable damage was caused, the bombing also served to strengthen the resolve of the population – not an unknown phenonemon.

Two years later, the B-52s were used to help lift the siege of the Khe Sanh combat base, dropping more than 100,000 tons of bombs! However, despite the massive use of air power, including specially equipped C-47 and C-130 gunships (one version of the latter mounting a 105 mm howitzer), the relentless pressure on the South continued.

The airlift which had built up American strength in Vietnam was put into reverse; the C-141A and C-5A transports were supplemented by chartered civil airliners. Finally, the last remaining Americans and some Vietnamese fled in helicopters as soldiers from the North occupied Saigon in 1975.

The latest technology in the art of air warfare had failed to stem the tide of the North Vietnamese advance, demonstrating the limits to the effectiveness of air power.

Brian Walters

One of the SPECAT Jaguar prototypes displays its full range of weapons, including bombs and overwing air defence missiles. Jaguar was sold to the RAF, French, Indian and Nigerian Air Forces but is being withdrawn from the first two.

Response time has always been an important element of ground-attack aircraft, and probably the most important ground-attack aircraft in this respect in recent years has been the STOVL Harrier. Its development was initiated in 1964, almost as a consolation prize to Hawkers following the cancellation of the supersonic P1154 VTOL fighter, specifically as a ground-attack aircraft for the RAF. During the mid-1960s, the VTOL option was thought to be the answer to the immediate response time demanded by troops on the ground requiring CAS, as they were perceived as being able to be operated from small sites close to the FLOT. As the Harrier evolved and the RAF gained experience following its service entry in 1969, it was seen that a larger payload could be lifted by using a short take-off and a vertical landing mode of operations – STOVL. The 'gimmickry' of vertical take-off, causing a payload penalty, did much to offset its operational flexibility and hence its marketability within NATO. It took many years of RAF operations, principally with their squadrons based in Germany, but also backed up by its creditable performance during the Falklands campaign (1982), to rid the ground-attack Harrier of its 'toy aeroplane' image.

Even the doyens of CAS – the US Marine Corps – had to admit that there was nothing like the Harrier and bought 102 AV-8A Harriers from the UK. As a replacement, American technology was adapted to the Harrier by McDonnell Douglas, including the redesigned composite-material wing, nose and cockpit together with current avionics, to produce the AV-8B Harrier II, and in 1981 the RAF selected the type as its own successor to their original Harriers.

The AV-8B is powered by the Rolls-Royce Pegasus 11-21 engine, rated at 9,979 kg (22,000 lb), fitted with zero scarf (non-slanted) front nozzles. The air intakes are enlarged and the lower fuselage features lift-improvement devices to direct the airflow more efficiently, especially when hovering within ground effect. The weapons capacity of the aircraft is 7,710 kg (17,000 lb) for a short take-off. The USMC AV-8B is armed with a single GAU-12/U 25 mm cannon, while the RAF's Harrier GR5 has a pair of 25 mm Aden cannon. From a short take-off of 305 m (1,000 ft), it can carry 12 Mk82 Snakeye bombs over a radius of 278 km (150 nm) on internal fuel with provision for a one-hour loiter over the battle area.

BRITISH AEROSPACE SEA HARRIER FRS2

Country of origin: Great Britain.
Role: Single-seat fighter, strike and reconnaissance aircraft.
Wing span: 8.3 m (27.25 ft).
Length: 14.1 m (46.25 ft).
Max T/O weight: 11,340 kg (25,000 lb).
Engine: One 9,752 kg (21,500 lb) Rolls-Royce Pegasus Mk104 vectored-thrust turbofan.
Max speed: 1,185 km/h (736 mph).
Max altitude: 15,240 m (50,000 ft).
Range: 1,500 km (920 mi).
Military load: Up to 3,630 kg (8,000 lb) of stores: typically (air defence) 4 × AIM-120 AMRAAMs, 2 × AIM-120s plus two 30 mm Aden cannon plus two 860 litre (190 Imp gal) combat drop tanks; (anti-ship) 2 × Sea Eagle missiles plus two 30 mm Aden cannon.

Below: arming a Harrier with unguided rockets for ground-attack strikes.
Bottom: the latest generation in the Harrier story is the AV-8B. It entered US Marine Corps service in late 1985, and will be followed by the Harrier GR5 in the RAF Strike Command service in 1987. Both aircraft retain the V/STOVL characteristics of the earlier models but have improved airframes and systems. Both are cannon-armed, the AV-8B with a 25 mm and the GR5 with a pair of Aden 25 mm types.

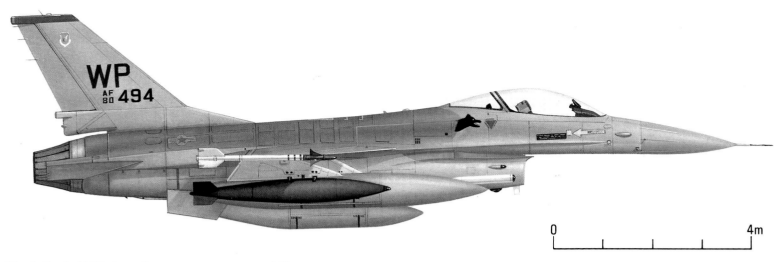

The dedicated CAS aircraft

In 1967, the USAF initiated the A-X programme to develop a dedicated CAS aircraft, with Northrop and Fairchild-Republic each building two prototypes for a competitive fly-off, the YA-9A and YA-10A respectively. The YA-10A was declared the winner in January 1973, and between 1975 and 1982, 707 A-10A Thunderbolt II aircraft were built for the USAF only.

The A-10A is of unusual configuration, having a twin tail, engines mounted on pods on the rear fuselage, a straight wing with four pylons each side (and three under the fuselage centre-section) capable of lifting a maximum external load of 7,258 kg (16,000 lb), plus a nose-mounted GAU-8/A 30 mm seven-barrel cannon. It is this gun which confers the 'tank buster' tag on the A-10. Configured for close air support, it has an operational radius of 463 km (250 nm) with 1.7 hours loiter capability.

Avionics now comprise an integral part of a CAS aircraft. The A-10 has a head-up display (HUD), a Pave Penny laser designator pod, the ALR-66 (V) radar homing and warning receiver, ECM pods, secure voice communications via a plethora of radio systems and space for growth in the fuselage. The current update programme for the A-10 is designed to offer night/adverse weather capa-

Although the Aérospatiale Epsilon was designed to meet a French Air Force requirement for a basic training aeroplane, it has been sold around the world as a combat trainer and cost-effective ground-attack aircraft. This one has underwing rocket pods.

bility under the LANTIRN (Low Altitude Navigation Targeting Infra Red for Night) programme. Apart from internal equipment and modifications, the system also comprises two sensor pods carried on the outer two fuselage pylons: one equipped with forward-looking infra-red (FLIR) equipment and the other equipped with a terrain-following radar.

The ability to survive hits from AAA (anti-aircraft artillery) is another factor in the A-10's design, and the pilot's cockpit is located in an armour 'bathtub' able to withstand direct hits from bullets up to 23 mm calibre. The location of the two 4,115 kg (9,065 lb) General Electric TF34-GE-100 turbofans offers maximum shielding by the airframe against acquisition by IR seekers on missiles, while the engines' high bypass ratio is used to cool the exhaust itself. Fuel tanks are self-sealing and tear resistant. All these design features enhance survivability in a hostile environment.

In the 1970s, the A-10 was the epitome of a CAS aircraft but, if a 1985 request for information to US industry from the USAF is anything to go by, its effectiveness for the late 1980s and early 1990s is not held in as high esteem. The USAF Aeronautical Systems Division has asked industry to provide information on a CAS/BAI aircraft, based on in-service or in-production types, as a means of keeping up with the threat until the Advanced Tactical Fighter of the 1990s is in service.

Of the four responses received, three – the F-16, AV-8B and Northrop F-20 Tigershark – are adaptations on in-production aircraft, while the fourth – the Vought A-7 Strike-

The Fighting Falcon is capable of carrying the Mk84 bomb or MRASM (Medium Range Air-to-Surface Missile) in a close air-support role. The General Dynamics F-16 is, however, better known as a fighter; this aircraft is based on Okinawa, Japan.

GENERAL DYNAMICS F-16C FIGHTING FALCON

Country of origin: USA.
Role: Single-seat, lightweight air combat fighter.
Wing span: 10.01 m (32.83 ft).
Length: 15.09 m (49.48 ft).
Max T/O weight: 16,057 kg (33,400 lb).
Engine: One 11,340 kg (25,000 lb) Pratt & Whitney F100-PW-200 afterburning turbofan, or one 13,164 kg (29,000 lb) General Electric F110 afterburning turbofan.
Max speed: Above Mach 2.0 at 12,200 m (40,000 ft).
Max altitude: In excess of 15,240 m (50,000 ft).
Range: In excess of 1,850 km (1,150 mi).
Military load: 1 × 20 mm M61A1 rotary cannon, 2 × AIM-9 Sidewinder AAMs on the wingtips, plus one fuselage centreline hardpoint and six underwing pylons for a combined total of 8,892 kg (19,600 lb).

fighter – is an extensively re-worked Corsair, using in-service airframes. Apart from the Strikefighter, which also proposes a new engine, all four solutions involve dedicated avionics and weapons appropriate to the role. With Pentagon budget reductions, whether this programme survives is open to debate.

Soviet CAS

Since 1960, Soviet ground-attack aircraft have been modified fighters, mainly from the MiG and Sukhoi stables, adapted in a similar manner to that previously described. The MiG-21 Fishbed series and the Sukhoi Su-17/20/22 Fitter series are typical examples: the combat radius of the Fishbed-J with four 250 kg (551 lb) bombs being 370 km (200 nm) in a hi-lo-hi profile; the Fitter-C with 2,000 kg (4,409 lb) of weapons has a radius of 630 km (340 nm) hi-lo-hi.

As a first step towards the dedicated CAS type, the Soviet MiG-23 Flogger, which first

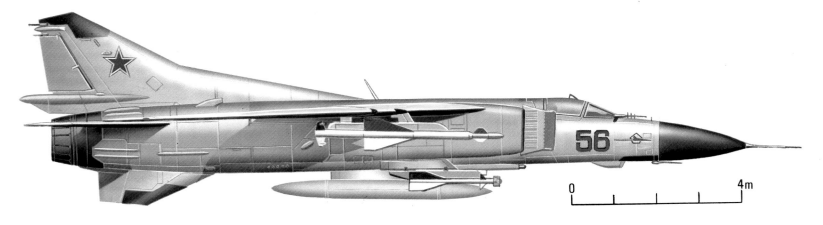

Demonstrating the MiG-23's swing-wing geometry, this aircraft was photographed shortly after take-off, with the wings fully extended. The Czech Air Force flies a hybrid version with a raised seat for the single pilot to fly attack sorties.

MIKOYAN-GUREVICH MiG-21MF FISHBED-J

Country of origin: USSR.
Role: Single-seat multi-role fighter.
Wing span: 7.15 m (23.46 ft).
Length: 15.76 m (51.71 ft) incl pitot boom.
Max T/O weight: 9,400 kg (20,725 lb).
Engine: One 6,577 kg (14,500 lb) Tumansky R-13-300 afterburning turbojet.
Max speed: Mach 2.1 above 11,000 m (36,000 ft).
Max altitude: 18,000 m (59,050 ft).
Range (internal fuel): 1,100 km (683 mi).
Military load: 1 × twin-barrel 23 mm GSh-23 gun in belly pack; plus four underwing pylons; (ground attack) 2 × 500 kg (1,102 lb) and 2 × 250 kg (551 lb) bombs; or (air defence) four K-13A Atoll AAMs.

appeared in 1967 as an air defence fighter, was modified as a ground-attack aircraft. The radar nose of the MiG-23 was replaced by a sloping window in a sharply tapered nose, behind which is housed a laser ranging and marked-target seeker. Two additional pylons are located under the rear fuselage, in addition to the existing five pylons (one centreline and two under each wing); while the GSh-23 cannon of the interceptor is replaced by a six-barrel 23 mm Gatling-type gun in a belly

pack. The initial MiG-27 Flogger-D entered service in 1971, while the improved Flogger-J was first identified in 1981.

Imitation, it is said, is the best form of flattery, and the Soviets have obviously been impressed with the A-10's dedication to the CAS role. Their response has been in the form of the Sukhoi Su-25 Frogfoot single-seat ground-attack aircraft, which looks more like the Northrop YA-9A than the A-10. Of single fin configuration with slightly swept wings, it has 10 pylons and a belly-mounted Gatling-type cannon of 'heavy calibre'. First photographs of Frogfoot were taken in Afghanistan in December 1982, and it was thought to have reached operational service in 1983-84.

Described as a single-engine multi-role fighter, the MiG-23MF Flogger serves with a number of client states, as well as with the Soviet Frontal Aviation. The aircraft illustrated is armed with AA-7 Apex and AA-8 Aphid air defence missiles.

CAS around the world

Around the world, the French Mirage series of fighters offers CAS as one of the fighter-bomber/attack options. The Mirage IIIE, Mirage F1-E and Mirage 2000 all have ground attack capability. However, it is the Mirage 5 and 50 series which are the ground attack dedicates: the Mirage 5 was originally built for Israel, but never delivered, and was then taken into French service, being basically a IIIE with simpler avionics for export; the 50 is powered by a more powerful SNECMA Atar 9K-50 afterburning turbojet and is described as a 'special mission' or multi-role aircraft. Its combat radius at low altitude with two 400 kg (882 lb) bombs is 630 km (340 nm). Both the Mirage 5 and 50 have been successfully exported.

During 1982, British forces had reason to be wary of the Argentine FMA IA-58 Pucara, and those pilots who encountered it were not derisive of its capabilities. Fortunately, it was not used to effect, and the level of threat it posed is exemplified by the fact that the commando raid on Pebble Island on 14 May 1982 was to destroy 11 Pucaras, their fuel and weapons.

Design work on the Pucara began in 1966

Although shown here in the air defence role configuration, the Mirage 2000 fighter can carry out short-range attack missions, armed with eight 250 kg (551 lb) bombs, and with two fuel tanks. It defends itself with the Matra Magic missile.

Top: the FMA Pucara has been used operationally in the Falklands and against 'rebel' indians in Argentina, proving itself to be both sturdy and robust.

Above: from Brazil (with 70% Italian origin) comes the Embraer AM-X tactical fighter. It will be procured for the Italian Air Force and for the Brazilian Air Force, in whose markings this prototype is shown. It may also be exported to non-aligned nations.

fuel tanks. Typically, its combat radius would be 370 km (200 nm) on a lo-lo-lo profile with a total military load of 1,088 kg (2,400 lb). Although its maximum speed at sea level is less than the AM-X it is a manoeuvrable aircraft, wherein lies its survivability option. In many situations, its lower speed is an asset for CAS operations, but not always. The success of its predecessor, the MB326K, is an indication of a future market.

Truculent tutors

The final category of CAS aircraft is the armed trainer, or trainer/light strike aircraft. In the 1960s, the weapons training phase of pilot training was carried out on two-seat versions of the operational type or single-seat obsolescent fighters; this was not cost-effective. With the advent of the third-generation advanced jet trainer, hardpoints for weapon pylons were designed into the aircraft from the start so that flying training and weapon training could be carried out on the same type, thus making the whole cost of ownership and training more effective. The availability of hardpoints on such aircraft also offers the export customer a more attractive product, as an armed, light strike aircraft.

Some of the second-generation trainers, typically the BAC 167 Strikemaster, Fouga Magister and Macchi MB326 series, were armed, but it was in the third generation that the concept was refined to a fine art. The Dassault-Breguet/Dornier Alpha Jet is a prime example, being initially developed as a jet trainer for the French and a CAS aircraft for the Germans. In the early 1980s, Dassault-Breguet developed the aircraft further to offer the *Nouvelle Génération pour l'Ecole et l'Appui*

by Fabrica Militar de Aviones (the Military Aircraft Factory) and the first prototype flew in 1969. The second prototype flew in 1970 powered by French Turbomeca Astazous XVI G engines, also adopted for the IA-58A and B production variants. This aircraft packs a powerful punch as the built-in armament comprises two 20 mm Hispano cannon in the lower nose and two 7.62 mm FN-Browning machine guns either side of the two-seat cockpit area. In addition one centreline and two wing pylons can accommodate a total of 1,500 kg (3,307 lb) of stores. Of the many examples of Pucara captured in the wake of the Falklands campaign, the RAF brought several back to the UK, made one airworthy and test-flew it. They confirmed the effectiveness of the type.

Italy, too, has a well-established aircraft industry and is currently developing, as a collaborative venture with Brazil, the AM-X attack aircraft, which will replace existing G91s and some Starfighters in Italian service. Both Aermacchi and Aeritalia of Italy are working with Embraer of Brazil. The AM-X is a single-engined attack aircraft, powered by a Rolls-Royce Spey Mk807 turbofan, and armed with a single 20 mm Italian rotary cannon or a pair of 30 mm DEFA 553 cannon for Brazil; it has four underwing pylons plus wingtip missile stations. Its attack radius with 2,720 kg (6,000 lb) of stores is 520 km (280 nm)

AERMACCHI MB326G

Country of origin: Italy.
Role: Tandem-seat training and light strike.
Wing span: 10.85 m (35.58 ft) – over tip tanks.
Length: 10.64 m (34.92 ft).
Max T/O weight: 5,217 kg (11,500 lb) (light attack).
Engine: One 1,547 kg (3,410 lb) Rolls-Royce Viper 20 Mk540 single-shaft turbojet.
Max speed: 867 km/h (539 mph).
Max altitude: 14,325 m (47,000 ft).
Range: 1,850 km (1,150 mi).
Military load: Up to 1,814 kg (4,000 lb) (underwing), to include bombs, rockets, drop tanks, recce and gun pods. Light attack variant (MB326K): 2 × 30 mm DEFA 553 cannon.

flying hi-lo-hi. Designed for a number of strike and attack roles, including CAS, the AM-X is a workmanlike state-of-the-art aircraft.

One of the partners in AM-X, Aermacchi, has its own breed of CAS aircraft in the MB339K Veltro II. Derived from the MB339A trainer, it features a redesigned nose housing a single-seat cockpit, avionics bay and a pair of 30 mm DEFA cannon, the remainder of the airframe and engine being common to the trainer. This includes six underwing pylons with a combined capacity of 1,815 kg (4,000 lb) of ordnance, plus integral wingtip

Developed for training but with tactical missions as a secondary role, the Aermacchi MB 339 Veltro already equips the Italian Air Force training force and the Aerobatic team, and will prove a worthy successor to the combat-proven MB326.

Following the success of the Hawker Hunter, the British Aerospace Hawk has sold well, both as a trainer and ground-attack aircraft. The company's demonstrator is shown here with underwing stores and an Aden cannon on the fuselage centreline.

(NGEA). At the 1985 Paris Air Show a further development known as Lancier was revealed as a low-cost strike aircraft with improved avionics and a wider variety of weapons options.

Other examples include the Aermacchi MB339 series, CASA C-101 Aviojet, SIAI Marchetti S.211, Aero L-39 Albatros, Fairchild Republic AT-46A and FMA IA-63 Pampa. The detailed evolutionary cycle in this range of aircraft can be examined using the BAe Hawk series as an example.

The Hawk began life in 1969 as the Hawker Siddeley HS1182 and by 1971 had been selected as the RAF's new advanced jet trainer. In 1972, it became the Hawk T1 and 176 were ordered for the RAF, powered by a single Rolls-Royce/Turbomeca Adour Mk151 turbofan. Fitted with a centreline and two underwing pylons the Hawk T1 carries a 30 mm Aden cannon pod, either rocket pods or practice bomb carriers, or even a Sea Eagle anti-ship missile.

From the Hawk T1 came the Mk50 series for export. Among the modifications were two extra underwing pylons, all five being cleared to carry 515 kg (1,135 lb) of stores, and an increased maximum take-off weight of 7,350 kg (16,200 lb). The Mk60 series followed with the uprated Adour Mk861 turbofan, which increased the maximum level speed from 991 km/h (616 mph) to 1,037 km/h (644 mph), an increase in pylon capacity to carry 900 kg (2,000 lb) of stores, a maximum take-off weight of 8,600 kg (18,960 lb), a wider variety of stores, including the option to carry up to four AIM-9 Sidewinder AAMs on twin-carriage racks on the outer wing pylons and drop tanks on the inboard pylons.

By mid-1984, two further versions were on the drawing board. The first was the Mk100, optimised for CAS and ground-attack missions. It can carry a wide variety of conven-

tional and guided air-to-ground weapons, plus a self-defence capability with 30 mm gun and AAMs, enhanced by whatever electronic warfare equipment the customer specifies. It can carry over 100 combinations of stores up to a maximum of 3,100 kg (6,800 lb). Powered by the Adour Mk861, its maximum level speed is 1,040 km/h (638 mph) and its maximum take-off weight is 8,570 kg (18,890 lb). The heart of the aircraft's effectiveness, however, lies in the adoption of an integrated inertial navigation system and HUD/weapons aiming computer allied to a multi-function cockpit display and HOTAS (hands on throttle and stick) controls to ease pilot workload. In short, the increased weapons capacity has been merged with a modern avionics suite that would do justice to a dedicated attack aircraft.

The second, more radical, development turned the Hawk from a trainer to a single-seat attack aircraft. The Hawk Mk200 series go-ahead was announced at the Farnborough Air Show in September 1984, and is aimed totally at the export market. It comprises a new forward fuselage and single-seat cockpit inte-

grated to the Mk100's avionics suite. In addition, there is provision for a pair of internal guns in the lower forward fuselage, while it is also proposed that a lightweight multi-mode radar or a FLIR sensor be located in the nose.

By the time Hawk 200 is being built in production quantities, the 2,767 kg (6,100 lb) Adour Mk871 will be available, and is the likely powerplant when it flies in late 1986. Carrying four 454 kg (1,000 lb) and four 227 kg (500 lb) bombs over a 250 km (135 nm) radius, the new Hawk 200 could well become the Hawker Hunter of the 1990s.

The eye in the sky

Ever since man began fighting man, he has had an overwhelming need to know what is on the other side of the hill. The first military balloons were for observation, and so were the first aeroplanes. The need for intelligence of the enemy is still with us today, and aviation's contribution spreads its net across a wide spectrum from cameras in aircraft through airborne radars looking down to spy satellites. As technology has progressed, so systems become more sophisticated, and the years 1960-85 have produced, perhaps, the greatest leap forward since the aeroplane itself took a man over the enemy lines for a 'look-see'. Today it is called reconnaissance or recce or recon in the parlance of military men.

Even in this age of technology, the camera,

America's 'spy in the sky' in the pre-satellite days was the Lockheed U-2, used for CIA-controlled reconnaissance flights over Communist territory, flying at altitudes of over 16,750 m (55,000 ft). It entered service in 1956 from European bases.

as a dependable and adaptable means of providing permanent imagery of enemy dispositions, remains the principal sensor for reconnaissance work. Today's cameras have become lighter, more rugged, easier to maintain and engineered to allow rapid film spool changes. Miniature cameras have been developed so as to fit into restricted spaces in aircraft. Automatic operation has become the rule rather than the exception. Computers are used to determine the best ratio of height to speed so that the exposure frequency (frame rate) is at an optimum value. High-speed focal-plane shutters are highly favoured, to reduce the effects of vibration in high-speed flight. If light conditions deteriorate and the lens aperture becomes fully open, the shutter speed is then automatically reduced in order to maintain the correct exposure.

Cameras may be set at varying angles pointing out of the aircraft: forward, vertical, oblique and panoramic. Oblique cameras are particularly favoured where politics or the tactical scenario place overflight restraints on the aircraft. The Lockheed U-2 during the Cuban missile crisis of 1962 was able to take photographs of areas some 22-111 km (12-60 nm) to the side of the flight path; a derivative of the U-2 known as the TR1 is still performing that task for the USAF today. A

The U-2's replacement, again from the Lockheed design team, is the TR-1. It uses various advanced reconnaissance techniques, including synthetic aperture radar and special cameras to view the Earth from altitudes of around 56 km (35 mi).

panoramic fan of between three and five cameras can provide 180° coverage below the aircraft for mapping purposes. These cameras use special lenses to ensure minimal distortion of the photographic images.

The past 25 years have seen the development of infra-red imagery, particularly as it offers a passive means to see through poor visibility and industrial haze. The principle is that it detects heat emissions, typically detecting whether engines have been, or are, running. It can tell whether an oil storage tank is full or empty, and can indicate whether particular areas of interest have been recently tenanted by vehicles or equipment by virtue of the thermal shadow that has been left behind.

These infra-red sensors operate by scanning narrow beams across its path to build up a total picture, hence the name Linescan has been adopted for the technique. The scanner is a mechanical device rotating at between 8,000-12,000 rpm, while the sensor has to be cooled and stabilised. The field of view is some 60° either side of the vertical, while the scanner works at 90° to the line of flight. Data can be either processed on board the aircraft or transmitted to a ground station via a data link.

The third major sensor for reconnaissance is radar. Several types are used but modern needs, apart from airborne early warning and control, are best met by a sideways-looking radar (SLR). As its name suggests, SLR offers the ability to see an area without having to overfly it in the conventional manner. Typical of such systems is Motorola's APS-94D system. Its main elements are a pair of slotted

The venerable Shackleton started life as a long-range maritime patrol aircraft with RAF Coastal Command but was later modified to an interim airborne early warning role, with AN/APS-20 radar taken from the Fairey Gannets scrapped by the RN.

wave-guide arrays, operating the I/J-band, which are gyro-stabilised and pivoted at the centre. It can operate in two modes – a fixed map, showing all detected targets in an area as if they were stationary; and a moving map which indicates all targets against a suppressed map. By using one array, on one side of an aircraft, targets up to 100 km (54 nm) from the aircraft can be detected: both arrays will give the picture on either side of the aircraft. More advanced systems offer a data link tied into a ground processor for real-time information. One such system, the UPD-4, was developed for the RF-4C Phantom, and has been adopted on other marks of the aircraft. Video equipment can be used to record images not transmitted by data link.

Airborne early warning (AEW) radar is something of a special case, as it is used as much for controlling aircraft already aloft as to gather information. AEW grew from the need to extend the horizon of ground-based radar. Sets were put in aircraft which then were able to see further over the horizon. Initial use was made of AEW by carrier-borne aircraft, and as recently as 1982, the lack of AEW by the Royal Navy led, indirectly, to the loss of four major ships in the South Atlantic.

In the early 1960s, AEW radar was becoming a useful tool for the military, because they had developed a way of eliminating permanent clutter such as waves and high ground from the radar screens and display only moving targets. This technique is known as moving-target indication (MTI).

The fourth major sensor for reconnaissance purposes is provided by refined television systems, which offer a good real-time surveillance capability. Typical of such systems is that developed by Israel Aircraft Industries and shown for the first time at the 1985 Paris Air Show. It is known as the Stabilised Long-range Observation System (or SLOS).

The platforms

Reconnaissance can be gleaned from the sensors mentioned, used individually or in combination. The sensors can be fitted internally to an aircraft or mounted in pods for carriage on a weapons pylon. The latter enables almost any type of aircraft to be adapted to the recce role.

At the smaller end of the spectrum come the fighter reconnaissance aircraft, usually a ground-attack type with slight modifications to carry between one and three cameras. The Hunter FR10 is typical of the genre for the early 1960s, being an FGA9 with three cameras mounted in the nose section. The Phantom FGR2, which replaced the Hunter in the attack role, is fitted with a single camera installation in a fairing under the fuselage, while a dedicated pod was also developed for the specialised role. The Harrier GR1/3 has a single F95 70mm oblique camera on the port side of the nose and an early modification of the Northrop F-5 offered a camera nose, as the RF-5, but this designation was later used in a dedicated version of the F-5E model, the RF-5E.

Dedicated fighter-reconnaissance aircraft first appeared in the 1950s with the RF-84 Thunderflash, but during the 1960s, both the F-101 and F-4 were produced with modifications. The ubiquitous Mirage III/5/50 and F1 series also had reconnaissance versions. For naval operation, the A-5 Vigilante was modified to RA-5C standard, as was the RF-8A/G Crusader. In recent years, the F/A-18 Hornet has also been developed into the RF-18, although only one development aircraft had been flown by 1986.

Among the types which have been fitted with reconnaissance pods as standard for the

Used extensively for maritime reconnaissance and intelligence-gathering, the Tupolev Tu-20 Bear-D was first seen in 1967. Since then aircraft have been detached to Somalia, Angola, Guinea, Vietnam and other Soviet client states.

ENGLISH ELECTRIC CANBERRA PR9

Country of origin: Great Britain.
Role: Two-seat photo-reconnaissance aircraft.
Wing span: 20.7 m (67.83 ft).
Length: 20.3 m (66.67 ft).
Max T/O weight: 26,082 kg (57,500 lb).
Engines: Two 5,103 kg (11,250 lb) Rolls-Royce Avon 206 non-afterburning turbojets.
Max speed: 870 km/h (540 mph).
Max altitude: 18,000 m (59,140 ft).
Range: 6,426 km (3,993 mi).
Military load: (High/medium altitude) 4 × F96 and 1 × F49 cameras, or 4 × F52 and 1 × F49 cameras; (high altitude night) 3 × F89 Mk3 cameras and 2 × photoelectric cells operated by 5 × 203 mm (8 in) or 3 × 419 mm (16.5 in) photo flashes.

role have been the RF-104G Starfighter, Phantom FGR2, Jaguar GR1 and some F-16A Fighting Falcons.

The medium bomber category also produced its reconnaissance versions of one sort or another. The Canberra was developed into three versions, PR3/7/9, for photo reconnaissance and mapping; the Soviet equivalent was the Yak-26 Mandrake. From America came the RB-47 Stratojet, while the UK's V-bombers were all modified at one stage for reconnaissance: the Valiant B(PR)1, the Victor SR2 and the Vulcan MRR2 – the latter designation standing for maritime radar reconnaissance. Several Soviet bombers – the Myasishchev M-4 Bison-B, the Tupolev

Used successfully around the world as a bomber, the English Electric/BAC Canberra is still employed in a reconnaissance role by many forces, including the Venezuelan Air Force. The Canberra mainly uses cameras for this role.

Tu-16 Badger-E and F, the Tu-20 Bear-D and F and the Tu-22 Blinder-A and C – have all been afforded maritime, reconnaissance, signals intelligence (SIGINT), electronics intelligence (ELINT) or surveillance roles.

The two most well-known reconnaissance aircraft in the world both come from the United States, and both from Lockheed-California: the U-2 (and its recent TR1 derivative) and the SR-71 Blackbird. The U-2 came to public prominence in May 1960, when one was shot down over Sverdlovsk in the USSR on a CIA spying mission. A graceful slim aircraft with a large span, 24.38 m (80 ft), it could reach an altitude of some 25,910 m (85,000 ft) flying at a maximum speed of 805 km/h (528 mph) with a maximum range of 6,437 km (4,000 mi). The Russian incident, followed by similar ones over China and Cuba, put an end to its clandestine role, but it remains in service today.

The TR1 single-seat tactical reconnaissance version is officially described by the US DoD as being 'equipped with a variety of electronic sensors to provide continuously available, day or night, high-altitude all-weather stand-off surveillance of the battle area in direct support of US and Allied ground and air forces during peace, crises and war situations'. Some 35

aircraft are on order, the first of which flew in 1981, and the last will be delivered during 1986. Its principal sensor is a sideways-looking airborne radar, while some aircraft will be fitted with the precision location strike system (PLSS) able to 'see' 55 km (30 nm) into enemy territory without overflying an actual or potential battle area. It is expected the main use will be over Europe.

When the stealth techniques of the mid-1950s were overcome, speed was seen as the counter for the strategic reconnaissance role. Mach 3 is the claimed speed, at 24,000 m (78,740 ft), for the SR-71A two-seat successor to the U-2, which has a variety of sophisticated sensors – photographic, infra-red and electronic – housed in the forward portions of the wing/body chine fairings of the aircraft. Exact details of its capability are classified, but it is known that the aircraft can survey an area of 155,400 km² (60,000 sq mi) in one hour from an altitude of 24,400 m (80,000 ft).

Maritime reconnaissance

Maritime reconnaissance (MR) is usually bracketed with anti-submarine warfare (ASW) missions and both roles are incorporated into the one aircraft. These are usually long-range missions, typically 2,500 km (1,350 nm), and in addition to air-sea surveillance radars such aircraft are equipped with a variety of ASW sensors, including expendable sonobuoys, sniffers, which detect diesel exhaust fumes from surfaced submarines, and magnetic anomaly detectors (MAD), which detect a change in the local magnetic fields around the aircraft, caused by the metallic bulk of the submarine.

Aircraft in this role range from the Lockheed P-2 Neptune, Avro Shackleton, Canadair Argus and the Soviet bomber types of the early 1960s, through the Breguet Atlantic, Hawker Siddeley Nimrod and Lockheed P-3 Orion series to the Dassault-Breguet Atlantique Nouvelle Génération and the current P-3 Orion Phase III aircraft. The Soviet bomber/MR types remain in service with updated systems, and purely in the MR/ASW role, the Beriev M-12 Mail is one of the two flying boat types still involved in such operations, the other being the Shin Meiwa SS-2 from Japan. Until 1966, New Zealand operated five Short

LOCKHEED P-3C ORION

Country of origin: USA.
Role: Maritime reconnaissance and ASW patrol.
Wing span: 30.37 m (99.67 ft).
Length: 35.61 m (116.83 ft).
Max T/O: 64,410 kg (142,000 lb).
Engines: Four 3,661 kW (4,910 ehp) Allison T56-A-14 turboprops.
Max speed: 761 km/h (473 mph).
Max altitude: 8,625 m (28,300 ft).
Range: 3,835 km (2,383 mi).
Military load: (bomb-bay) 1 × 907 kg (2,000 lb) Mk25/39/55/56 mine, 3 × 454 kg (1,000 lb) Mk36/52 mines, 3 × Mk57 depth bombs, 8 × Mk54 depth bombs, 8 × Mk43/44/46 torpedoes or a combination of two Mk101 nuclear depth bombs and four Mk43/44/46 torpedoes, total weight of 3,290 kg (7,252 lb); (underwing) up to 5,443 kg (12,000 lb) stores, including 2 × AGM-84 Harpoon anti-ship missiles.

Sunderland GR5s of Second World War vintage in this role.

During the 1970s, the establishment of the 370 km (200 nm) Exclusive Economic Zone for resources off-shore led to the appearance of surveillance derivatives of medium-range

Highly successful and cost-effective, the Lockheed P-3 Orion maritime reconnaissance aircraft is in service with several NATO and friendly nations, including Australia, New Zealand, Japan and the Netherlands, as well as with the US Navy.

turboprop airliners, the most successful being the Fokker F27 Maritime. The advent of micro-electronics meant that surveillance radars and associated data processing allowed such systems to be installed in smaller aircraft. A whole new range of what have become known as coastal patrol types were then launched, derived from business jet types and twin-engined utility aircraft. The Britten-Norman Maritime Defender, Dornier Do 28 and Dassault-Breguet HU-25G Guardian are but examples from a whole range of aircraft adapted for this role.

At opposite ends of the maritime surveillance league are the Embraer Bandeirante *(below)* for EEZ surveillance and anti-smuggling type operations (Brazilian Air Force version shown), and the jet-powered British Aerospace Nimrod long-range maritime patrol aircraft *(bottom)*, which is reportedly the best-equipped aircraft of its type. Nimrod carries various stores and sensors, including the Searchwater radar and a magnetic anomaly detector designed to detect submerged submarines from the air.

0   4m

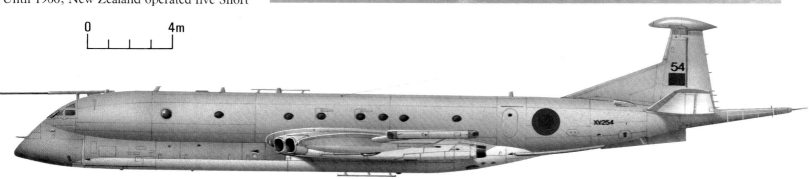

Remotely piloted vehicles

There is no doubt that to overfly an important tactical or strategic target in war is a dangerous business. The use of remotely piloted vehicles (RPVs) with reconnaissance sensors has been pursued with varying degrees of intensity over the last 25 years.

During the Vietnam War, the United States adopted several RPVs derived from drone targets for this role, and achieved a measure of success. This led the US Army to formulate a requirement for a dedicated RPV for such purposes, and in the mid-1970s, the Lockheed YMQM-105 Aquila project was born. It has been plagued by technological and financial problems; it was still not operational in 1986. Beechcraft and Northrop Ventura Division are other companies which are involved with this field, while LSI and Boeing also offer such systems.

Belgium flirted with an RPV known as Asmodée or Epervier during the 1970s, but apart from a Belgian Army order, it did not make an impact. The Canadair CL-89 achieved more success, and was ordered by several NATO nations, including the US and UK. As a benchmark for this type of system, it

The Hercules transport is used by the US military to carry remotely-piloted vehicles (drones) both for target practice and for the gathering of intelligence in battlefield or clandestine situations where a pilot cannot be risked.

is equipped with a camera and infra-red linescan equipment, has a maximum launch weight of 156 kg (343 lb), a maximum speed of 741 km/h (400 kt) and a maximum altitude of 3,050 m (10,000 ft).

Germany, France and Italy have all developed RPVs of varying configurations, both fixed and rotary winged. Some have looked like mini fast jet aircraft, while others have looked like glorified flying model aircraft. The UK has followed a similar pattern, and Westland was well into the development of a rotary-wing Wisp/Wideye for the British Army, when it was cancelled in the late 1970s. The replacement is known as Phoenix, and in 1985, the UK MoD announced it had selected the GEC Avionics/Flight Refuelling bid for development. The latter company are producing the air vehicle while the former will produce the sensor fit.

It has been Israeli experience with RPVs, notably in the Bekaa Valley in Lebanon during 1984, that has made most impact on the growing acceptance of the type. Two types, the IAI Scout and the Tadiran Mastiff, were used in action, not only for reconnaissance purposes, but also as decoys for SAM units. The Bekaa experience has had two major results: IAI and Tadiran have pooled their resources to form Mazlat, and develop their RPVs further and the US Navy bought a number of Scouts off the shelf for evaluating the concept.

Airborne early warning

In the early 1960s, the United States and Great Britain were the only two nations with a viable AEW force: the Grumman WF-2 Tracker of the US Navy, with a mushroom antenna configuration above the fuselage; the Lockheed EC-121H Constellation of the USAF and the Fairey Gannet AEW3 of the Royal Navy. With the demise of conventional air groups in the mid-1970s, a number of Gannet APS-20 radars were fitted into Shackleton airframes to give the AEW2 version.

In the US Navy, the Grumman E-2 Hawkeye series replaced the Tracker in three consecutive variants, the E-2A, B and C models, the latter being the current version. The USAF began looking for an EC-121 replacement in the 1960s under the Airborne Warning and Control System (AWACS) programme. By 1970, Boeing had been selected to provide the platform, a modified 707 airliner, and in late 1972, Westinghouse was selected to provide the radar. The resulting system was designated the E-3A Sentry, but is still known as AWACS. It can provide comprehensive surveillance to a range in excess of 370 km (200 nm) against low-flying targets, and farther for targets at higher altitudes.

As early as 1974, NATO had identified the E-3 Sentry as the type to equip a NATO force. However, considerable procrastination within the member countries at the political and industrial level, resulted in Great Britain developing her own AEW technology as well as being involved in the NATO discussions. The UK RAF launched its own AEW project using the Nimrod airframe, and NATO decided on an 18-aircraft force of E-3A Sentries. Cost overruns on the mission system avionics for the Nimrod AEW3, causing a minor political/industrial scandal in 1985, have meant that it will be 1987 before it comes into service, if at all, and even then it will not be up to the original specification. The NATO force, however, took delivery of the last E-3 in mid-1985. Despite the problems with the British AEW radar and avionics system, it is interesting to record that Lockheed, in association with GEC Avionics, is offering an AEW version of the C-130 Hercules transport equipped with the British radar.

In the last five years or so, several other nations have ordered and received AEW aircraft. Saudi Arabia has opted for five E-3 Sentry derivatives, while the E-2C Hawkeye has been ordered by Israel, Japan, Egypt and Singapore.

During the Falklands campaign a crash programme was launched to fit Thorn EMI Searchwater radar to Westland Sea King helicopters to provide naval AEW. The Sea King AEW 2 was too late to be used operationally, but the system has now been

Above: among the various platforms now being offered for airborne early warning (AEW) is the Orion in a concept known as the airborne early warning and control system (AWACS). This has smaller capital and running costs than jet-powered aircraft.

developed and procured for the air groups of the 'invincible' class aircraft carriers. The system has also been sold to Spain, Italy and India. Another variant of the same radar, now known as Skymaster, has been fitted into the nose of a Pilatus Britten-Norman Defender, and is being offered as a low-cost AEW system for nations not requiring the sophistication of an E-2, E-3 or Nimrod type of aircraft.

If the Western nations have decided that AEW is a major requirement, then its adoption by the Soviet Union was but a matter of time. However, the when and how of Soviet AEW technology has not been determined. What is known is that they do have such a capability in the Tupolev Tu-126 Moss, fitted with a mushroom rotordome system above the fuselage, similar to the E-2/E-3 systems. It is also understood that there is an AEW version of the Ilyushin Il-76 Candid in existence, and that the Il-86 airliner is also being modified to provide an AEW type.

Inside the NATO E-3A Sentry AWACS aircraft, controllers for the Continental European nations, Canada and the United States keep watch of the movements of Warsaw Pact and other aircraft. The AWACS concept is designed around the Boeing 707.

After the loss of *Sheffield* during the early stages of the Falklands conflict, the Royal Navy acquired a number of Thorn EMI Searchwater radar sets and fitted them to the Sea King Mk2 helicopter, as on this Sea King from the 849A flight.

The airborne truck

Unless personnel are specifically involved in the mission, transport aircraft in a military context tend to be taken for granted. On the face of it, their role has none of the excitement of the fast combat jets, but transports do have their moments – ask any C-130 Hercules pilot who flew in and out of Khe Sahn during the Vietnam War.

Transport aircraft fall neatly into two categories: strategic and tactical. The air transport of passengers and cargo over long distances, or from country to country, is

considered to be strategic. Once within the theatre of operations all air movement of passengers and cargo becomes tactical air transport.

The strategic side of the category is, perhaps, the easiest to cover, comprising as it does mainly civil airliners in military guise and cargo aircraft designed specifically for military loads and operations.

Of the five major types of military freighters in use since 1960, one is British, two are American and two Russian. Weighing in at 104,462 kg (230,300 lb), with a payload of 35,181 kg (77,500 lb), the Shorts Belfast C1 was the first British aircraft designed from the outset as a military transport. Powered by four 4,274 kW (5,730 ehp) Rolls-Royce Tyne Mk101 turboprops, the Belfast could fly some 5,794 km (3,128 nm) with a 9,979 kg (22,000 lb) payload. It entered service with the RAF in 1966 and was withdrawn in 1976, following a round of defence cuts. Five of the 10 aircraft built eventually went to a civilian freight airline, and three of these found themselves hauling military cargo, under contract to the UK MoD, during the Falklands campaign.

Lockheed-Georgia produce both the USAF's long-range strategic transport aircraft, the C-141 Starlifter and the C-5 Galaxy. The C-141 was designed to USAF Military Airlift Command (MAC) requirements, and can haul a maximum payload of 32,136 kg (70,847 lb) a distance of 6,565 km (3,899 nm). Comparable to the Belfast in terms of size and cargo/passenger capacity, it can carry 153 troops. It entered service as the C-141A in 1964 and during the late 1970s was stretched in the fuselage by some 7.1 m (23.3 ft) to produce the C-141B, which also features an in-flight refuelling capability. Although supposed to be able to carry '90 per cent of all air portable items in the (US) Army and Air Force' its flexibility was restricted by its body cross-section of 3.3 m (10.75 ft) – the same as the C-130 Hercules.

One of the world's most widely used transport aircraft, civil or military, the C-130 Hercules has been used in both conflicts and relief operations, such as the RAF operations in Ethiopia and Nepal.

ANTONOV AN-124 RUSLAN (NATO CONDOR)

Country of origin: USSR.
Role: Strategic transport/cargo aircraft.
Wing span: c.73.3 m (240 ft).
Length: c.69.5 m (228 ft).
Max T/O weight: 405,000 kg (893,000 lb).
Engines: Four 23,430 kg (51,655 lb) Lotarev D-18T turbofan engines.
Max speed: c.850 km/h (528 mph).
Max altitude: c.12,000 m (39,360 ft).
Range: c.16,500 km (10,253 mi).
Military load: Maximum payload of 150,000 kg (330,750 lb). Up to 85 troops in upper cabin, with cargo in main hold.

Providing the US military with worldwide transportation, the C-5A Galaxy is one of the largest aircraft in the world. From the high flight deck *(top)*, the crew are able manoeuvre the aircraft on confined airfields. Powered by four General Electric TF39-1 turbofans, the Galaxy *(above)* is capable of operations from rough field locations thanks to the 28 main landing wheels. Entering service in 1986, the C-5B Galaxy has updated systems for ease of maintenance.

propellers, it has a maximum cruise speed of 679 km/h (403 kt) and with its maximum payload has a range of 5,000 km (2,700 nm), while with a lighter payload and maximum fuel it can fly 10,950 km (5,913 nm). Beaver-tail doors in the rear allow loading into the cargo hold of 4.4 m (14.4 ft) square section. The Cock is in military and Aeroflot service.

The 1985 Paris Air Show allowed the West its first public look at the most recent Soviet strategic transport, the An-124 Condor. Looking like a Russian Galaxy, the Condor has a maximum take-off weight of 405,000 kg (893,000 lb), and a maximum payload of 150,000 kg (330,750 lb). It is powered by four Lotarev D-18T turbofans and has a cruise speed between 800-850 km/h (432-459 kt). Range with maximum payload is 4,500 km (2,430 nm) and with a smaller, unspecified payload, maximum range is 16,500 km (8,900 nm). The nose door can be raised, visor-style, and there are beaver-tail doors, so simultaneous loading/unloading is possible, while the undercarriage has special features to lower the floor height and adjust inclination. The first flight of the Condor was 26 December 1982, and production deliveries began in 1986.

Tactical transport

Four major tactical types of transport operations can be identified: airborne and assault landings operations; air logistic support operations; special missions and aeromedical evacuation.

The fixed-wing aircraft used in the tactical role can vary from the medium-sized transport/cargo aircraft, which in some air forces fulfil a semi-strategic role, earning themselves the unofficial classification of 'tactical 'tweenies', through modified medium-range airliners and commuter airliners, down to utility types and versions of standard light aircraft. These aircraft come, therefore, in all shapes and sizes, and at prices to meet a variety of defence budgets. When new air forces or air

This restriction, together with other needs, led MAC to formulate a requirement that resulted in the C-5 Galaxy. This high-wing transport has a fuselage cross-section of 6.8 m (19 ft) and an interior height of 4.1 m (13.5 ft), and can fly a maximum payload of 100,228 kg (220,967 lb) a distance of 6,033 km (3,258 nm). To meet a MAC requirement it should be able to operate with heavy loads out of rough short airstrips. It has a high flotation undercarriage with 28 wheels. It entered service in 1969, and the need for additional heavy airlift capability was recognised in 1972, when a programme to build a further 50 Galaxies, designated C-5B, was initiated. It incorporates the experience of C-5A operations, and is virtually identical. Among the improvements are digital avionics, improvements in construction materials and better reliability, maintainability and availability. The first aircraft flew in September 1985, and first deliveries to the USAF were scheduled for December that year. It will be able to haul a similar payload in excess of 5,370 km (2,900 nm). Both versions of the Galaxy and the C-141B will remain in USAF service for the foreseeable future.

The first of the two Soviet giants was first seen in public in 1967, when three Antonov An-22 Cock transports were demonstrated at an air show. After the Boeing 747 and C-5 Galaxy, it was the largest aircraft in the world. Its maximum loaded weight was 250,000 kg (551,160 lb), with a maximum payload of 80,000 kg (176,350 lb). Powered by four 15,000 shp Kuznetsov NK-12MA turboprops, each driving two four-blade contra-rotating

Built by a Franco-German-Dutch consortium, the Transall C-160 serves with the Federal German, French, Turkish and South African Air Forces as a tactical transport aircraft. The aircraft was first delivered in 1968 and will remain active until 2000.

TRANSALL C-160 (SECOND SERIES)

Country of origin: France/Germany.
Role: Tactical transport aircraft.
Wing span: 40.00 m (131.25 ft).
Length: 32.40 m (106.29 ft).
Max T/O weight: 51,000 kg (112,435 lb).
Engines: Two 4,549 kW (6,100 ehp) Rolls-Royce RTy20 Mk22 turboprops.
Max speed: 593 km/h (368 mph).
Max altitude: 8,230 m (27,000 ft).
Range: 5,095 km (3,166 mi).
Military load: 93 troops; 61-88 paratroops; 62 stretchers plus four attendants; or 8,000 kg (17,637 lb) of air-droppable cargo.

arms are being established within emerging nations, it is tactical transport aircraft which form the nucleus of the force, well before any combat aircraft are discussed.

Among the many types which may be included in the former bracket are the Franco-German Transall C-160, Aeritalia G222, Antonov An-12 Cub and de Havilland Canada DHC-5D Buffalo. The C-160 is a European collaborative project of the 1960s,

first flying in 1963. Powered by two Rolls-Royce Tyne 22 turboprops, it can carry a maximum payload of 16,000 kg (35,274 lb) a distance of 5,200 km (2,808 nm). Although the initial production run ended in 1972, the first of a second series of 25 C-160s for France was delivered in 1979.

Collaboration for NATO-standard aircraft was popular in the early 1960s and a competition resulted in a Fiat design, the G222 with two Rolls-Royce Dart turboprops and eight lift jets. Further drawing board development traded off the lift jets for fuel and substituted General Electric T64 turboprops for the Darts. Although the aircraft had grown in empty weight by two tons, it was ordered by the Italian Air Force in 1974, and nearly 80 aircraft were built for home and export, although none of the export sales were achieved within NATO. Although it can carry up to 9,000 kg (19,840 lb), its range with a 5,000 kg (11,025 lb) payload is 2,950 km (1,593 nm).

The An-12 Cub entered Soviet service in the early 1960s. Powered by four Ivchenko A1-20K turboprops, it can carry a maximum payload of 20,000 kg (44,090 lb) some 3,600 km (1,942 nm). This aircraft is widely used within the Soviet Bloc and sympathetic countries. One particularly interesting design feature is a power-operated tail gun turret, housing a pair of 23 mm cannon.

A gun turret is also a feature of the Cub

replacement, the Ilyushin I1-76M Candid. Similar in concept and configuration to the C-141A, the Candid has a nominal range of 5,000 km (2,700 nm) with a maximum payload of 40,000 kg (88,185 lb). It is powered by four Soloviev D-30KP turbofan engines, and can cruise at between 750-800 km/h (405-432 kt). The Candid is in wide use with Aeroflot and the Soviet military, and has been exported. It made its public debut at the 1971 Paris Air Show, although series production did not start until 1975.

The ubiquitous Hercules

Perhaps the most well-known of military transports after the Douglas C-47/DC-3 Dakota must be the Lockheed C-130 Hercules. Production of this four-engined tactical transport was begun in 1952 and at the time of writing, Lockheed are confident it will still be in production at the turn of the century. At the end of October 1985, some 1,802 examples had been ordered, of which 1,755 had been delivered. In all, 27 different military variants have been developed. Apart from tactical transport variants, special mission, tanker, SAR, EW and drone-launch aircraft have been produced, as well as three civil cargo variants. As might be expected, the aircraft has sold all around the world in 57 countries.

LOCKHEED C-130H Hercules

Country of origin: USA.
Role: Tactical transport aircraft.
Wing span: 40.41 m (132.58 ft).
Length: 29.79 m (97.75 ft).
Max T/O weight: 79,380 kg (175,000 lb).
Engines: Four 3,362 kW (4,508 ehp) Allison T56-A-15 turboprop engines.
Max speed: 602 km/h (374 mph).
Max altitude: 10,060 m (33,000 ft).
Range: 7,876 km (4,894 mi).
Military load: 92 troops; 64 paratroops; 74 stretchers plus two attendants; or cargo up to 12,080 kg (26,640 lb).

Used on the Israeli rescue mission to Entebbe (Uganda) in 1976, this C-130H Hercules is fitted with two 5,146 litre (1,132 US gal) underwing tanks. As many as 92 troops can be carried in the standard version, 128 in the stretched type.

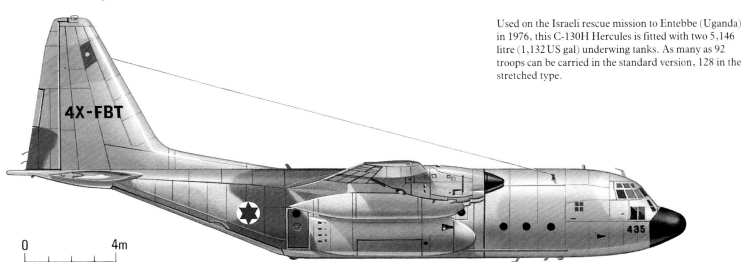

4X-FBT

0   4m

Air War South Atlantic

The conflict in the South Atlantic began on 2 April 1982 and ended on 14 June 1982, again demonstrating the indivisibility of air power. Aviation played a major role in both sides' campaigns, and was there from the first landings by Argentine Naval Sea Kings around Port Stanley to the RAF Harrier GR3s on their way to deliver Paveway laser-guided bombs against ground targets when the mission was aborted at the 'eleventh hour'. It saw the operational debut – and proved the concept of – the Harrier/Sea Harrier V/STOL fighters and the modern anti-ship missile, particularly the AM39 Exocet and Sea Skua. It also saw, in the very twilight of its career, the only operational use of the Vulcan bomber – in its conventional role.

The air war of 1982 did for the strategists and tacticians of the 1980s what the Six Day War of 1967 did for their predecessors in the late 1960s: it proved many new weapons and aircraft, it taught new lessons and reinforced basic points which had been almost forgotten.

The campaign proved to the world that the STOVL (Short Take-Off, Vertical Landing) concept worked. The UK's Chief of Naval Staff encapsulated the importance of the concept: 'Without the Sea Harrier there could have been no Task Force.' It was the major combat aircraft deployed by British forces in both its air defence (Sea Harrier FRS1) and ground-attack (Harrier GR3) versions because it was the only type of fixed-wing combat aircraft deployable with existing resources and access.

The Falklands campaign confirmed the operational flexibility by safe launch and recovery without an aircraft carrier having to turn into wind. This was achieved in poor weather and in high sea states that would have prevented conventional carriers flying-off and recovering their aircraft. It could deploy to ships other than carriers, such as assault ships and merchant vessels, even ashore. The Sea Harrier showed a high combat agility, using vectored thrust, particularly at low level.

With effective airborne early warning (AEW) capability, as provided by the Gannet AEW3 prior to the withdrawal of the last conventional British carrier in 1978, the Sea Harriers' use would have been even more effective. With AEW, since restored with the adaption of Searchwater radar on a Sea King helicopter, the Super Etendard strike

Top: providing air defence and some ground-attack support, the Sea Harrier was outstanding during the Falklands conflict. Using the AIM-9L Sidewinder, Fleet Air Arm pilots were able to destroy 16 enemy aircraft without loss.

Above: daily attacks on the British task force in San Carlos were carried out by Argentine strike aircraft including Mirage III and Israeli Aircraft Industries Dagger. This Dagger is seen attacking a troop transport (LSL) of the Royal Fleet Auxiliary.

fighters, which launched the Exocet anti-ship missiles which sank the *Sheffield* and *Atlantic Conveyor*, would never have reached a launch position.

The success of anti-ship missiles launched from aircraft and helicopters proved that major naval units were far from invulnerable to destruction by a relatively small threat. On the naval side, this had led to the adoption of more effective ESM, ECM and point defence weapons. Bombing proved to be less effective; despite continuous attacks against the runway at Port Stanley with high-explosive and cluster bombs, it was never taken completely out of action. The

one major hit on the first of the Vulcan Black Buck raids did, however, cut its effective length in half and deny its use to the high-performance jet fighters such as the Skyhawk and Dagger. In one respect the longest-ever RAF bombing raid of 11,260 km (7,000 mi) round trip was effective because it also proved the value of air-to-air refuelling (AAR).

Air-to-air refuelling proved to be a 'force multiplier' which, apart from putting one Vulcan at a time over Port Stanley, expanded the range of Nimrod maritime patrol and Hercules transport aircraft. Initially it fell to the Victor tankers of the RAF, but later they were joined by a number of hastily modified Hercules. Argentina, too, used AAR to effect despite only having two KC-130 Hercules tankers. Many strike aircraft were relieved to rendezvous with them on the return from San Carlos Water.

Again both sides confirmed the vital place of the helicopter over the modern battlefield. The conflict saw its effective use in many roles, such as ASW, ASVW, assault, transport, resupply, casevac and SAR. The loss of Wessex and Chinook helicopters in *Atlantic Conveyor* almost cost Britain the land battle and precipitated the disaster at Bluff Cove.

Bluff Cove, together with San Carlos Water, illustrated the continuing vulnerability of ships to hits from both high-explosive bombs and unguided rockets. The latter weapons, fired from an MB339 trainer, were responsible for the loss of the frigate *Ardent*. Fortunately for the British, many Argentine bombs were dropped too low for their fuses to take effect. Had they detonated, British shipping losses would have been greater.

The principal reason for the ineffectiveness of low-level attacks over British ships was the effectiveness of their SAM defences, both ship-based and, after the landings, ground-based Rapier and Blowpipe. Guns too, played their part, both sides deploying weapons from 40 mm down to small arms against aircraft.

Two types of air-to-air missiles were used: Matra Magic (Argentina) and AIM-9L Sidewinder (UK). Although some Magics were launched in the early stages by Argentine Mirage IIIEAs to no effect, it was the Sea Harrier's AIM-9L Sidewinders which swept the sky with 16 confirmed kills.

The United Kingdom actually fought the air war with distinct disadvantages, while Argentina held most of the trump cards. There is no easy answer to why Britain won, but in the context of the air war, the superior training level of British air and ground crews must have had its effect. Anyone who saw the air attacks over San Carlos Water cannot doubt the courage of the Argentine pilots, but they had never been trained for such warfare or operational conditions. For the British, too, the Falklands scenario was never foreseen, but their NATO role and training gave them the edge to press home what advantages they possessed and make use of Argentine inexperience. It was, indeed, 'a close-run thing'.

Michael J. Gething

The aircraft's specification seems to have covered the role envisaged in the 1950s, yet left room for growth, improved engines and new methods of construction. Typical of its growth is the fuselage stretch option, similar to the C-141B modification, which was initially developed for the RAF. Their C-130K, basically an H model with avionics and instruments of British manufacture, was delivered during the late 1960s. In 1978, it was announced that 30 of the 66 C1 versions bought would have their fuselages stretched by 2.54 m (8.3 ft) forward of the wing and 2.03 m (6.67 ft) aft of the wing to produce an equivalent of the L100-30 commercial model, designated C3 by the RAF. As with most cargo, the volume bulks out before the weight, and so the capacity of the Hercules C3's hold was increased to 171.5 m³ (6,057 cu ft) or over 25 per cent.

The flexibility of the Hercules is well illustrated by the modifications done by Marshalls of Cambridge (UK) during the Falklands campaign of 1982. In order to extend the aircraft's range, to keep up with the progress of the Task Force as it made its way south, a two-tier fuel tank system was installed in the cargo hold. This was followed by the modification of several aircraft with an inflight refuelling probe, and later by a ramp-mounted Mk17 hose drum unit in order that the aircraft could act as a tanker, as well as a receiver of fuel, easing the pressure on the RAF's overworked Victor tanker force. During the campaign, one RAF Hercules flew an operational mission lasting 28 hours, and set a new Hercules duration record.

A Hercules replacement?

During the 1970s, the United States initiated a programme to develop an Advanced Medium STOL Transport aircraft, which resulted in two manufacturers, Boeing and McDonnell Douglas, building two development aircraft for evaluation – the YC-14 and YC-15 respectively. This programme was not proceeded with because the best replacement for an old Hercules is a new one!

In 1980, a slightly different requirement was formulated, under the designation C-X. Conceived as a long-range, heavy-lift, air-refuellable cargo transport, it was intended primarily to provide intra-theatre airlift of outsize loads, including the US Army's M1 Abrams tank and the M2/3 Bradley infantry fighting vehicle, directly into airfields in potential conflict areas. It would have the outsize load capability of the C-5 Galaxy combined with the STOL performance of the C-130 Hercules.

McDonnell Douglas's C-17 design was selected in August 1981, but full-scale development was not authorised until February 1985. An indication of its capability shows a maximum payload of 78,110 kg (172,200 lb), take-off in 2,320 m (7,600 ft), landing in 915 m (3,000 ft) and a range of 4,445 km (2,400 nm). Current plans call for a first flight in 1989, and initial operating capability in 1991.

The modified medium-range airliner sector

Supporting ground forces around the world, and to many people a replacement for the Dakota, the DHC-5 Buffalo is good for short and rough field work. It has also been a substitute for the larger Hercules and it has served with the US Army.

of the spectrum covers a variety of types, of which the Fokker F27 and the Hawker Siddeley 748 series were the world leaders during 1960-85. Among the utility types and commuter airliners adopted to the role are the CASA C-212 Aviocar, de Havilland Canada DHC-6 Twin Otter, Dornier Do 28 and 228, Embraer EMB-110 Bandierante, Pilatus Britten-Norman Defender series, Shorts Skyvan and its most recent development the C-23 Sherpa, for the USAF.

Training the fledglings

In the 1980s, flying training patterns are different to those 20 years ago. More aircraft types were available and operating costs were lower in the 1960s, but today costs of acquisition and operation are higher, with the needs of air forces more demanding. In 1960, for a relatively sophisticated NATO-level air force, pilots could expect to go through an elementary phase, basic phase and advanced phase of training before being streamed into fast jet, multi-engine or helicopter categories, where there would be a second level of advanced training and operational conversion on to type. Other types of aircrew could expect a similar process up to operational conversion.

If one takes the UK RAF as an example, in 1960 elementary training was carried out in the de Havilland Canada DHC-1 Chipmunk T1, a piston-engined aircraft, with basic training on either the last of the Hunting Provost T1s or the Hunting BAC/BAe Jet Provost T1/3s. Advanced flying training at this time was on the de Havilland Vampire T11, with the Hawker Siddeley (Folland) Gnat T1 taking over from 1965 and supplemented from 1967 with Hawker Hunter T7s. Apart from the Chipmunk and Gnat, which were tandem

The mount of the famed Red Arrows formation display team for many years, the Folland Gnat was highly successful in British service; it was also built under licence in India, serving as a fighter and receiving the nickname 'Sabre Slayer'.

configuration, the other types had side-by-side seating, as favoured by the RAF at that time. As the Jet Provost T1s were phased out, the streamlined and pressurised T5 version took over, while the elementary Chipmunk phase was eliminated.

From 1977, a rationalisation programme for training was initiated, with the BAe Hawk T1 taking over from the Gnat and Hunter (for basic weapons training), plus some Jet Provost roles. As the programme stands in 1985, direct-entry student pilots, who have not gained elementary flying training experience on the Scottish Aviation Bulldog T1 at University Air Squadrons, spend some 93 hours on the Jet Provost T3, 57 hours on the Jet Provost T5 and 75 hours on the Hawk T1.

During the 1960s-70s, RAF thinking preferred the side-by-side seating arrangement, but in the early 1980s the tandem seat configuration was gaining prominence. So when, in 1983, the RAF began formulating its requirement for a Jet Provost replacement this

configuration was preferred. At the same time, it indicated that jet-powered and turboprop types would be considered. This was a result of ever rising fuel costs. In the event 15 possible types were narrowed down to four contenders, all turboprops with tandem seating. The final decision was announced in March 1985 in favour of the Shorts-developed version of the Embraer EMB-312 Tucano from Brazil. The other contenders were the Pilatus/BAe PC-9, the NDN Aircraft NDN-1T Turbo-Firecracker and the Australian A20 Wamira, bid in collaboration with Westland. Some 130 Tucano T1s are to be supplied to the RAF up to 1990 to cover between 100-140 hours presently flown by Jet Provosts. The difference in flying hours will be taken up by extra Hawk time or increased use of simulators.

In America, the USAF has preferred the side-by-side arrangement in basic training on the Cessna T-37. Its presumed replacement, the Fairchild Republic T-46A, has a similar arrangement. Pilots then move to the Northrop T-38 Talon for advanced training before operational conversion. In France, after elementary training on the Mudry CAP 10B, pilots will move on to the Aérospatiale Epsilon and thence to the Dassault-Breguet/Dornier Alpha Jet. Italy starts its pilots on the SIAI-Marchetti SF260AM piston-engine basic trainer before progressing to the advanced Aermacchi MB 339A jet trainer, which replaced the MB326 from 1981. After operating a multitude of older types, Spain is now bringing in a syllabus which uses the Chilean Enaer T-35 Pillan, built under licence in Spain, for basic training and the CASA C-101 Aviojet for advanced training.

What has happened in the last 25 years of pilot training is a condensation of training types in service as a result of cost-effectiveness studies into pilot training. Advanced trainers are more akin to the fast combat aircraft flown on operations today than they ever were in 1960. As witness of this fact is the success of the development of the Hawk, Alpha Jet and

Capturing an interesting segment of the 1980s training market, the Embraer Tucano was selected for the Royal Air Force (with licence production by Shorts) in 1985. These five, first-generation Tucanoes, are from the Brazilian Air Force.

Combining a training role with light ground-attack and anti-helicopter operations, the Dassault-Breguet/ Dornier Alpha Jet has been successful in the European and export market places. The aircraft is flown by the French aerobatic team.

MB339 types as light strike/ground-attack aircraft. Training of non-pilot aircrew, navigators, weapon operators and observers, begins with multi-seat trainers such as the HS125 Dominie, Boeing T-43A (737 airliner) or equivalent, while flight engineers and electronic warfare operators follow a similar course on specially modified types similar to those mentioned or in service in the role.

Electronic warfare

Despite popular contentions to the contrary, electronic warfare is as old as the application of electronics in the military environment. Although the concept was employed during the First and Second World Wars, its widespread use in all types of combat aircraft did not make its impact until the mid-1960s when the volume constraint of components was solved. In 1960, electronic warfare was the prerogative of medium to large aircraft configured specially for the role. Such aircraft still exist today. The Grumman EA-6B Prowler and EF-111As are prime examples, and reduction in system size has meant an increase in EW capabilities limited only by the power

generation capability of the aircraft. Today, by a variety of means and systems detailed below, an aircraft such as the F-16 has the ability to detect and jam hostile radar emissions, and deploy decoys against both infra-red and radar-guided missiles: something the Hunter did not possess in 1960.

The simplest form of EW suite for any military aircraft is a radar warning receiver (RWR) and a decoy dispenser. The RWR is the electronic 'ears' of a combat aircraft, which can detect whether the aircraft is being 'painted' by a radar, and in more sophisticated systems, the direction and type of radar used can also be displayed. Even without a counter to the threat, the knowledge it gives can help the aircrew survive.

The simplest counter to radar detection is decoys: chaff to spoof a radar or radar-guided missile; and flares to provide a 'hot spot' in the sky to convince an IR-guided missile there is a better target to home in on. These chaff/flare dispensers can be fitted to aircraft in three ways: internally, with only their exit ports showing, on the fuselage side or on an

A new generation of internally mounted electronic countermeasures equipment has been developed by Thomson CSF for the Mirage 2000 series. Details are classified but it is thought that the housing on the fin top contains a radar warning receiver.

underwing pylon. They are simple, relatively cheap and remarkably effective when used in conjunction with RWR.

Having detected enemy radar emissions directed at an aircraft, prevention being better than cure, a means to jam the transmission is useful. The need to find a means of jamming North Vietnamese SA-2 SAM radars tracking US combat aircraft in 1965 led to the development of the jamming or electronic counter-measures (ECM) pod, which could be carried on a weapons pylon and connected to the aircraft power supply.

Jamming theory is simple, but in practice it is somewhat more complicated. Assuming an illuminating radar can be identified and classified, then the jammer can transmit large amounts of RF energy over a wide frequency waveband – a technique known as broadband or barrage jamming. This, however, is wasteful of energy, and leaves the aircraft open to illumination on a narrow frequency. If the detecting sensors can acquire the waveband of the transmission classified accurately, as is now possible in many systems, then the available jamming energy can be used on the narrow frequency in a technique known as spot jamming. Early ECM pods were pre-calibrated to deal with specific threat radars such as the SA-2 missile, but over 20 years, techniques and equipment have vastly improved. Today, it is possible to re-programme ECM pods on the ground while the aircraft is being rearmed and refuelled.

This same sophistication has also led to the development of internal ECM fits within aircraft, thus freeing valuable weapons pylons. The need for internal ECM is now being addressed at the design stage of aircraft, rather than as an afterthought. The value of pods today, however, is that they can be readily 'hung' on any aircraft with hardpoints. Tactical flexibility is all, but eventually the majority of aircraft over the battlefield, fixed wing or helicopters will have internal ECM fits. Airborne electronic warfare is here to stay.

Described as the best packaged aircraft in history, the Lockheed S-3 Viking is a carrier-based anti-submarine warfare aircraft. The package includes a host of electronic systems, including magnetic anomaly detection in the extended boom.

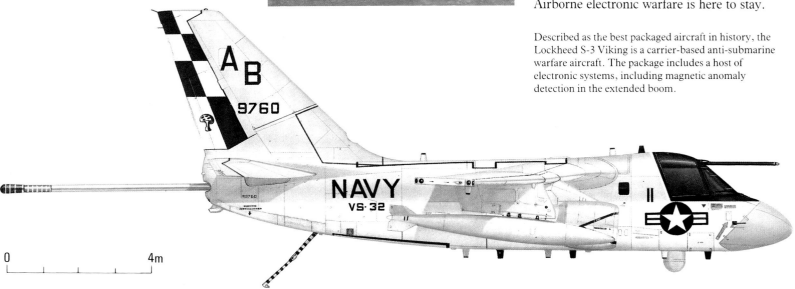

0   4m

4

AIRLINES AND AIRLINERS

Today's jumbo jets, high-technology commuter aircraft, and high-speed airliners owe their development, in part at least, to the legacy of surplus military aircraft from the First World War. These aircraft saw service in the 1920s and 1930s in developing the first air routes. So widespread are these now that today there is not a single nation without an airline or internationally served airport.

Technology has allowed the flight deck to become simpler yet more efficient, reducing the number of crew needed without abandoning the safety priority. In engine technology developments have been made in noise reduction, fuel economy and size. The prop-fan and turbo-prop are particularly good examples.

When on 17 December 1903 Orville Wright successfully completed the first controlled man-carrying mechanical flight in history a new world was born.

The aeroplane offered the potent promise of a form of transport that is not limited by earth-bound conventions, problems or obstacles. Under the wings of the aeroplane national boundaries disappear; rivers and hills have little relevance to travel and even the widest oceans can be bridged. It seemed that the aeroplane not only offered a mechanical solution to the chore of travel but also created a new mental attitude that had little time for the artificial constraints of national limits or exclusive areas.

It was an attractive dream and one that did not have time to develop, yet much of that early promise still remains and the intrinsic freedom of aviation may still be experienced today and noted in the demeanour and attitudes of the people involved with all aspects of the operation of aircraft.

.The triumph of the Wright Brothers' aircraft set the future course of heavier-than-air flight. Prior to their experiments most of the current aviation technology had been concerned with lighter-than-air flight. Balloons and airships were relatively common and seemed to offer a more practical solution than the mechanical complexities inherent in the emerging aeroplanes.

Certainly the early aeroplanes were flimsy,

Flying over New York's skyline, the German Zeppelin Hindenburg, seen in 1936, the year of its completion. The following year the craft burned out at Lakehurst, New Jersey in a devastating hydrogen fire.

prone to failure and usually underpowered. In contrast, some of the airship designs suggested large machines capable of sustained and, some would say, safer flight.

As the aeroplane struggled towards improved capabilities and performance, considerable work was being undertaken on both sides of the Atlantic with various forms of airship. Certainly most of the inventors concerned had an eye to the commercial possibilities of their ideas and several designs underlined this intention.

Airships were evolving in three basic categories – non-rigid, semi-rigid and rigid. The rigid airship was to prove the more persistent design and it began with the launching of just such an airship through the design genius of Count Ferdinand von Zeppelin. His first ship, the LZ-1, was launched on Lake Constance in July 1900. It featured an aluminium structure and was covered by cotton fabric. It was a remarkable machine as it established the basic pattern for all of the company airships to follow, up to and including the LZ-129, which was more popularly known as the *Hindenburg*.

Despite setbacks, other airships were built, each improving on the previous ship in the light of hard-earned experience. They began to be more successful and Zeppelin created a company called the Deutsche Luftschiffahrts-Aktien-Gesellschaft, which became known as the rather more manageable Delag. In 1910 the company began to carry passengers for hire and by the outbreak of the First World War more than 34,000 passengers had been carried without injury, establishing the concept of airline operations.

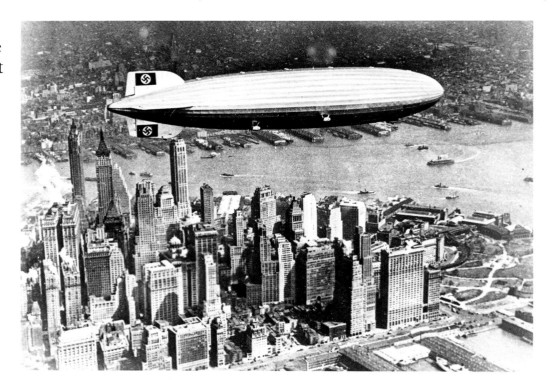

Heavier-than-air operations

As the First World War led to an upsurge in technology, heavier-than-air flight became more advanced and by the end of the conflict large aeroplanes that could fly long distances and carry a useful weight or payload had been developed.

Meanwhile, in the United States similar developments were leading to a number of new aircraft and some work had also been carried out on semi-rigid airships, but this work was suspended in 1912 following a fatal accident.

Nevertheless it is to the United States that the distinction of the first scheduled, heavier-than-air airline operation must be awarded. In December 1913, the city of St Petersburg in Florida agreed a contract with Thomas Benoist to operate a local air service. Benoist was an aircraft manufacturer who had previously made good in the motor car industry and wished to demonstrate the practical value of the aeroplane. The air service was to be provided by a recently formed company, known as the St Petersburg-Tampa Airboat Line, formed by an electrical engineer named Paul E Fansler and supported by local businessmen.

The service began on 1 January 1914 and took just 23 minutes to cross the bay from St Petersburg to Tampa. It quickly established routine operations of two services a day. Initially a Benoist Type XIV flying boat was used but as demand increased a second and larger aircraft was used. At first the airline achieved financial viability, but after some five months passenger loads declined, reflecting seasonal fluctuations in tourism, still a bane of airline life, and the service was terminated having carried 1,204 passengers safely and largely on time.

Other possible ventures of this kind on both sides of the Atlantic went into suspension during the four years of the war. During that time considerable technological advances were made and engineers developed practical aircraft of considerable capability in Germany, France, Great Britain, Italy, Russia and the United States. Large aircraft appeared in Germany, Great Britain and Russia. These were used in the bombing role, but it is

Starting the world's first scheduled air service, between St Petersburg and Tampa, Florida, this Benoist Flying Boat leaves the St Petersburg port on 1 January 1914. The route was the only pre-First World War service to be flown.

significant that the Russian candidate was derived from a design by Igor Sikorsky, of later fame as a helicopter designer, which was the world's first successful multi-engined aircraft and carried 16 passengers in an enclosed cabin.

After the war

With the end of the war the aircraft industry, which had expanded at an enormous pace, was suddenly faced with the problems of just as rapid a contraction. This meant that there was a surplus of aeroplanes and suitably trained personnel and this combination led to a variety of aeronautical enterprises.

Although this appeared to be an advantage initially, the large numbers of readily available cheap aircraft inhibited the introduction of new and better types, though it did not entirely rule out innovation, with Fokker and Junkers introducing new manufacturing techniques and materials and the monoplane configurations which laid down the foundations for the modern airliner.

In 1919 scheduled daily services were introduced in Europe on the Berlin–Weimar; Paris–Brussels and London–Paris routes. Lessons were quickly learned, but it soon became apparent that the economics of air transport are complex and often confusing.

Other forms of activity were also evolving with the advent of barnstorming operations, which had the effect of creating and sustaining aviation-mindedness among the general public. Cheap joy rides at seaside and carnival venues awakened an appetite for aviation in many minds and helped to sell some of the

Delivering the post became the first task of converted wartime bombers, in this case US-built de Havilland DH4s. The aircraft are pictured at the Omaha, Nebraska, staging post of the trans-continental US Air Mail. Note the open cockpits.

airline seats that would be available in the future.

Similar activities in the United States gained a degree of additional status by the development of the airmail routes, which were the true trail-blazers of today's airways. Many of the returning members of the Commonwealth forces elected to attempt to return home by air. Most failed even to start, but some achieved remarkable flights to South Africa and Australia, and were justly honoured. At the same time these early exploratory flights carried the message of aviation to remote areas and offered the basis of route surveys that were to prove invaluable over the next decade.

It was probably at this time that the seeds of a design philosophy were sown that led to a certain type of outlook by designers on either side of the Atlantic. In general terms the Americans were looking for range to suit the need for a 'coast-to-coast' capability. In Europe, and Britain in particular, the worry was to suit Empire routes with few facilities and primitive airfields.

This seems to have influenced designs ever since, with the American designers happy to accept increased wing loadings while British designs often emphasised short-field performance. It is interesting to see this philosophy borne out in later years by comparing the operational characteristics of the Boeing 707 and the BAC VC-10.

Perhaps it was the inevitable euphoria that always follows the cessation of war that now makes 1919 look something of a golden year in

Above: at Paris, an early post-War picture of a Liore &
Olivier 213, named 'Rayon d'Or' ('Golden Ray'). This
was one of the aircraft which gave France a lead in air
transport in the 1920s, although it was not long before
other nations caught up.

retrospect. Records were being set and reck-
less claims made about the future of aviation.
In July a British pilot and a navigator – Alcock
and Brown – made the first successful
non-stop crossing of the Atlantic from west to
east, thereby taking advantage of the prevail-
ing winds.

It was a considerable achievement but in
the public mind it had seemed relatively easy
and once more raised public expectation over
practical reality.

A more potent indication of the future of
civil aviation was to be noted by the holding of
a meeting in Paris now known as the Convention of Paris, which laid down the first rules for
European air traffic. Similar rules were ratified
by the United States in 1928 following the
Havana Air Convention.

Perhaps such rulings were inevitable but
they were the first of many blows against that
carefree promise of the freedom of flight.
These rules were couched in terms that had
already been used for the regulation of other
forms of transport on national lines – aviation
was now constrained by the declaration of
complete and exclusive sovereignty over the
airspace above a nation's territory.

With the coming of such ideas the concept
of using air transport for national purposes
inevitably followed and many of Europe's
emergent airlines were happy to co-operate,
for without subsidy there seemed to be little
chance of growth or even existence.

This background resulted in the formation
of several airlines, leading to mergers and
collaboration, of which a typical product was
Air Union in France. Meanwhile the glamorous operations with airmails across the
Sahara and down to West Africa, carried out
by an airline known simply as The Line, have
been preserved in the writings of Saint-
Exupéry.

In Britain the early days were marked by
the emergence of several operators but the
question of subsidies was vexed: they were not

attractive to central government. Sadly the
politicians seemed to lack the vision that was
required of those who would be part of the new
industry. British aviators could only watch the
growing strength of their European competitors and despite several attempts to gain
official support the government of the day was
adamant that British airlines must 'fly by
themselves'. It was less than satisfactory and
for one brief period there was no British airline
in operation.

Thankfully a new Secretary for Air – Sir
Samuel Hoare – was appointed and he had a
true belief in aviation, even if it was generated
by a desire to unite the many scattered
countries of the Empire. In true British spirit a
committee was set up under the chairmanship

HANDLEY PAGE HP 42

Engines: 4 × 555 hp Bristol Jupiter XFBM
driving 3.7 m (12 ft) diameter, four-bladed
propellers (W).
4 × 490 hp Bristol Jupiter XIF driving 3.7 m
(12 ft) diameter, four-bladed propellers (E).
Wing span: 39.6 m (130 ft).
Length: 27,35 m (89.75 ft).
Height: 8.2 m (27 ft).
Weights: 13,381 kg (29,500 lb) (W); 12,701 kg
(28,000 lb) (E).
Max speeds: 204 km/h (127 mph). Cruising:
161 km/h (100 mph).
Accommodation: Crew of 3 and 38 passengers
(W); 24 passengers (E).
Remarks: Following its first flight in November
1930 a total of eight HP 42s were built. These were
in two versions of which (W) stood for Western
and (E) for Eastern, reflecting the conflicting
passenger configurations for operations in Western
Europe and the emerging Middle Eastern routes.

of a banker – Sir Herbert Hambling. After the
necessary deliberations it was considered that
the best approach would be the creation of a
single airline operating with the benefit of a
subsidy of a million pounds over ten years.
This was accepted and Imperial Airways was
created in 1924. Students of airline history will
recognise that an inevitable pattern affecting
the air transport industry had emerged from
the very earliest days.

Post-war technical developments
During this time of deliberation, the engineers
were bringing along new ideas and improved
technical answers to persistent questions. It
was realised that multi-engine configurations
bestowed a greater degree of reliability and
improved performance. The move towards
longer sectors, the increasing frequency of
over-water flights and the need to overfly
mountainous terrain all underlined the need
for reliability and this was being increasingly
emphasised by the growing amount of night
flying now becoming necessary under the goad
of commercial competition.

The day of the old wartime engines was
clearly passing. The rotary engines had not
been widely adopted for civil use and the
demand for increased power led to a dependence upon liquid-cooled engines. These were
inevitably more complex and their systems of
radiators and pipes were fertile areas for leaks
and failures. Now a new generation of
air-cooled engines appeared. These offered the
required power without the complexities of
liquid cooling. Typical of these was the Bristol
Jupiter, which soon became the subject of
foreign manufacture licence arrangements.

By the late 1920s the air routes were
spreading rapidly in many parts of the world.
Much of this activity was prompted by
European nations anxious to improve communication, and particularly mail links, with
their various overseas dependencies. Shortage
of suitable airfield sites in parts of the northern

In order to speed up the trans-Atlantic mail service, the
Germans put floatplanes on fast passenger liners and
launched them when in range of America. This was the
beginning of the intense trans-Atlantic competition still
much in evidence today.

A remarkable period piece: a night shot of the Imperial Airways Handley Page HP 42 Hannibal at Baghdad (Iraq) on its way to India in the late 1920s, taking 38 passengers in a state of luxury to be envied by modern airline passengers.

countries and Africa combined with the natural barrier of the Mediterranean had resulted in the development of a number of seaplanes and flying boats.

In other areas of Africa and the Middle East new land airfields were being developed, often in co-operation with military forces garrisoned in the area. So began the creation of the 'rest houses' that were beloved of both crew and passengers and provided a name which still lingers as a euphemism for night-stop accommodation!

By the beginning of the 1930s an increasing degree of sophistication was to be found in both operational and marketing techniques. Enclosed cockpits increased crew comfort and efficiency, while wireless was offering added advantages in some areas.

Navigation techniques were still relatively simple, with a heavy reliance on basic map

reading. This could sometimes be alleviated by night flying, when a greater use could be made of astro navigation, but this option was not always widely available as the airfields often lacked the most rudimentary night landing facilities. Aircraft were sometimes fitted with underwing flares which were electrically actuated from the cockpit at a late stage in the approach. Surprisingly this system operated quite well, though on at least one occasion it led to tragedy when a flare set fire to the wing. Perhaps the real flavour of the times is best appreciated by the fact that one of the best known navigation aids in the Middle East was a long furrow ploughed in the sand by an obliging RAF team!

The wide diversity of types was also being superseded by a degree of standardisation in the name of efficiency. In practice this tended to favour types built in the country of the

Britain commenced serious commercial airship development in 1924 with the construction of the R100, designed by the great aircraft thinker, Barnes Wallis. The R100 first flew in 1929 and in 1930, journeying to Canada and back.

airline, which was to be expected, particularly when national subsidies were being used to maintain the airlines. Of course, not all countries enjoyed indigenous aircraft industries; those without tended to purchase Fokker and Junkers products, recognising the enhanced technology of the countries where these aircraft were built.

Meanwhile, from the Netherlands, KLM was operating the longest scheduled flight in the world to its possessions in the East Indies, and the German airline, Lufthansa, was developing a South Atlantic mail service using a flying boat which was catapulted from a ship in order to achieve the required range.

Airship developments

Airships had still been making an impression on the commercial scene. Although their main use since the war had been military, there were a number of projects aimed at the civil market. In Britain the competitive programmes of the R100 and R101 had unfortunately evolved into a squabble between the relative merits of free enterprise and state control.

The R100 was seen as the free enterprise product and achieved a good performance, including a flight to Canada. The R101 seemed to be lagging and beset by problems. In fact it was altogether a more advanced concept and was a creditable attempt to develop safer, new technology. Typically both design teams had considered diesel engines and the less volatile fuel used by such engines. The R100 designers had decided that the choice was wrong and returned to the petrol engine, while the R101 team stayed with engines that were basically too heavy for the

nature of their task.

With hindsight both sides of the story may be appreciated and understood, though at the time it all seemed to represent a much more doctrinal issue.

Certainly there were considerable pressures on all concerned as it was the intention of the British to develop commercial airship services that would rival the standards of passenger comfort then associated with the ocean liners. The whole venture foundered with the loss of the R101 at Beauvais as it set out on a journey to India. In the general confusion of national shock that followed the whole programme was cancelled and despite some protest the R100 was scrapped.

Great preparations had been taken to develop the route, including the erection of a truly enormous hangar at Karachi. This latter offered ample service to Imperial Airways for its landplanes and existed as a dark sober reminder of tragedy until it was finally demolished in the late 1960s.

In Germany the airship had fared better and the famous *Graf Zeppelin* had initiated mail services to South America. Passenger services followed and between 1932 and 1937 the *Graf Zeppelin*, later joined by the *Hindenburg* established a good record of trans-oceanic operations from Frankfurt to Lakehurst, New Jersey, and Recife in South America. When the *Hindenburg* was destroyed by fire at Lakehurst in 1937 the hopes and aspirations of commercial airship operations died as well. One can only ponder the outcome if helium had been more readily available.

The 1930s – electronic and mechanical advances

For the aeroplane the future was now assured. Aircraft were adopting a multi-engined monoplane configuration with perhaps the exception of the Handley Page HP42, which sedately crossed the Channel between Croydon and Paris at a modest 160 km/h (100 mph) but achieved a reputation for comfort and punctuality that still lingers today.

While some of the shorter European routes could still allow this type of luxury the demands of increasing range were creating a design revolution. All-weather operations were now a pressing requirement and although true all-weather operations were still many years away much of the foundation work had already been accomplished.

In 1929 Lieutenant (later Lieutenant-General) J H Doolittle had made history by taking off seated in a fully hooded cockpit, flying a predetermined course and returning to his point of departure simply by reference to on-board instrumentation. It was a pointer to the future and work was being carried out by a number of air forces on all aspects of 'blind flying'. Electronics were becoming increasingly established and as new and better instrumentation came along it was being increasingly supplemented by the use of new communication and navigation aids.

Aircraft were sprouting the ubiquitous circular antennae of direction-finding receivers and a greater element of control from the ground was being introduced. Significantly, as

Croydon (south of London) was the scene of British pre-Second World War commercial aviation; pictured here on the tarmac is the de Havilland Frobisher all-wood monoplane and above it a Douglas DC-3 airliner on finals to the runway.

the decade progressed much work was being undertaken on all aspects of blind flying and advanced navigation techniques in Germany. The emergent Luftwaffe had a school of instrument flying at Berlin that was attracting students from all over Europe and the first glimmerings of electronic beam assistance for landing were to be seen.

In the United States the emphasis was on domestic services, which had evolved from the mail plane operations. Airway structures had evolved and even used illuminated beacons for night operations and had reliable radio links all along the way.

Even so the evolution was not achieved without difficulties, and following an alleged infringement of standards applied to the award of airmail contracts in 1934, the US Government introduced a Civil Aeronautics Act in 1938 that paved the way for regulation and control of airline economic and safety functions.

The almost frenetic air activity in Europe had long since been seen as a need for some form of collaboration between airlines in order to standardise elements such as fares, safety and timetables. This had been recognised as long ago as 1919 and led to the creation of the often vilified International Air Traffic Association (IATA). By 1930 this organisation com-

The rather primitive conditions of the DC-3 cockpit, although the basic flight instrument displays have been standard until very recently. Very little attention was paid to crew comfort.

prised 23 members and had its headquarters at The Hague. Perhaps at least it was a sign of increasing sophistication among airlines, at worst it was regarded as a restrictive cartel.

Certainly the 1930s was a decade of great change. Technology was changing fast, and in addition to electronic innovation, the mechanical engineers were developing improved high-lift devices, variable-pitch and even reverse-thrust propellers, retractable undercarriages and more efficient all-metal airframes. A new era was about to begin.

The first modern airliner

In 1931 the Boeing company built a clean, modern-looking bomber which it named the B-9 and offered it in competition for a requirement for the US Army Air Corps. It was not lucky on that occasion, but the company used it as the basis for a new airliner and built the Boeing 247.

Its configuration allows this famous aeroplane to be justly called the first modern airliner. It was a sleek, twin-engined aircraft featuring many of the new manufacturing techniques. It could carry 10 passengers and outperform all existing competition. The new aircraft made its first flight in February 1933 and entered service with United Airlines in the following month.

The Boeing 247 proved to be a sensation and by the summer of that year had established a record schedule of 19 hours and 45 minutes coast to coast. It proved popular with passengers for the new standard of comfort and speed that it offered. At first this enthusiasm was not shared equally by the pilots. The aircraft performance proved disappointing when operating at the higher airports such as those in the Rocky Mountains.

This was quickly alleviated by the adoption of variable-pitch propellers, which created a new designation of 247D. United Airlines was extremely pleased with the aeroplane and ordered 70, which provided Boeing with an enviable backlog, though it squeezed out competitor airlines from ordering the revolutionary aeroplane in the short term. Eventually 75 B 247s were built, of which three were ordered by DLH, the German airline.

With hindsight it is easy to suggest that Boeing's design team should have built a larger version using the available Pratt & Whitney Hornet engine of 700 hp instead of the 525 hp Wasp. Such are the pivot points of history and the cause of this aircraft's short-lived ascendancy.

The challenge from Douglas

Concerned at the success of the new Boeing, a major competitor of United were the thrusting Trans World Airlines (TWA), who were unable to gain an early entry into the production line. Instead the airline decided to go it alone and approach another manufacturer and ask it to build a completely new design to a TWA specification.

Even by today's standards the specification compiled by TWA was a formidable technical challenge. It called for a three-engined, all-metal aircraft capable of carrying 12 passengers at a cruising speed of 233 km/h (145 mph).

It went on to call for the most up-to-date electronics for communication and navigation; for the aircraft to be capable of climbing initially at 6 m/sec (1,200 ft/min) and to have a ceiling of 6,400 m (21,000 ft) and a range of at least 1,610 km (1,000 mi). It was then requested that all this should be provided in an

The forerunner of the modern airliner and great rival of the DC-2/DC-3, the Boeing 247, which was a streamlined, low-wing monoplane powered by two Pratt & Whitney radial engines. Although overtaken by the Douglas DC-2/3 series the 247 was ahead of its time.

The Douglas family of airliners made a major impact on Europe in the late 1930s, which continued into the 1940s, despite the Second World War. The Swiss (illustrated here) and the Swedish, both neutral countries in the conflict, were active users.

aeroplane with a weight of 6,440 kg (14,200 lb).

After all these years it is interesting to speculate just how this formidable requirement was formulated, but it is hard not to imagine a major input from TWA's technical adviser, Charles Lindbergh, a stickler for aerodynamic excellence, who had made the famous solo crossing of the Atlantic in 1927.

Among the manufacturers being canvassed was Douglas of Santa Monica. A strong engineering team was assembled and succeeded in getting TWA to accept a twin-engined design. In turn, TWA insisted that the aircraft could take off, with full load, on one engine from any airfield on the company network. It was also insisted that the aircraft should be able to maintain height on one engine over the highest *en route* ground.

Douglas agreed that it could meet the specification and received an order for the Douglas Commercial No. One, or as the series was to become more familiarly known the DC-1. Perhaps the really significant point about this new aeroplane was that to achieve the stringent demands of the specification an element of 'over-engineering' had to be em-

ployed. This resulted in the remarkable longevity of the series and an apparent inability even today to duplicate the concept by an aeroplane of comparable cost-efficiency.

The DC-1 made its first flight in July 1933 and following some modifications became established as the DC-2, the first production version of which made its maiden flight on 11 May 1934. It was an instant success and by August of that year had established a coast-to-coast service with a scheduled time now cut to only 18 hours.

There was an element of consternation in the Boeing and United Airlines camp and the airline converted all of its 247s to 'D' standard, with better engines and redesigned interiors.

The die was well and truly cast, and airlines on both sides of the Atlantic made their way to Douglas. Sales were further stimulated by the success of a KLM DC-2 in the England to Australia air race held in October 1934. The race was won by a specially designed racer from de Havilland and named the Comet. The real message for the future was to be noted in the inescapable fact that the second place was

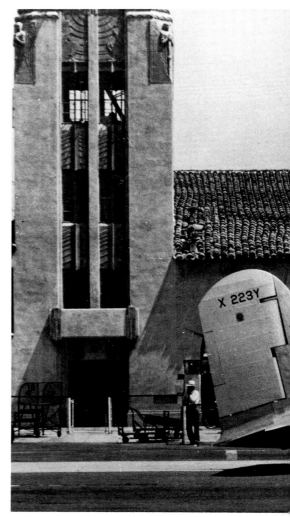

To answer the original Boeing challenge, Douglas Aircraft built the DC-1, seen here in Trans World Airlines colours, probably during the early route-proving flights across the United States. US air travel has never looked back.

DOUGLAS DC-3

Engines: 2 × 900 hp Wright Cyclone GR-1820-G102A radials driving 3.7 m (12 ft) diameter, three-bladed propellers.
Wing span: 29 m (95 ft).
Length: 19.7 m (64.5 ft).
Height: 5.2 m (16.92 ft).
Max weight: 11,431 kg (25,200 lb).
Speed: Normal cruise – 290 km/h at 3,048 m (180 mph at 10,000 ft).
Max range: 2,092 km (1,300 mi).
Accommodation: Crew of 2 and 14 passengers first class or 32 tourist class.
Remarks: Arguably the most famous airliner of all time with a working life span now in excess of 50 years. This aircraft has appeared in all guises ranging from an engineless 'glider' to a sea plane. Licence to manufacture the DC-3 outside the United States was granted to Japan, the Soviet Union and the Netherlands, though the latter country (Fokker) did not exercise this right, becoming instead the Douglas agent in Europe.

taken by the DC-2 and the third place went to the Boeing 247. American supremacy in the airline manufacturing business had begun.

Perhaps the next step for Douglas was inevitable. Although at first the company was reluctant to comply, pressure from the airlines suggested that the DC-2 should be stretched to provide increased space and allow a sleeper configuration to be adopted. The design evolved as a 14-berth aeroplane and was given the designation DST, which simply stands for Douglas Sleeper Transport. A daylight version was on offer with 21 seats, and this was known as the DC-3. A significant milestone had been reached.

The flying boats

For long-haul intercontinental operations it was the general belief in the 1930s that the flying boat provided the best answer. Both in the United States and Europe a number of flying boats were either in service or being developed. Dornier introduced a veritable leviathan, the Do X. It was powered by 12 engines and was the world's largest flying boat. Astonishingly it could carry 170 passengers and its design had begun as early as 1925.

Initially the aircraft was fitted with Siemens Jupiter engines of 525 hp. These proved unreliable and were changed to Curtiss Conquerors of 640 hp. This was an undoubted triumph in engineering terms for the time and the aircraft completed a somewhat protracted flight to the United States and back. Basically it was too far ahead of its time and lacked the right amount of power, and in 1934 it was placed in the aviation museum in Berlin. Sadly it was destroyed during air raids in 1945. Two other Do Xs were built for the Italians. These were fitted with Fiat engines and seemed to be more successful in terms of performance. Their fate is something of a mystery.

France had a Latecoere 300 flying boat service across the South Atlantic, while Pan American World Airways and Imperial Airways were establishing major flying boat networks.

Competition was hotting up on many of the new routes, with KLM operating DC-3s and Imperial Airways predominantly using flying boats to the Far East and Australia. In 1934 Imperial Airways and Qantas of Australia co-operated on a route to Sydney via Singapore. Air France had developed services to South America and through the east to Hong Kong and Lufthansa was also operating a service to South America and to Bangkok via the Middle East.

Politically Europe was less stable, there were clear indications of emerging trouble, and the situation was slowly polarising into factions that would eventually lead to war. For the airlines the problems of the present were

Boeing 314 flying boat, c.1942.

0   5m

One of the early German seaplane mail and passenger services, flown by Lufthansa immediately prior to the Second World War, using the Dornier Do 26. Surprisingly, limited services continued into the war years.

disconcerting enough and in particular the United States in the shape of Pan American Airways and Britain through Imperial Airways were eyeing the prospects of transatlantic operations.

Crossing the Atlantic

The Atlantic tended to be looked upon as something special, a true blue riband prospect. Communication links across this ocean join the two largest industrially developed areas in the world. Traffic generated in this area is both commercial and personal, reflecting the economic considerations of trade and the considerable ethnic links between the two areas. This is not a new phenomenon, as sea traffic across the Atlantic in the 19th century was considerably greater than long-distance traffic between other continents.

Little wonder therefore that the airlines could not help pondering the market and establishing the 'special relationship' that has since characterised the air transport in this area.

Crossing the 'pond' posed a formidable challenge. Westbound flight would be faced with headwinds most of the time and the weather conditions, especially in winter, can be notoriously fickle and unpredictable. In order to achieve success the aircraft must be capable of long range with a good payload and have an element of self-protection against icing. Powerful, reliable engines of modest fuel consumption were probably the real key and certainly by the late 1930s this element of technology was available.

First discussions on an Atlantic air route actually took place as early as 1928 and 1929.

These were between Pan American, Imperial Airways and the Compagnie Générale Aéropostale.

In both geographic and meteorological terms a mid-Atlantic route offered the best compromise, using the Azores and Bermuda as refuelling stations. In 1930 Pan American and Imperial Airways obtained permission from the Bermudan authorities to begin a service to the island from the United States, but this lapsed in 1932. Meanwhile Aéropostale had made inroads into the South American market and negotiated exclusive landing rights in the Azores on the way.

True to the tenor of the times, the French company then suffered financial problems and failed. Pan American and Imperial Airways then attempted a joint enterprise and regained the concession to use the Azores. Other efforts were made to gain additional landing rights in Newfoundland and Greenland, which emphasizes the political uncertainty of those days.

By 1935 Pan American had made a formal announcement of its intention to start a transatlantic service and flying boat bases were established on Long Island and later at Baltimore. There was still some way to go and political bickering conspired to confound plans until finally, in February 1937, the United Kingdom Director General of Civil Aviation authorised Pan Am to start a civil air transport operation between the United States and the United Kingdom, on a reciprocal basis with Imperial Airways.

During the long years of negotiation many of the engineers and visionaries had not been idle. Technology had provided some of the answers, additional and invaluable route surveys had been undertaken in the northern latitudes and across Greenland, and it had even been suggested that large, floating platforms be stationed at convenient points along the way to allow aircraft to land and refuel.

The two airlines now began preparations for what everyone realised could be an extremely marginal operation. Pan Am initially opted for landplanes and ordered three

Above: the Italians made a series of successful flying boats between the wars, including the Savoia-Marchetti S66, powered by three Fiat aeroengines and pictured taking off from the River Tiber, near Rome, on a proving flight.

of the new Boeing S-307 airliners, which had pressurised cabins to enable them to fly high and so keep clear of much of the heavy, turbulent and ice-inducing cloud that is such a feature of winter North Atlantic flying.

It was recognised that the aircraft lacked the range to carry an economical passenger load on the route but it was felt that it could be used for trial mail-carrying operations. In the event survey flights were carried out during 1937 using the Sikorsky S-42B and by the end of the year the airline had created an autonomous Atlantic Divison based at Baltimore.

Certainly by this time it would seem that a more collaborative political element was coming into being and which was very necessary in order to develop properly integrated long-range air services. Typical of this new realism was an invitation from the Canadian Government in early 1938 to Pan Am, British and Irish representatives to a joint meeting. The intention was to establish standard procedures for communication and weather monitoring. Another similar meeting took place in Dublin in 1939 and led to the creation of an organisation known as the Transatlantic Air Service Safety Organisation (TASSO).

As the various European landing rights were negotiated, Pan Am was also preparing to receive the aeroplane that would prove to be the eventual key to their transatlantic future –

the Boeing 314 flying boat.

Back in 1935 Pan Am project staff had realised that there would soon be a need for a large, long-range aeroplane to cater for their oceanic links in the Pacific and Atlantic. Boeing won the competitive design contract and undertook to develop an aeroplane that incorporated the many lessons already learned in the development of long-range routes in the Pacific.

The result was a massive 38,556 kg (82,500 lb) flying boat which scored a notable first in transport circles by using 100 octane fuel. Its maximum passenger load was 74, but this was reduced to some 30 passengers on long sectors.

Europe was also undergoing a flurry of activity in anticipation of an Atlantic air service. During 1937 and 1938 the British government called for investigations into the establishment of a mail service across the ocean. For this purpose two of the C Class flying boats were fitted with additional fuel tanks within the hull. This provided the necessary range and the first flight was made during the night of 5/6 July 1937. Other trial flights followed but the operational success was degraded by the fact that no payload was

Pan American World Airways began to open up international routes with the Boeing 314 Clipper series of flying boats, which were especially important on the Pacific Ocean services, where they island hopped.

In an effort to give quicker mail delivery to the customer, the British manufacturer Shorts developed the Mayo combination. This involved the S21 flying boat 'Maia' carrying the smaller S22 'Mercury', piggy-back style, for inflight launching.

carried in order to achieve the required range.

This basic compromise of range versus payload will always be a factor in aviation, and two other attempts were then made to overcome the problem. The first of these was the concept of a piggy-back operation put forward by Major R H Mayo, who was Technical General Manager of Imperial Airways. The idea had first been mooted in 1932 and, following an evaluation by Shorts, the Mayo Composite Aircraft Company Limited was formed in 1935.

The idea was to use a large flying boat as a carrier for a smaller aircraft which could be loaded to its limit. The smaller aircraft could be carried up to a safe height and speed and then released and allowed to continue under its own power. The lower element was an existing flying boat, while the upper element was a specially designed four-engined seaplane.

The pair were ready for flight by 1937 and

In another effort to extend range, some of the first air-to-air refuelling tests were carried out over Southampton in 1939 using a Handley Page Harrow (top) and a Short S30 C-Class flying boat. Air-to-air refuelling is not necessary today.

quickly became known as the Maia and Mercury, these being the names of the flying boat and seaplane respectively. The first combined flight took place in 1938 followed shortly afterwards by the first separation. The first Atlantic crossing took place in July. Several record-breaking flights followed, including a mail service to Egypt.

The other method of achieving the optimum range/payload compromise was to carry out aerial refuelling. This project was undertaken by Flight Refuelling Limited using a modified flying boat carrying a hose winch and receiver in the tail. Initially the trials were conducted using a modified Armstrong Whitworth 23 as a tanker. Later four Handley Page Harrows were converted to the tanker role and eight flying boats were adapted for the purpose. Two of the tankers were shipped to Montreal, reassembled and flown to a base at Gander. The first crossing using this technique took place on 5 August 1939 and was the first of 16 successful operations.

Although time was running out for Europe's peace once more and all the careful plans would soon be disrupted, there still remained one significant pointer to the future.

Below: Taken at Berlin's Tempelhof Airport, just before the Second World War, this picture of the Focke Wulf Condor shows the airliner's clean lines. The Condor was soon developed as a long-range naval patrol aircraft for the North Atlantic.

Early in 1936 the design team of Focke-Wulf began work on the design of a new four-engined airliner for Deutsche Lufthansa. Its first flight took place in 1937 and the aeroplane was designated the Fw 200 Condor. Unlike the rather bulbous flying boats that currently dominated the long-haul scene the new German airliner was a sleek, land monoplane that demonstrated its capabilities in 1938 by flying non-stop from Berlin to New York. Later in the same year it flew from Berlin to Tokyo in just over 46 hours with refuelling stops at Basra, Karachi and Hanoi.

Perhaps it was a sign of the times that these flights were not fully recognised for what they were. The aircraft is now remembered more as a warplane over the Atlantic rather than an airliner, but it certainly established the long-term trend and marked the beginning of the demise of the flying boat.

Events were now moving quickly. On 20 May 1939 the world's first transatlantic scheduled airmail service was inaugurated by Pan Am and on 28 June the first passenger service began using the new Boeing flying boats.

In spite of the war, civil air transport continued to a limited degree in Europe, especially where there was no other means of getting there; this included the Scottish Airways DH Dragon Rapide service to the Outer Hebrides.

The Second World War

On 3 September 1939 war was declared in Europe and the plans and hopes of all the airlines began to be confounded by fate and a gradual widening of the conflict.

As so often happens the goad of war is an important element in the speed of development of new technology. Long-haul air transport played an important role in the conflict and laid the foundations for many of the operational techniques that would be used to regenerate air transport later in the 1940s and 1950s.

In Europe, and Britain in particular, development of specialised air-transport aircraft virtually ceased. The infrastructure of commercial operations was largely lost, while in the United States both the production of transports and the expansion of domestic airline operations developed at great speed. The American airline systems and airways structures developed along with new techniques in navigation and air traffic control. Europe meanwhile lagged in these technologies and began to rely increasingly upon the United States for air transport needs. This has since bred an element of resentment in many European minds which clouded many judgements that were made in the first two or three

Whilst the Swedish airlines, AB Aerotransport, continued to operate its Douglas DC-3s during the Second World War (*top left*), it was not until the immediate post-war period that the newly formed British European Airways (later part of British Airways) (*bottom left*) regularly flew the aircraft. Today, over 50 years after its first flight, about 2,000 of these Douglas transports are still flying all over the world on regular services, with many in military service.

decades after the end of the war.

Although the United States was not initially involved in the war the projected services to Europe were curtailed. By this time many of the operational problems of transatlantic operations could be more clearly understood but it still remained a formidable challenge, particularly in winter.

All the more reason therefore to recognise the achievement of the North Atlantic Return Ferry service, which was introduced to shuttle aircraft and, eventually, crews from the United States and Canada to the UK. This was taken over by BOAC and became a year-round operation. There can now be little doubt that this service created the real foundations of future commercial operations on this route and many BOAC crews who cut their Atlantic teeth at this time later came to operate Stratocruisers, Britannias and Boeing 707s over these hostile waters.

The war in Europe was later complicated by the Far Eastern hostilities and air links were extended via the famous horseshoe route from South Africa to Australia.

Despite the preoccupation of many Europeans simply to survive there were some far-sighted people who could look ahead to a future. In Britain many people felt that the future of Britain's civil aviation industry would be severely inhibited unless urgent action was taken. A committee was set up to consider the needs of the future and it was put under the chairmanship of the redoubtable Lord Brabazon of Tara. The report of the Brabazon Committee was presented in 1943 and it called for some revolutionary thinking. Perhaps the most exciting of the findings was the call for a very large airliner of more than 100 tons that could carry up to 150 passengers.

This project was handed over to the Bristol Aeroplane Company and eventually appeared as the Brabazon. De Havilland received a commission to develop a smaller multi-jet mailplane, possibly the most significant pointer to the future of all of these designs. Other suggestions covered a medium-range airliner that evolved into the Britannia and two other smaller twin-engined types. Whatever the outcome of these designs it was a remarkable testament of faith in relation to the times.

Despite the smothering cloak of war some airline activity had continued within Europe. The amalgamation of Imperial Airways and British Airways had created BOAC in 1940 and in the same year Spain started its own airline – Iberia. Portugal had a small airline operation and Lisbon served as a neutral meeting ground for the crews of BOAC, Pan Am, KLM and Lufthansa. Swissair was in virtual isolation, while Sweden had the addi-

tional elbow room among the combatant areas to actually form a new airline. This was known as SILA and it ordered the new Douglas DC-4 four-engined airliner.

The early days of peace
When peace returned once more in 1945 the airline industry of the United States was particularly strongly placed in terms of expertise, equipment and route structures. The DC-3, an advanced aeroplane to many of the world's hopeful airlines, was already relegated to a secondary role as new four-engined aircraft appeared. The DC-4 and the emerging Lockheed Constellation offered faster, nonstop services that eliminated unnecessary refuelling stops and created genuine coast-to-coast, first-class service. Navigation techniques were much improved, *en route* navigation aids allowed for a better standard of control, while radar was making airfield management more efficient and safer. The problems of weather were now being addressed in a realistic manner and improved landing aids reduced many of the problems created by fog or low visibility.

As in the aftermath of the First World War the rebirth of civil aviation was founded upon a wide range of different types, many based upon wartime designs and bomber aircraft. The Lancaster and Liberator became airliners, and derivatives such as Avro's York were

In the 1940s the international airline business mushroomed, and types from the Douglas stable, like the DC-4 and DC-6, were in widespread use. They had been developed from the DC-3 experience.

On long flights it became necessary to offer sleeping accommodation for passengers, as on this Boeing 377 Stratocruiser. Many of the long-range aircraft had flying boat style luxury. This idea has been revived recently for Pacific flights.

created by adding a new commodious fuselage to a set of wings and empennage of the Lancaster. Boeing applied a similar philosophy to create the distinctive lines of the Stratocruiser.

To the idealistic the early days of peace seemed full of bright hope for aviation. Unfortunately that is rarely the nature of reality and typically as the tide of war seemed to turn in the favour of the Allies the first political divisions began to appear. In operational terms, and particularly under the flag of Pan Am in international terms, the Americans were supreme. They were also realistic enough to sense hardening attitudes over questions of national sovereignty as peace drew near.

In particular the Americans recognised the potential commercial threat that could be created by an integrated policy being adopted by all the members of the British Commonwealth. A brief glance at a map of that period will confirm that these countries could control the greater part of the world's air traffic by judicious manipulation of landing rights.

As early as 1943 a conference of British Dominions was called in London, and although its overall achievements are debatable it created a number of definitions which later became known as the Five Freedoms of the Air and are still crucial to airline marketing and route applications, as they are the basic rights accorded to an international airline.

These freedoms are now stated as:
1 To fly over a foreign territory;
2 To land for refuelling, maintenance or for non-commercial purposes;
3 To carry traffic from the home country to a foreign country;

South African Airways

Scheduled air links within South Africa have played an important role in the country's transport facilities since 1929, when Union Airways began operating regular services. In 1934, the government took over the assets of Union Airways and formed South African Airways, which in turn, a year later, acquired South West African Airways.

Domestic and international services have been the responsibility of SAA ever since, the Springbok route to London being the most important of its overseas links.

The company has long maintained a policy of regularly updating its fleet of aircraft, exchanging its DC-4s for Constellations, which in turn gave way to DC-7Bs. SAA took delivery of Viscounts in 1958 to serve on domestic routes, and in 1960, Boeing 707s began to operate on long-distance routes.

As more and more black African

countries achieved their independence, South Africa found itself increasingly isolated and therefore dependent upon air transport links to Europe.

SAA has been able to overcome attempts to isolate South Africa by operating non-stop services to European and other cities.

A sizeable fleet of Boeing 747s (including the very long-range SP version) maintains services to North and South America, the Far East and Australia as well as to Israel and many cities in Europe.

Domestic routes are served by a fleet of Airbus A300s backed up by a large number of Boeing 737s. These aircraft play an important role in maintaining daily links between widely separated cities.

Early services throughout South Africa were carried out in Junkers Ju52-3m aircraft bought from Germany prior to the Second World War, but the modern SAA has the latest high-technology airliners, including the Airbus and Boeing 737.

The Boeing Stratocruiser, with its characteristic double-bubble fuselage, seen in service with American Overseas Airlines (later merged with Pan Am). It was a classic airliner and was greatly loved by its crews.

4 To pick up traffic from a foreign country destined for the home country;

5 To pick up traffic from one foreign country and carry it to another foreign country.

Such simple phrases were to be the cause of much future debate as national and commercial interests once again reasserted their influences over the simplistic wartime doctrine of mutual co-operation. The Americans, in particular, were now arguing for a policy of open skies, which certainly suited their strength in air transport.

Fortunately an element of moderation was also abroad and in November 1943 President Roosevelt held a meeting at the White House to consider post-war aviation policy. Certainly there was some input from Britain at this meeting and a number of points emerged, including the notions that none of the Axis Powers should participate in airline operations and that each country should control its own domestic routes.

The pattern for future political domination of commercial aviation was beginning to emerge and several discussions took place between a number of interested countries. Britain started to draw up plans for a major international conference to be held in London, but the United States stole a march on the British by convening a conference in Chicago and issued invitations to 54 countries.

Held in late 1944 this conference led to the creation of a permanent International Civil Aviation Organisation (ICAO), which exists today and currently comprises 33 member states.

A second conference was held in Bermuda in 1946 in which many of the disagreements between the UK and the United States were resolved to some mutual satisfaction. It also resolved that fares and rates were to be regulated by governments after agreement with IATA. In latter years this decision has often caused criticism of a cartel that ensures artificially high fares, a weapon much used by the champions of open competition.

Although there were still many flying boats available, indeed in the UK design work was progressing on a leviathan that would be named the Princess, the shadow of the Condor now lay across this scene and the growing number of sleek, new landplanes was being encouraged by the legacy of war in the form of a world-wide distribution of long runways. Landplanes were easier to handle on the ground, while flying boats required considerable assistance while on the water.

Air India was one of the newly formed international airlines that acquired the long-range Lockheed Constellation airliner for its international routes to Europe and in Asia. Later the aircraft were transferred to the Indian Navy.

TWA

Formed as Transcontinental & Western Air in July 1930 as the result of a merger between two American domestic airlines, TWA changed its name to Trans World Airlines in May 1950.

In the 20 years or so between these historic dates, the airline grew from one of America's Big Four domestic carriers to become a powerful international operator with a network extending from the United States to Europe and the Middle East.

Born as a result of the US Postmaster General's insistence that mail subsidies would only be granted to strong airlines, TWA devoted much of its energies to the improvement of its services on transcontinental routes. Two of the other Big Four airlines – United and American – provided strong competition, the former by ordering a fleet of Boeing 247 airliners.

TWA's president, Jack Frye, countered by backing the Douglas Commercial No 1, an aircraft which was developed in 1935 to become the DC-3, the most successful airliner ever built and one which was carrying 75 per cent of American domestic passengers by 1939.

TWA continued its pioneering role by introducing the Lockheed Constellation into service in 1946, the first pressurised transport aircraft to be produced in significant numbers. The larger L1049G Super Constellation was used by TWA to open non-stop services from coast to coast in the 1950s. Astute management has helped TWA to become the largest transatlantic carrier, supported by an extensive network of domestic services. To operate these services, the airline has a large fleet made up of Boeing 727s, 767s and 747s, Lockheed TriStars and McDonnell Douglas MD-80s.

Trans World Airlines (TWA) was an early customer for the Lockheed L1011 Tristar wide-bodied jet. It uses it on both domestic and international services, from its St Louis and New York hubs.

The Atlantic still retained the aura of glamour and was the primary blue riband route. As the post-war era dawned the United States carried 90 per cent of this traffic. This superiority was further emphasised by the introduction of the first scheduled operation by a landplane, the DC-4, and in 1946 TWA launched the pressurised Constellation on the route.

The British renaissance

In Europe the renaissance of commercial aviation was, not unexpectedly, led by the UK using a wide variety of aeroplanes, not all of which were suited in operationally economic terms to the task. Converted bombers such as the Handley Page Halifax were used with a protruding belly container to increase volumetric capacity and captured enemy aircraft, such as the Ju 52, took over on some domestic routes.

In 1946 BOAC split off its European routes and these were taken over by a newly formed British European Airways Corporation or BEA. BEA also absorbed some of the small UK domestic operators and quickly estab-

British South American Airways Corporation flew the Avro Tudor, a development of the Lancaster and Lincoln bombers, and later BOAC flew the Tudor 2 on medium-range routes to the Commonwealth and other international destinations.

lished itself as the leading European airline. In time its domain spread beyond the boundaries of Europe and extended to North Africa and the eastern Mediterranean.

BEA also had the distinction of introducing the first new post-war British-built airliner, the Vickers Viking. This aircraft was a direct descendant of the famous Wellington bomber and shared its inherent strength and ruggedness. This was amply proven in 1950 when a BEA Viking *en route* to Paris suffered a bomb

explosion in the rear toilet. At first the incident was ascribed to a lightning strike until the full extent of the damage was appreciated.

By comparison, BOAC was less well placed to use indigenous designs. Long-range aeroplanes were its pressing requirement and its initial reliance on a variety of flying boats and converted bombers was lacking in competitiveness and market appeal. New designs were on the drawing board but these were to be delayed by various technical problems.

Although this was a period of severe economic restraint the Government was forced, in the face of considerable political opposition, to part with precious American dollars and purchase a number of Lockheed Constellations. At the time it was thought to be a stop-gap operation but in fact it was to be the beginning of a long association between BOAC and the American aircraft manufacturing industry.

BOAC had been preparing for the advent of the new Avro Tudor, which was to be the first British pressurised airliner. Originally intended as a derivation of the Lincoln bomber, the design had been constantly refined to create an entirely new aeroplane. It first flew in 1945 and was joined by two production models for further trials in 1946.

These showed the need for considerable modifications and it has to be suggested that further delays were created by somewhat artificially high expectations by BOAC. Shortfall in the performance finally caused the airline to reject the aeroplane at the end of that year. Development work continued and the aircraft entered service with British South American Airways Corporation until two of the aircraft were lost in mysterious circumstances.

The remainder of this early version were then either scrapped or relegated to freight duties, a role in which two of the aircraft made a valuable contribution to the Berlin Air Lift.

In 1952 Aviation Traders Limited purchased the available airframes and carried out extensive modifications, including stripping

Based on the wartime Wellington bomber (another Barnes Wallis design), the Vickers Viking was the first British post-war airliner. It is seen here at Copenhagen in the colours of its chief customer, British European Airways.

out the pressurisation system, and started to use them in a 42-seat configuration on a Colonial coach service between Stansted, Tripoli and Lagos. Others were used as freighters and did valiant service, particularly to Australia and Christmas Island, remaining in service until 1959.

Although the Tudor 1 and its variants were intended as long-haul aircraft the Tudor 2 was intended for the medium-range Empire routes. BOAC retained an interest in this configuration and their increasing requirements resulted in a fuselage stretch to 32 m (105 ft), making it the largest aircraft produced in Britain to date.

The airline ordered 30 of these aircraft in late 1944 and Qantas, together with South African Airways, decided to opt for the same aircraft, the order being increased to 79. The first flight took place in 1946, but the earlier pattern began to be repeated and performance once again fell below what was expected. Not surprisingly Qantas and SAA cancelled their orders and purchased Constellations and DC-4s instead.

Although the aeroplane operated in various freighting roles and charter duties the type was never destined to operate the major routes and it passed into history.

In retrospect the sad tale of the Tudor can be seen as typical of the situation faced by

Lockheed's Constellation, seen here in early Air France livery, was one of the finest-looking airliners ever built, at almost the limit of the era of piston-engine technological development. A handful remained in the early 1980s.

LOCKHEED 049E & 749A CONSTELLATION

Engines: 4 × 2,200 hp Wright Duplex Cyclone R-3350-C18-BA1 radials driving 4.6 m (15.17 ft) diameter three-bladed propellers.
Wing span: 37.5 m (123 ft).
Length: 29 m (95.17 ft).
Height: 7 m (23 ft).
Weight: 47,628 kg (105,000 lb) (749).
Max speeds: 558 km/h at 5,500 m (347 mph at 18,000 ft). Cruise at 60% power – 480 km/h (298 mph).
Range: 4,827 km (3,000 mi).
Accommodation: Crew of 7 and 43 passengers.
Remarks: First flown in 1943, the Constellation was to be continuously developed through a number of variants until it had reached the limits of contemporary aerodynamic and piston engine technology. It had been designed to maintain a 7,620 m (25,000 ft) altitude on three engines and could fly at 4,572 m (15,000 ft) on only two engines, an excellent safety factor. First model in service was the 049, followed by developed versions of which the 749 had increased capacity and strengthened undercarriage.

Seen in South African Airways colours, the Douglas DC-7B, was used for internal flights and international routes to neighbouring countries, like Mozambique, Northern Rhodesia (Zambia) and Southern Rhodesia (Zimbabwe).

British industry in its attempts to catch up the lost years of the war. Caught between the twin problems of long range and performance limits of hot and high airfields the overall result was one of false hopes, overoptimism and a degree of vacillation by customer airlines over their precise requirements.

In contrast, French air transport returned to the international arena quite quickly after Air France was re-formed in early 1946. Its European routes were re-established and a transatlantic service began. Later that year a welcome taste of true international co-operation was demonstrated when the airlines of Sweden, Denmark and Norway formed a consortium under the name of Scandinavian Airlines System (SAS). It was a significant step as it demonstrated a willingness to transcend narrow national attitudes in order to exploit the real potential of air transport while

In Latin America, these airlines began serving the communities of fast-growing nations after the war in a social as well as commercial capacity. This is an early Mexicana Airlines Douglas DC-3.

enabling three relatively small countries to play a major part in the future development of the industry.

Airlines were growing in other parts of the world also and the aeroplane was making an increasing contribution to the development of the less wealthy areas of the world, which were to become known collectively in future years as the Third World.

The glamorous years

Perhaps this was the most glamorous period of the air transport business. Air travel, especially over long distances, was still a relatively uncommon experience. Airline advertising heavily emphasised the luxury aspect and its social selectivity. Cabin staff had greater resources and air hostessing was now one of the most prestigious of occupations for young ladies of gentle birth.

Airlines vied with each other over the standards of cabin service with vast amounts of food and drink freely available. A variety of ploys was used to amuse the passengers during long sectors, ranging from visits to the flight deck, through raffles, lucky draws based on navigational features to films. Smart overnight bags were given on boarding and these rapidly assumed a considerable social significance when the trip was finished, passengers flaunting them as a mark of self-esteem.

It all added up to an advertising man's dream and with IATA holding sway over fares, the airlines could only compete on the exotic and superficial aspects of what should have been presented in a much more serious manner. Public relations departments also fortified the image and used heavy defensive PR to overcome the fairly common accident or incident and to retain the euphoric hold on the

Air France

Formed in August 1933 after a series of mergers between smaller airlines, Air France was subsequently nationalised in 1948 and it remains state-owned to this day.

As well as forging links with French colonies, Air France carried on the tradition of pioneering long-distance routes, but having inherited about 35 different aircraft types in a fleet totalling 260, the airline set about replacing slow aircraft with more modern designs.

After the Second World War, Air France very quickly began to rebuild its operation, ordering DC-4s and Constellations to supplement its fleet of indigenous designs. Its network grew steadily and at one time Air France boasted the largest in the world.

It kept in the vanguard of technical advancement too, being one of the first to order the Viscount and pioneering the use of jets on short/medium routes with the Caravelle. This helped Air France to become the largest European airline by the end of the 1950s.

It shared with British Airways (BOAC) the honour of inaugurating transatlantic supersonic air services with Concorde. The Airbus A320 was launched by a substantial order from Air France, which also operates the Boeing 727, 737 and 747, along with the Airbus A300/A310 and some smaller types.

Air France has long been a champion of French aerospace industries and was one of the first customers for the Airbus Industrie A300/310 series; the airliner is used on high-density inter-capital routes in Europe and the Middle East.

In Italy the newly re-formed Alitalia flew the Convair CV 440 and the Douglas DC-6Bs, photographed here at Ciampino Airport in Rome. They were a common sight in the skies of Europe in the 1950s and 1960s.

One of the last piston-engined airliners in British Overseas Airways Corporation (BOAC) service was the Douglas DC-7C (*top right*), used for longer-range Commonwealth flights. In Canada, there was a considerable early demand for air transport to cover the long internal routes, especially between British Columbia and the Atlantic coast, and this led to the purchase of types like the Constellation (*below right*), seen here over Toronto.

passengers' perceptions. Often newspapers carrying photographs of accidents would 'not be available' on board and in any incident that left the company name in prominence some-one would attempt to paint it out prior to general media attention.

It is an aspect of aviation that dies hard and even today many echoes of this market treatment are still to be found. Yet survey after survey over the years have all indicated that the main feature of air travel that the passenger considers important is able flight crews fol-lowed by good maintenance and safety re-cords. Then comes the desire for fast service at check-in, then ease of reservations. Concern about pretty air hostesses and in-flight food are well down in the list.

All of these developments and perceptions were founded in the maturing of the post-war air transport industry between 1946 and 1950. Large, four-engined landplanes had become established on all major routes, while the DC-3 still ruled on the short haul, though increasing-ly common were the new Convair CV-240s and Martin 4-0-4s.

Terms like 'jetstream' and 'cobble-stone' entered the vocabulary and the benefits of pressurisation encouraged the use of 'over the weather' flight to reassure the nervous passen-ger. Many of the hard-learned lessons of the war were being implemented through emer-gent technology, with the beginnings of avionic systems taking over from the old, unreliable aircraft electrics.

Already the first signs of a real social revolution could be seen. The benefits of time-saving were obvious despite the many delays caused by meteorological or mechanical problems. In terms of Atlantic operations it can be accepted that a sea voyage can take between five and ten days. In the early 1950s a trip to North America took around 20 hours, while today the flight between London and New York is less than four hours by Concorde.

In 1947 the total number of passengers crossing the Atlantic by sea was more than 400,000 per year. From that point sea traffic slowly increased to a maximum in 1957, from which time it decreased every year and led to the demise of the prestigious ocean liners.

Product improvement

As the airlines settled down to developing the routes, and negotiations over traffic rights became ever more problematical, the aircraft manufacturers kept up a steady flow of developments. The American manufacturers, in particular, have always demonstrated a considerable flair in product improvement and stretching of the original design. Perhaps Lockheed's Constellation was one of the best examples, as this evolved in both size and wing shape and required ever-increasing amounts of power from its piston engines.

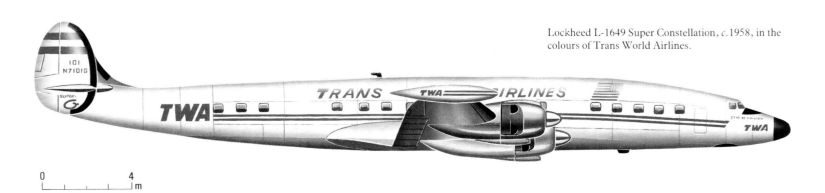

Lockheed L-1649 Super Constellation, *c.*1958, in the colours of Trans World Airlines.

0   4
|___|___|___|___| m

The great success story of the British airliner industry was the Vickers Viscount, ordered by many international airlines around the world, including Lufthansa. Many remain in service today, especially in Africa and the Far East.

This indicates an area which had always been something of a restrictive factor. Aeroplanes demand a great deal of power for take-off in relation to size. By the late 1940s designers could, and sometimes did, design very large airframes. In order that they might fly, several engines had to be mounted, leading to increased mechanical complexity and weight which then required more power.

The piston engine had proved to be a remarkably adaptable mechanism and through processes of refined fuel and supercharging power outputs had steadily increased. It was becoming apparent that the quest for additional power had to be found by other means. The piston engine had reached such complexity that overhaul life was actually beginning to decrease. It was time to embrace the gas turbine.

During the early post-war years the American commercial aircraft manufacturing industry consolidated their lead in the industry, which was considerably benefited by a large and booming domestic market.

The turbine age

To the European manufacturers it seemed that in piston-powered aircraft they held an impregnable position, even though the technology was declining. Perhaps the answer was to move forward more quickly and develop turbine-powered aeroplanes.

Although the use of gas-turbine techniques to develop jet propulsion was a radical departure in commercial aircraft design the concept of such an engine dated back to the late 1920s. At that time a young English inventor named Frank Whittle wrote a thesis on the use of gas turbines for aircraft propulsion and in 1930 he consolidated his position by taking out the first patents.

Inevitably, as with all technology, the basic idea was available to other methods of interpretation and in Germany a scientist called Hans von Ahain was also working on a similar project, with patents dating from 1935. The first flight of a jet-propelled aeroplane took place in Germany just before the outbreak of war and the first British engine took to the air in 1941. Later in the same year one of the Whittle engines was flown to the United States, where it was copied by a US manufacturer and used to power the first American-built jet aircraft in 1942.

These achievements were of considerable significance. They offered a simpler form of propulsion, largely devoid of vibration and with the promise of virtually boundless increases in power in the future. It seemed to be the answer to the airframe designer's dream.

By the end of hostilities there were several jet engines either in being or on the drawing board. Germany ceased to be a contender and it seemed that Britain was well ahead in the new technology. Visitors to the Rolls-Royce test airfield were somewhat surprised to see a Gloster Meteor apparently fitted with two propeller engines. This installation paved the way for the Dart engine, which has now become one of the most ubiquitous engines of the turbine age.

This was the powerplant for the innovative Vickers Viscount, which was the first turboprop airliner to enter production. In retrospect it is strange to recall that this aircraft, which originated in the Type 630 of 1948, was not readily accepted by the airlines at first and it was almost five years before the developed version entered service.

Once it began operations it quickly attracted the attention of the world's airlines. In a remarkably short time it was ordered by Air France, Aer Lingus, TAA, TCA (now Air

Certainly the turboprop is now a well-established powerplant and is the virtual mainstay of the short- to medium-range sector. Its use in long haul was less successful.

One of the aircraft proposed by the Brabazon Committee began to take shape as a four-engined piston-powered machine that evolved into the first big turboprop, the Bristol Britannia 100. The engine was the Bristol Proteus, which featured a reverse inlet flow for the airstream and which carried the seed of future problems, though the configuration resulted in a powerful engine of modest overall length.

Although the prototype first flew in 1952 it did not enter service with BOAC until 1957. The aircraft certainly had its share of teething troubles and with hindsight it can be confidently suggested that in many ways it suffered from an attempt at over-sophistication in technology.

But it was a remarkable aeroplane and extremely comfortable from the point of view of the passenger. For the maintenance engineer it could often prove to be a nightmare and offered the worst of the 'old electrics' and little of the promise of emerging electronics. A longer-range Series 300 was developed and entered transatlantic

Below: although not an outstanding success, the Lockheed L-188 Electra, seen here in QANTAS livery, remains in limited service in South-East Asia and Latin America. The highly successful Orion maritime aircraft was developed from it.

BRISTOL BRITANNIA TYPE 175-102

Engines: 4 × 3,900 hp Bristol Proteus 705s driving four-bladed, fully reversing propellers.
Wing span: 55.5 m (182.25 ft).
Length: 34.7 m (114 ft).
Height: 11.2 m (36.67 ft).
Max weight: 70,308 kg (155,000 lb).
Speed: Cruise – 582 km/h (362 mph).
Range: 5,550 km (3,450 mi) (max payload). Max range 7,369 km (4,580 mi).
Accommodation: Crew of 3 and up to 90 tourist passengers. The long-range version used by BOAC on the Atlantic route could carry a maximum of 139 passengers.
Remarks: Representing the 'interlude' between piston and jet economics, the Britannia appeared in a number of versions and acted as the base for future freighter and maritime reconnaissance aircraft. It introduced innovative engineering and was considered to be an all-electrical aeroplane. Its introduction into service was delayed by engine intake icing problems, but it proved to be a comfortable and forgiving aeroplane and was generally considered to be easy to fly.

service in 1957.

Sadly it was not the overall success that was expected and ice-ingestion problems with the engine, exacerbated by the intake design, often inhibited the theoretical performance. Yet the type had considerable passenger appeal, being described in the media as the Whispering Giant, and it had an enviable safety record.

The United States also looked at the long-haul turboprop market and Lockheed

Above: the 'whistling giant', the Bristol Britannia 102, was very popular because of its low noise levels and unusually high level of comfort. It also enjoyed an important role in military transport to link the UK with the Far East.

Canada) and Capital Airlines of the United States. It seemed to offer the chance of gaining a valuable foothold in the lucrative American market and re-establishing a European manufacturer in world terms.

Meanwhile the massive Brabazon had flown, though the developed version using turboprop propulsion already seemed doomed and the whole programme was finally concluded in 1953 when the sole example was scrapped.

With a maximum speed of 870 km/h (470 mph), the Tupolev TU 114 is one of the fastest propellor-driven aircraft ever (it was initially designed for military service). It was used extensively for long-range services by Aeroflot.

developed the Electra, powered by four Allison 501 engines driving distinctive paddle-blade propellers. The aircraft was realistically a medium-haul aircraft and it began to gain sales in Australia, New Zealand and the Netherlands in addition to its domestic market.

KLM used the aircraft on African routes and its speed in the cruise was often noticed by the Britannia crews, who found that they were regularly being overhauled on the long desert sectors.

Unfortunately the aircraft suffered from an inherent problem created by a phenomenon of vibration. Accidents followed and the aircraft came under restrictions on operational speeds and height. Eventually extensive modifications were made and the airframe is now mainly seen as the basis of the Company's Orion maritime reconnaissance aeroplane.

In many respects the advent of the longer-range turboprop aeroplane was only an interlude in the progress of commercial aviation. Many major airlines did not acquire this type of equipment, choosing instead to move directly from the piston engine to the pure jet form of propulsion.

In one respect the type has survived thanks to the USSR. This massive country entered an intense period of technological development in the aftermath of the war by continuing to develop military systems at a time when many of the Western nations were only concerned with domestic interests. The Russian aviation programme was aided by the purchase of gas-turbine aero engines from the West.

Both jet-propelled and turboprop aircraft appeared in the military inventory and the large Tu-16 Badger bomber later appeared as the civil Tu-114, which could carry up to 220 passengers. Another Russian turboprop,

Pan Am

On 19 October 1927, a seaplane crammed with mail took off from Key West, Florida, bound for Havana, Cuba. It was the first flight operated by Pan American Airways – an airline founded by Juan T Trippe, one of the giants in an industry which was just getting under way in the United States.

Although America was the father of the first successful aeroplane, it was slow to exploit its potential and Europe forged ahead in the development of air transport after the First World War. However, while many airlines clamoured for domestic routes in the United States, Trippe could foresee the greater potential of international air services.

The 28-year-old president of the fledgling airline was a man of many talents and a shrewd negotiator who quickly built up his airline to become the unofficial but widely accepted national carrier of the United States. Supported by valuable contracts to carry airmail (by which means all major airlines became established in America at that time), Pan American pioneered routes to Latin America and across the Pacific.

Trippe equipped his airline with large flying boats – the Jumbos of the 1930s – but as landplanes took over most air transport roles after the Second World War, Pan Am did not lose its pioneering spirit. The first airline to order Boeing 707s, Douglas DC-8s and, later, the mighty Boeing 747, Pan Am maintained its position as a world leader.

However, the unofficial title of America's overseas airline ultimately proved to be a liability rather than an advantage. Lacking an extensive network of domestic services to feed its worldwide network, Pan Am dominance of overseas routes was gradually eroded by carriers such as TWA, and Trippe tried vainly to get the permission of government authorities to start domestic operations.

Its opportunity came in 1979, when permission was obtained to acquire a majority holding in National Airlines. Still a large airline by any yardstick, Pan Am has decided to rationalise its fleet so that it is made up of Boeing 747s along with Airbus A300/A310 and A320 aircraft.

Pan Am is as synonymous worldwide with American air transport as is the Boeing 707 with American airliner design. Pan Am was the first operator to order it, as well as the DC-8 and the 747 wide-bodied jet.

the Il-18, entered service with Aeroflot and was supplied to a number of overseas countries, including Air Guinée, Air Mali, CAAC, CSA, Cubana, Ghana Airways, Interflug, LOT, Malev, Tabso and Tarom. Although popular among many of the emerging-nations' airlines, the life of the engines was severely restricted, which inhibited its sale to more commercially orientated countries.

The Comet

Although the propeller turbine period was relatively short it did give BOAC a better position in the Atlantic stakes and created an upswing in traffic that was later maintained and increased when the Comet IV entered service. The problem of the large turboprop was the inherent restriction in speed and altitude, which was inevitable with the use of propellers at that time. By comparison, the pure jet offered the eventual goals of massive power increase, higher speeds and increased operational heights without loss of propulsive efficiency. The future trend was obvious and inevitable.

Once again it is necessary to return to the wartime deliberations and recall the conclusions of the Brabazon Committee. The type IV aeroplane suggested by that far-sighted body was to be a high-speed aeroplane using pure jet engines. Although it was acknowledged that at that time the engine power available would still reflect the early stage of development of the new technology, it was felt that only by making a major step from the obviously fading piston engine to jet propulsion could British industry

regain the initiative from the United States in building viable commercial aircraft. De Havilland accepted the challenge and began design work on the project in 1944. Initial performance calculations excited BOAC and led to an agreement in December 1945 to purchase 10 of the new aircraft, which was designated DH 106 Comet.

Several different configurations were considered including a twin-boomed aircraft and a tailless design, which led to the development of experimental DH 108s to assess the aerodynamic feasibility. These exotic designs were rejected as being impractical for passenger aircraft and instead the design evolved as a more conventional, though sleek aeroplane.

The first flight was in July 1949 and it seemed as if a new era in air transport had dawned. The new aeroplane quickly demonstrated its astonishing capability in speed and height and it seemed as if the timetables would all have to be rewritten. Extensive flight trials and route-proving operations by the manufacturer and BOAC developed new techniques in cruise control and flight planning and finally the first scheduled passenger service left London for South Africa on 2 May 1952.

It all appeared to be a breathtaking success and it seemed as if Britain had once more asserted itself in the vanguard of civil aircraft manufacture. In retrospect it is easy to recognise some of the inbuilt problems. Engine power was limited and the aeroplane had to be carefully designed for maximum performance. Range of the Comet 1 precluded transatlantic operation and it was decided to

Surely one of the most beautiful jet airliners ever designed, the de Havilland Comet, but it was sadly plagued with problems in its early development which allowed American makers to come to dominate the world airliner market.

develop a Mark 2, which would have more powerful engines and increased fuel capacity.

Just as it seemed that a major commercial success was in the making, fate was already at work. A non-fatal take-off accident at Rome was later echoed in a more serious disaster at Karachi. Then there was a total loss of life when a Comet crashed during climb-out from Calcutta in bad weather. At that point it seemed that the accidents could be explained in the light of extenuating circumstances. Then early in 1954 two aircraft were lost and despite several modifications introduced between these two events the aircraft was grounded.

The blow was doubly severe – in morale the whole industry was shaken while the bright promise of massive commercial success was lost almost overnight.

A major inquiry was held and metal fatigue was much cited. So too was the fact that the aeroplane used powered controls and innovative bonding techniques in fabrication. Despite this almost mortal blow De Havilland built a Mark 4, which evolved into the Comet 4. This eventually entered service with British Overseas Airways and a number of other airlines with considerable success. Nevertheless the British had by now lost their long lead in jet-propelled airliners.

Above: a Boeing 707-441 of the Brazilian national airline, Varig, seen on its delivery flight from the Boeing factory near Seattle, Washington. The 747 has been a revolutionary airliner, making mass air travel possible for the first time.

Boeing 707-437 of Air India, c.1961.

American jets

Of course the American industry had not been sitting idly by while the Comet rose to glory. Much of the experience gained in the wartime developments in technology and aerodynamics was already being built into military aeroplanes. Boeing in the United States had built the B-47 bomber and was developing an even larger machine, the B-52. To the observers of the time these aeroplanes represented radical ideas such as thin, highly swept-back wings and the use of podded engines to offer relieving forces on the structure.

These ideas were now being translated into a new design intended as an aerial refuelling tanker, though with obvious transport possibilities. A prototype was built under the

Boeing designation Type 367-80, more commonly called the Dash 80. This was the dawning of the real jet age and spawned a series of aeroplanes which were then dubbed the 'big jets'.

Despite the apparent confidence of all concerned the project was a considerable gamble, but the tension was greatly relieved when the United States Air Force ordered a number of tankers in March 1955. The type was then put on offer to the airlines, Pan Am in particular, and the Boeing 707 was on its way.

Douglas had also been working on jet designs, and mindful of the way that it had held on to a leading manufacturing role since the DC-3 was certainly not going to miss out at this stage. Its design, which was outwardly similar to the 707 configuration, was introduced

BOEING 707-320C

Engines: 4 × Pratt & Whitney JT3Ds of 8,165 kg (18,000 lb) thrust.
Wing span: 46.7 m (153 ft).
Length: 44.5 m (146 ft).
Weight: 152,410 kg (336,000 lb).
Speed: Cruise – 965 km/h (600 mph).
Range: 6,436 km (4,000 mi).
Accommodation: A crew of 4 and up to 202 passengers when entering service in 1963.
Remarks: Originally conceived as a military tanker aircraft, the basic design soon became the basis of a number of different variants of what was the first of the so-called 'big jets'. Different airframe sizes and engines offered the operator a choice of domestic or long-haul aircraft and gave Boeing the dominant position it now holds in civil air transport.

Although not as successful as the Boeing 707, which it was designed to rival, the Douglas DC-8 has been a popular aircraft. It is seen here in the colours of the independent Texas-based Braniff airline.

as the DC-8.

The commercial battle now began in earnest, but Boeing had an enviable two-year lead. For both companies the investment was massive and although Boeing could offset some of the development costs on the military programme both companies were taking a hefty commercial risk.

The first Boeing 707-120 flew in December 1957 and it entered service with Pan Am in October 1958. This was the inauguration of American jet operations over the Atlantic, but the laurels for being first had been narrowly snatched just three weeks earlier by BOAC with the Comet 4. It was a hollow victory however as the world's commercial aviation market was to be revolutionised and turned into a true mass travel operation, thanks to the big jets.

In the light of so much activity it is interesting that the first jet airliner to enter sustained commercial operation was the Russian-built Tu-104, which began operations in September 1956.

In terms of passenger capacity Pan Am was now supreme on the North Atlantic, while TWA actually suffered a decline as it could not command a large market share while using the suddenly dated Lockheed 1649As. The only match for Pan Am at this time was BOAC with the Comet, and it is significant that for a time the British airline rose to second place in numbers of passengers carried on the route.

In American minds the requirement was to offer direct flights from North America to the main cities of Europe. It was the old coast-to-coast philosophy of the American designer, and Boeing developed the -320 version, which was the first truly intercontinental jet airliner.

Meanwhile a third US jet airliner entered the ring when Convair flew the CV-880. It was a fast aeroplane that was expected to have even greater passenger appeal and give a first-class service. The manufacturers felt that its block-to-block times would eclipse the larger Boeing and Douglas aircraft. No doubt they were right in theory; but in practice the expected performance fell short and extensive modifications had to be incorporated. This hurt the sales programme, and although the aircraft eventually operated well, it was no longer a threat to the market leaders. Instead Convair faced a heavy financial loss said to be the heaviest ever incurred in the business.

Operating the Moscow-London service in the late 1960s, the Tupolev Tu 104 jet transport (*left above*) was the first Soviet turbojet transport aircraft and saw service with Aeroflot, the Soviet national airline. By the time that the Convair CV 880 jet (*left*) had been designed and built, the market was in the hands of the 707 and DC-8. Less than 100 were delivered, mainly to US domestic airlines, but also, as illustrated here, to Japan Air Lines.

Crews and operating procedures

They were heady times for airlines and their staff. Passenger reaction to the jet age had been good and air travel was increasingly being made available to a whole new mass market.

The scars of war were now healing fast and the airlines of Germany, Italy and Japan were rapidly gaining an international reputation. Confidence was high and to be an airline employee was a cachet in itself. Pay was generally far higher than the average earnings, and female staff members were treated to regular and expensive re-issues of uniforms, emphasising the glamour image of the industry.

This high standard of expectation also had the effect of generating a certain amount of industrial unrest and fragmentation among the various groups of employees. The rapidly evolving technology also played a part in the unrest. In the early post-war days the use of voice links over HF had led to the demise of the specialist wireless operator. Instead the pilots conducted all communications and received a financial allowance for this additional duty.

Such a trend could not be ignored and during the 1960s many of the industrial problems of some airlines centred on the complexities of crew complement. Inevitably the emotive subject of safety was brought into the discussion, but basically it was all in the effort to raise salaries.

The new 'big jets' had also had quite an impact upon the crews and operating procedures. They required a somewhat different approach in handling and anticipation of events. They had to be operated 'by the book' and were less forgiving than their more stately predecessors. Performance could be closely judged and crews became used to the large number of graphs used to obtain the very last ounce of payload out of any particular airfield.

In return the aeroplane could deliver optimum results in every application and offer the airlines the chance to make money. In fact many airlines had been losing money for years

The ultra-modern cockpit of the Fokker 100 airliner. It is designed to reduce pilot workload and thus make flying safer in the ever-more crowded skies. The aircraft was ordered by Swissair even before its first flight in 1986.

and it has to be said that the turnaround in their fortunes began in the early 1960s, leading to an exciting boom which foundered some 13 years later in the oil-price crisis.

Rear-mounted engines

While the secret of success was based upon large capacity, there was still an element of influential opinion which considered that the jet engine would be uneconomic on the short- to medium-range routes. It was to the credit of the French that, as the big jet philosophy

AEROSPATIALE SE 210 CARAVELLE I

Engines: 2 × 4,763 kg (10,500 lb) thrust Rolls-Royce Avon RA 29 Mk 522 turbojets.
Wing span: 34.3 m (112.5 ft).
Length: 32 m (105 ft).
Max weight: 43,500 kg (95,901 lb).
Speed: Cruise – 740 km/h (460 mph).
Range: 1,665 km (1,035 mi) (max payload).
Accommodation: Crew of 3 and up to 80 tourist class passengers.
Remarks: Originally conceived as a fast transport for the operation of services between France and North Africa, the Caravelle was to appear in a number of variants and achieved sales in many areas, including the United States. Later in its development it was fitted with both Pratt & Whitney and General Electric engines. It is primarily remembered as the first French jet transport to enter production, the world's first short-haul jet and the aircraft which introduced the fashion of rear-mounted engines.

Sud Aviation Caravelle VI-N of Alitalia, c.1961.

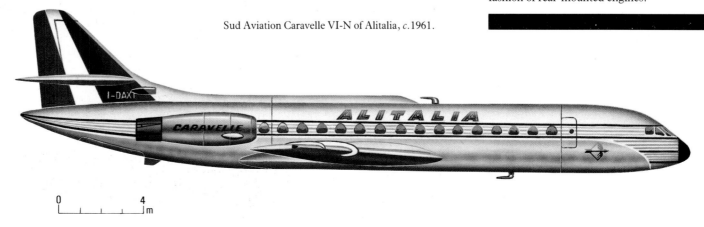

Surprisingly, the French captured a share of the US domestic airliner market with the rear-engined Caravelle, especially with domestic giant, United Airlines, despite tremendous competition from Douglas and Boeing. Deliveries began in 1961.

evolved, their Sud-Est company looked at the design of a medium-range jet aeroplane. The result was the Caravelle, a remarkably clean airliner which introduced the trend towards rear engine mountings.

The design received strong support from Air France and it ordered 12 aircraft in 1956. The type entered service with Air France in 1959 and quickly established a reputation for comfort and reliability. Several European airlines ordered the Caravelle and in July 1961 it received the accolade of entering service in the United States following the purchase of the type by United Airlines.

Economic factors that had previously been expected to militate against jet aircraft were proving to have been somewhat less than correct. The reliability of the new engines was far in excess of the piston engines and overhaul lives originally calculated in hundreds of hours were soon being noted in thousands of hours. Low vibration levels resulted in improved maintenance of airframes and associated systems, while aviation jet fuel was very cheap, often as much as half the price of high-octane petrol.

The timely arrival of the Caravelle proved that the jet was of equal utility in short- and medium-haul routes while the ploy of rear-mounted engines left the wing clear of all protuberances, allowing it to operate at maximum aerodynamic efficiency. While European operators ordered the Caravelle, Britain's BEA attempted to face up to the competition by introducing a short-range variant of the Comet under the designation 4B, which proved to be remarkably effective in the role.

The airline's long-term answer was the

Alitalia

Formed in 1946 with the assistance of British European Airways, Alitalia has grown to become a major carrier with a network of passenger and cargo services radiating from Italy to the rest of Europe, the Americas, Africa, the Middle and Far East and Australia.

Most long-haul services are operated by Boeing 747s or DC-10s, while short- and medium-range routes are served mainly by Airbus A300 and McDonnell Douglas MD-80 aircraft. The airline owns ATI, which operates a fleet of DC-9s on an extensive network of domestic services.

Alitalia operates its own flying school equipped with four SIAI Marchetti SF-260 single-engined trainers and two Piaggio P166-DL3 navigation trainers. The airline also operates a number of flight simulators, including an MD-80 and a DC-10 on which, in addition to its own pilots, the air crews of some other airlines receive training.

Having received help from both BEA and TWA in the early post-war years, Alitalia has since provided management and maintenance services for the airlines of some developing countries. In particular, Zambia Airways began its operations with the help of Alitalia in the late 1960s and subsequently received further assistance when it introduced the DC-10 into service.

The extensive engine and airframe maintenance facilities of the airline are devoted not only to the service of the Alitalia fleet but provide some services to other members of the ATLAS group. Originally founded by Air France, Alitalia, Lufthansa and Sabena (later joined by Iberia), ATLAS provides mutual aid in the form of maintenance and training facilities for member airlines.

The Italian flag carrier, Alitalia, has long been a user of the Douglas (now McDonnell Douglas) range of airliners, like the DC-9/MD-80 series. These aircraft are used for high-density domestic and international services to Europe and North Africa.

adoption of a totally new design which resulted in an order in 1959, for 24 de Havilland Trident airliners. The design featured a three-engined configuration and all the engines were rear-mounted. The final layout and size of the new aeroplane reflected current BEA requirements and perhaps the implicit lesson at this stage in the story is that narrow domestic needs of a particular operator rarely suit the wider demands of the international market. The aircraft finally appeared in three different versions to suit conflicting route demands, but was not an outstanding market success in international terms.

Meanwhile, in the United States it seemed that Boeing was becoming unstoppable. Although the intercontinental 707 was reaping rich benefits for the airlines the company was also anxious to address the needs of the medium-haul operators. The initial move in this direction was a shortened version of the basic 707 airframe to create the 720, which entered service in 1960. Perhaps this marked the real beginning of the Boeing 'family' concept, which stressed commonality of layout, components and systems, and a philosophy that offered airlines an attractive package with minimal complication.

Top: Ansett Airlines of Australia have a comprehensive domestic airline route structure which corresponds to many of the international European routes. Amongst the airliners operated by the Australian company is the successful Boeing 727.

Above: Laker was a highly successful charter and scheduled service independent airline in the United Kingdom but it suffered from direct competition in the North American route sector which eventually led to its demise. This is a BAC One Eleven.

Below: Swissair has gained a reputation for good time-keeping and for flying the more advanced designs on international service worldwide, including the use of the latest McDonnell Douglas MD-80 and DC-9 airliners on routes in Europe.

It was a move that was widely accepted and further strengthened Boeing's grip on the market. Thus encouraged, Boeing took up the three-engine, rear-mounted fashion and introduced the 727 in 1960. To many eyes it was a Trident look-alike, though in terms of performance it was an exceptional aeroplane. This attribute was largely due to to a clean, highly efficient wing which boasted an astonishing array of high-lift devices on both the leading and trailing edges.

A number of incidents and accidents were experienced at first, but following a review of techniques and flap settings the new aeroplane then settled down to establish itself as the best-selling airliner.

The rear engine mounting fad had now really caught on. The BAC 1-11 was introduced and represented a design which had first been mooted in the mid 1950s by the Hunting company. This nippy little jet was an instant success and achieved the much-desired penetration of the American market. It was also time for Douglas to make another move.

That company had originally teamed up with the French manufacturer to sell the Caravelle in the United States. This arrangement had proved less than satisfactory, so Douglas made the decision to develop and sell a product of its own. The result was the revival of an earlier project which now evolved as a neat, rear-engined twin jet, similar to the BAC 1-11, and named the DC-9. American airline customer loyalty to the indigenous industry was quickly re-established and the design

Many of the world's developing nations, especially in Africa and Asia, have used airliners to link the centres of population in their countries. This is a Boeing 737 in the livery of Air Zaire which operates on domestic and short-haul routes.

became a major success.

Although advertising hyperbole is often discounted it must be recalled that much of the sales thrust of the DC-9 centred on claims for much-reduced maintenance costs. Perhaps this was the first time that such stress had been placed upon cost of ownership rather than cost of acquisition. If so it marked the beginning of another significant trend in the history of air transport.

Not to be outclassed in any area, Boeing then introduced the 737 twin jet. This time the engines returned to a wing-mounted configuration. Boeing was aided in its conception by the sponsorship of Lufthansa, though the initial reactions from other airlines were cool. Then United Airlines became the first US

operator and the aircraft was on its way to becoming one of the most widely used of all the types.

There now followed a period of astonishing technical innovation. Much of it was lost upon the casual traveller because there was little external evidence of change. The thrust of the engines was improved by by-pass and turbofan techniques. Constant-speed generators offered better standards of electrical power on board and improved electrics, now firmly known as avionics, enhanced navigation and communication.

Increasing capacity

All of the models underwent varying degrees of stretch and these were usually dictated by every increase of power that was made available by the engine manufacturers. Douglas had perhaps been rather more far-sighted than was initially thought, for as the 707 fuselage seemed to stretch or shrink at frequent intervals, the DC-8 had remained constant, while the wing span was increased and extra fuel capacity provided.

Finally Douglas went in for a stretch, which, thanks to the geometry of the airframe, was no less than an extra 11.2 m (37 ft). This gave the aeroplane a distinctive profile in the

One of the world's largest airlines is Delta, based at the Atlanta International Airport regional hub. Amongst the aircraft flown by the airline in the last twenty years is the Douglas DC-8, once an intercontinental jet, now a regional transport.

air. Termed the DC-8-61 the new aircraft made its first flight in 1966 and offered the airlines a capacity of some 252 passengers.

In fact there was a great deal of capacity now available and in order to ensure commercial loads, considerable advertising and special offers began to become apparent. In the social sense the advertising emphasised cabin service and quality of meals. More significantly the term 'air hostess' was gradually being replaced by 'stewardess', surely a sign of things to come.

Although IATA members were restricted in price wars, other operators could be more adventurous though within certain limits. Inevitably many of the charters touched on the slightly dubious and there were cases of special interest charter flights being halted at the last moment because it had been proved that not a sufficient number of passengers were judged to be of an affinity group.

Clubs sprang up like weeds in spring as attempts were made to provide cheap flights around the world. Significantly not all of the guilty were small, upstart airlines; on occasion even the largest operator was a transgressor.

There were other pointers to future problems. The glamour image of the airline business was now under the scrutiny of more travellers than ever before. As demand boomed so did the need for more facilities and longer runways. Airports have a gargantuan appetite for land. Delays on the ground and in the air tended to create irritation out of all proportion with actuality, for the airlines were becoming victims of their own propaganda campaign.

Environmental factors

Other factors were also at work. In the years after the Second World War many Western countries strove to rebuild in an ordered and structured manner. Many laws were enacted to ensure good standards of construction and regard for the surrounding area. Such legislation applied equally to domestic and commercial activities.

This was not the case in other parts of the world, and in Japan, where regeneration and expansion of industry had been conducted to the virtual exclusion of all other activity, there had been a marked reduction in the quality of life in many areas.

From such beginnings had sprung the growth of a movement which liberally sprinkled its conversation with words like 'ecology' and 'environment'. Pollution of any kind had to be combated, although the term was not always used in the same meaning by all the converts. Nevertheless it became a potent force and much of its ire now fell upon aeroplanes and airports.

Most of the world's major airports are

towns within their own right. In turn they generate the growth of local dormitory suburbs often largely inhabited by airport workers or others closely associated with the airport through the provision of service industries.

These areas were liable to considerable disturbance as the new jets became more numerous and soon any attempt to develop airports met with hostile reaction.

Noise monitoring became frequent, and although the manufacturers vied with each other to develop rather rudimentary sound suppressors it was inevitable that the main burden of noise reduction should be placed on the shoulders of the flight crews.

Various noise abatement procedures were developed to suit each airport and runway. Noise monitors placed at strategic positions maintained a check on these methods and errant crews were quickly taken to task. Of course, it was a botched approach, but the only one possible at the time. The real answer lay in the hands of the designers, but new equipment was still some years away. Meanwhile restrictive curfews became common, with resulting complexities in airline schedules.

Short-range developments

As the large new long-range jet fleets opened up more and more areas of the world to tourism there was something of a revolution taking place at the lower end of the airline operational scale. The Caravelle and Viscount were now familiar shapes on many routes, though they were only viable on the relatively high-density routes. Low-density routes were still largely serviced by the DC-3 and not for the first time the cry was heard for a 'DC-3 replacement'.

Several attempts had been made to design this elusive aeroplane and the relative lack of success had led one aviation magazine to

Below: Fokker F27 Friendship of Air UK, *c.* 1984. This is used on domestic services in the United Kingdom and for scheduled services to the capital cities of the near Continent, on so-called commuter services. About 50 passengers can be carried.

Bottom: the British Aerospace HS 748 airliner has been successful in penetrating specially selected markets around the world, including the British Airways Highlands & Islands Service, which links the Scottish mainland with the Hebrides, Orkney and Shetland. A replacement type is known as the ATP (Advanced Turbo-Prop).

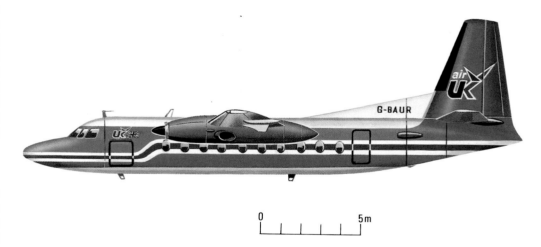

publish a cartoon depicting a line of new aeroplanes in production which looked just like DC-3s. The caption suggested that this was the elusive replacement!

The breakthrough seemed to coincide with the availability of the Rolls-Royce Dart and this in turn made possible the Fokker F27 and the Hawker Siddeley HS 748. These were rugged aeroplanes with good field-length performance and they opened up many new areas, often in remote locations, to the benefits of air travel.

Despite all the activity it is arguable that one of the most significant events in terms of future service to short, high-density routes was the introduction of a shuttle operation in the United States. This was basically a no-frills, no-reservation, frequent service between two points offering a guaranteed seat.

Typically Eastern Airlines operated such a service between New York and Boston, where there was a very heavy commuter traffic pattern. It was an instant success. Passengers merely turned up at the desk, filled in a form with name and address and walked out to the aircraft. When it was full the engines were started and it departed. Fares were collected *en*

'Try a little VC10derness' the advert used to say. Not as successful as it could have been because of programme costs and engine problems, the Vickers (BAC) VC10 was the first four-jet airliner with aft-mounted engines.

route. If more passengers arrived, a back-up aeroplane was available. This was always a good idea, and although it has been copied to some degree in other areas the original concept now tends to be manifested as an unbookable scheduled service, though back-up aircraft are often claimed. Perhaps more significantly it has still to be implemented on true international commuter routes, such as London to Paris.

By the mid 1960s the new aircraft were well entrenched and freight was also tipped to be a major growth industry. Side loading doors appeared and freight 'villages' flourished at many major airports. Although air freight seems to be an ideal venture there is always the slightly niggling doubt that it has never quite made it – yet!

In 1962 the last of the new generation of big jets appeared. This was the Vickers VC-10. This was to be the last aeroplane designed and built solely by Vickers, since the British industry was undergoing a regular spasm of rationalisation and the noble name was soon to be lost within the emergent British Aircraft Corporation (BAC).

The VC-10 had been designed with the hot and high airfields in mind and perhaps it can be said, not unkindly, that this was the lingering Empire Routes philosophy that had so long inhibited many British aeroplanes. Powered by four Rolls-Royce Conway engines in a rear-mounted configuration, the aero-

plane represented considerable aerodynamic sophistication. A longer-range version was developed and BOAC operated both versions. Other growth designs never got off the drawing board and despite its superb passenger comfort it could not compete with the scale and head start of the American manufacturers.

Supersonic flight

It now seemed that American manufacturers were in an unassailable lead and the only way in which they could now be challenged would be by means of a massive leap in technology. The early days of the Comet programme evoked euphoric memories in Britain and it was felt that the same advantage could be obtained once more, and, in the light of evolving aeronautical science, earlier mistakes would not be repeated.

The one basic lesson of the evolution of air transport technology to that date had been the simple fact that speed spelled success and commercial gain.

At a time when there seemed to be a growing trend towards high capacity and lower fares there seemed to be an equally growing disregard for the 'high-price' traveller who was prepared to pay for speed and priority. Suppose that this passenger could then be accommodated in an aeroplane that would offer twice the speed of existing aircraft and halved journey times.

Waiting for passengers at New York's J. F. Kennedy airport is an Air France Concorde which, along with its British Airways counterpart, celebrated ten years of service in 1986. French-built Concordes have also operated to South America.

The argument seemed seductive, and as many people began to experience 'jet lag' the idea of an aeroplane which would reduce this effect and overcome the common disorientation of long-haul travel seemed to offer an exciting new prospect and revolutionise business or priority travel. Once this argument was accepted, supersonic transport (SST) was accepted.

Military aircraft had been achieving supersonic flight for some time, but the idea of building a passenger aircraft capable of such a performance was daunting. It had to be big enough to be commercially viable yet capable of being light enough for acceleration by existing-technology engines.

The programme implied high cost and from the very early days it was felt that, so far as Europe was concerned, it would prove too expensive for a single country. As the publicity campaign began it was obvious that several companies were eyeing the problem and many were the paper aeroplanes that began to adorn the pages of aviation magazines.

As in so many other areas of aeronautics much of the thinking was based upon German efforts during the war. In 1956 a significant event occurred: the British government, in an uncharacteristically far-sighted move, formed the Supersonic Transport Advisory Committee. The intention of this body was to initiate and monitor a co-operative programme of research to pave the way for an SST.

It was the beginning of a long programme of research, though it was soon realised that the best compromise speed would be in the region of Mach 2 and allow the use of metals the characteristics of which were understood.

Below: BAC/Sud Aviation Concorde in the livery of Air France, *c.* 1980. Seven supersonic transports are still operated by the airline, mainly on services to the United States.

Bottom: the Soviets, always ready to copy Western designs, developed their own supersonic transport, the Tu 144, the first SST to fly. It is seen here at the Paris Air Show. However, the aircraft was very crude and it is no longer in airline service in the USSR; none were delivered other than to Aeroflot for high-speed freight work. No further Soviet SST work has been reported.

AEROSPATIALE/BAC CONCORDE

Engines: 4 × Rolls-Royce/SNECMA Olympus 593 Mk 602 turbojets with partial re-heat.
Wing span: 25.6 m (83.83 ft).
Length: 62.1 m (203.75 ft).
Height: 11.3 m (37.08 ft).
Weight: 176,450 kg (389,000 lb) max take-off.
Speed: Mach 2 at 15,545 m (51,000 ft).
Range: 6,940 km (4,313 mi).
Accommodation: Crew of 3 and up to 144 passengers.
Remarks: As the world's first supersonic transport to enter regular passenger service the Concorde is assured of its place in aviation history. Various modifications have helped to improve its performance and it has proved to be a particularly popular vehicle for specialist charters. Despite initial promise the aircraft has only been operated by Air France and British Airways.

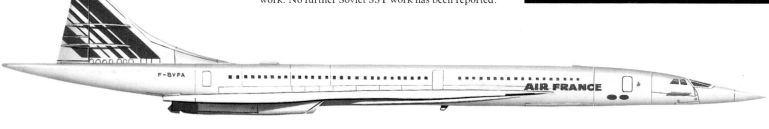

F-BVFA   AIR FRANCE

0   6 m

Mach 3 would create problems of structural heating and the adoption of exotic materials and further complex investigation.

Work on very high speed aircraft was also underway in the United States, France and Russia. The Americans were working on the awesome B-70 Valkyrie Mach 3 bomber, which flew in 1964 and demonstrated so many problems that it foundered in its own complexities. In Britain much work had been done by the Bristol Aeroplane Company and this eventually crystallised into a project known as the Bristol Type 223, which was powered by four Olympus engines. Coincidentally the French had come to similar conclusions, and at the 1961 Paris Air Show the stand of Sud-Aviation carried a model of the Super Caravelle, which bore a remarkable similarity to the Bristol project.

Bristol had now been absorbed in the new BAC organisation and it was that body which began negotiations with Sud-Aviation. Sixteen months later agreement was reached and a formal political and industrial treaty was concluded. It was all very exciting and at the time it did not seem very noteworthy that the wording also included a 'no cancellation' clause.

In the years that followed the SST, which became Concorde, was the cause of much debate, with a particular emphasis on the environmental effects. Much was made of the sonic boom and the possible effects of exhaust emission in the upper atmosphere. The cost also escalated supersonically and the 'no cancel' clause was invoked time and again.

No matter what the ultimate fate of supersonic travel, the arguments, particularly those with the benefit of hindsight, will continue to rage. And no matter how successful the programme in technological terms, it certainly drained a large part of European industry of research, resources and high-grade manpower for too long a period.

The first Concorde took to the air in 1969, but it had been preceded by the first flight of a Russian SST, the Tu-144, just two months earlier. While credit must be given for the first flight it must also be admitted that the Tu-144 in its initial form was very much an aerodynamic test bed. Performance relied almost entirely on brute power and there was little sophistication in the aerodynamics or engine intake systems.

Whatever the detractors of Concorde suggest it must be equally admitted that the Concorde wing and matching of the intake are major achievements.

In the United States Boeing, Douglas and Lockheed all looked at various configurations and tended to opt for large and even faster (Mach 3) performance. Boeing received a government contract for the construction of such an aeroplane with an expected in-service date of 1972. It was a truly complex aircraft and although the company enjoyed showing off a magnificent mock-up of the venture there did not seem to be any real regret when all government funding was suspended in 1971.

Meanwhile the Concorde development programme continued and moved towards the introduction into service. Controversy still

surrounded the whole programme and as certification came near much was made of the fact that the aircraft would be banned from US airspace because of noise. Supersonic passenger flight finally arrived when an Air France and a British Airways aircraft made simultaneous departures from Paris and London to South America and the Middle East respectively in January 1976. Once the operation got underway many of the objections vanished and eventually services were extended to the North Atlantic.

Non-scheduled operations

As the scheduled airlines developed their route structures in the post-war years there were continual efforts by other operators to gain a share of the overall market. The growth of the so-called non-scheduled operations stemmed from this desire. Initially these operators came into being to serve military troop and freight contracts, but by the 1950s this type of operation was beginning to enter the tourist market.

It proved popular and in an effort to protect their own market from attrition the scheduled operators introduced a new tourist-class fare. The result was a rapid growth in tourist-class passengers, and when a further new fare was

Several nations have in recent years become highly successful in marketing light transport aircraft. The Embraer Bandeirante from Brazil has managed to penetrate the European and American markets.

introduced in 1958 with economy class, a further stimulation was given to the scheduled carriers and to the inevitable loss of the charter operators.

This is typical of the cyclical nature of the business and this type of commercial warfare was to have many victims over the years. By the early 1960s a number of American operators failed and in the UK smaller operators grouped together to provide a more united front. Governmental aid tended to reflect the current complexion of politics with the consequent result that charter operators could not experience any degree of stability or the long-term prospects that are so necessary to the development of healthy airlines.

But by the 1960s many of these companies began to enter the jet market and as the Inclusive Tour Charter market became established the real boom in air tourism began to make itself felt. In retrospect it seems it then became a good time for all airlines. With relatively low costs and a booming demand, there really was more than enough trade for all involved.

At about the same time the United States Air Force had been considering the requirement for a very large strategic freighter which could carry large numbers of personnel or equipment over very long distances. It seemed an attractive prospect to the manufacturers and Boeing, Douglas and Lockheed all vied for the contract. It went to Lockheed and was developed as the C 5. Boeing looked over its design and offered Pan Am the chance of its second big step in jet transport with an aeroplane that could carry more than 350 passengers. The Boeing 747, popularly referred to as the Jumbo, had arrived.

The design and construction of the leviathan was a triumph of production engineering and administration. It was built and test flown in a remarkably short time scale and was ready to begin scheduled services by January 1970.

The depression of the 1970s
Unfortunately the 1970s did not begin very auspiciously in financial terms. Inflation was causing concern and one of the cyclical downbeats that so characterise air transport was beginning. The problems were first felt by the scheduled operators, who reduced scheduled services while maintaining the charter or tourist aspect. Significantly many of the operators began to enter the hotel arena by investing in new developments or joining consortia to develop new enterprises.

This was typical of the diversification that many of the larger operators felt necessary at that time. It was also a period when computers really began to make their presence felt in all aspects of the business, from monitoring the payroll to providing flight plans. Some airlines seemed more set on entering the computer business than operating aeroplanes and sold their peripheral services to other operators.

There was still an element of euphoria left over from the expansive 1960s, but to the far-sighted there were several ominous financial indicators that suggested harder times ahead. Rates of traffic growth slackened and companies began, reluctantly, to reassess many of their routes while at the same time

attempting to accept the new generation of wide bodies and their massive capacity.

The greatest blow fell in 1973, when the oil embargo was followed by a staggering quadrupling of the price of oil. Almost overnight the least significant aspect of operating costs now became one of the most expensive. Many airline managements felt at a loss as to how to cope with this dramatic turn of events.

Suddenly fuel shortages struck in areas where such an event would have been unthinkable just a matter of weeks before. Flights were cancelled or merged, and as the fuel bills rose inexorably the growing economic recession was inhibiting travel, leading to falling demand. The airlines now faced the most serious crisis in their history.

The Boeing 747 had now been joined by other wide-bodied jets – the Douglas DC-10 and the Lockheed 1011 Tristar. Soon these aircraft were flying with only minimal passenger loads and many of the most prestigious names in the airline business faced grave economic consequences.

The recession was to last for years and the business was never quite the same again. As the proud flag carriers of the Western World faced up to the new realities, other emergent airlines began to make headway. They were not restricted by custom and could more easily cope with the changing situation. Staff numbers tended to be relatively low and highly efficient. Restrictive practices were unknown and operational flexibility was always a management option.

Other ideas were now abroad. In the United States the political climate was moving

Boeing 747SP

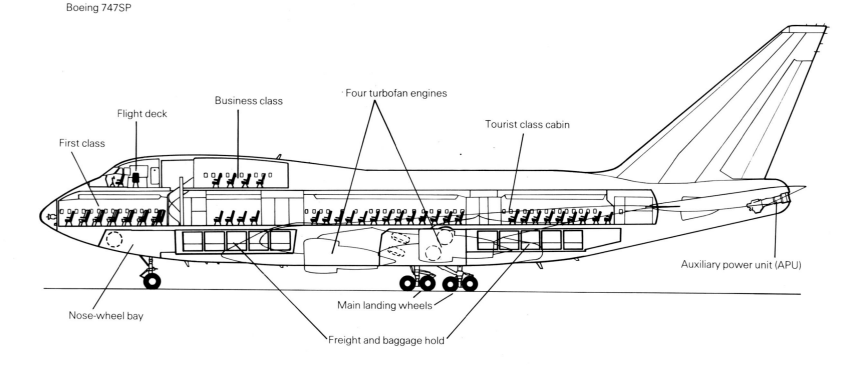

First class · Flight deck · Business class · Four turbofan engines · Tourist class cabin · Auxiliary power unit (APU) · Nose-wheel bay · Main landing wheels · Freight and baggage hold

Boeing 747 of Wardair Canada, *c.*1981.

0   5m

0   4m

McDonnell Douglas DC-10 of British Caledonian, *c.*1983.

strongly in favour of consumer interest and there was an increasing demand for deregulation. In retrospect it now seems that the American airline industry was an obvious target. For some 40 years the Civil Aeronautics Board (CAB) had ensured that United, American, TWA and Eastern maintained supremacy on the profitable transcontinental routes. All applications from other operators to fly major routes were summarily refused.

The new generation of short/medium-haul jet airliners from the Long Beach plant of McDonnell Douglas includes this model of MD 80 for American Airlines. The polished metal finish is characteristic of the airline's aircraft.

Deregulation

As the 1970s progressed there was an increasing campaign in all areas of US political life to allow a greater degree of competition to emerge. The result was the Airline Deregulation Act of 1978, which created a six-year programme of implementation which would be completed in January 1985 with the abolition of the CAB.

The results were dramatic. New operators entered the market and in the flurry of competition fares fell and the market was considerably stimulated. New operators became darlings of the media and then often fell from grace when their financial structures collapsed under the strain of the pure market forces now let loose.

Perhaps it is still too early to draw any firm conclusions. Certainly there are many good bargains to be had, some destinations are better served than ever before, while others have seen a decline in frequency and even an increase in fares. Certainly, what had become a slack market has been stimulated and revived.

Some critics suggest that it was too sudden and comprehensive a change and point to the instability of the situation with mergers and bankruptcies almost a weekly event. Others suggest that this is merely the shakeout that is to be expected in the cause of good economic order and eventual well-being.

As the concept of deregulation spread to other countries it was obvious that many of the major airlines were in serious trouble. The recession was continuing and long-haul routes showed continuing losses. The staff of the major operators began to see deregulation as a form of 'union bashing' and as the free market forces began to take hold, employees' positions, with high salaries and stable conditions, seemed to come under increasing threat. Perhaps this period marked the end of the glamorisation of the airline career and its substitution by an awareness of it being just another job with a negotiable price tag.

As the effects of deregulation began to be felt more deeply the average annual pay for US airline employees actually began to drop. This was the first time that such a decrease was ever experienced in the business. Some airlines

Boeing 737-200 from US Air, one of the regional carriers that started to compete with the established airlines when the US deregulated air transportation. The 737 has proved an ideal aircraft for commuter services.

adopted a two-tier structure which was aimed at lowering the pay of new recruits. The long recession aided the introduction of such decrees as many crew members had been furloughed for years and were often willing to accept the change of climate.

By the early part of 1985 it was noted that the hiring rate in the United States for flight crew and cabin attendants rose sharply but that the average wage for both of these groups of employees was down.

Implicit in the Deregulation Act was the assertion that a continuous and convenient system of air transportation to the nation's small and medium-sized communities must be maintained. This ensured that the commuter or regional airlines had a major stake in the future development of the US air transport system.

The Act also allowed regionals to operate aircraft with up to 60 seats, which permitted optimum suiting of aircraft to market and has spurred a major development in new aircraft to suit this need. This has generated a considerable growth in this type of operation in the United States which in 1984 represented a 17 per cent growth rate.

In turn this has led to a number of new aeroplanes being designed for this market and these include the Beechcraft 99, de Havilland Canada Dash 8, Aérospatiale/Aeritalia ATR 42, CASA/Nurtanio 235, Embraer 120, Saab SF 340 and the British Aerospace ATP.

Meanwhile the initial frenetic scramble is resolving itself as commuters increasingly ally themselves with a major carrier. In recent times there has been a high turnover of commuter operators but in spite of these complications this part of the air transport industry has continued to grow at an unprecedented rate and has been considerably aided by the upturn in the US economy during 1984/85. This trend is expected to persist over the next decade as existing markets expand and new routes are developed.

Average stage length is around 240 km (150 mi) and turboprops are the chosen powerplants since they best match power demands and operational efficiency on such routes. Despite the best of intentions airport congestion continues to pose a problem, though it is possible in some cases to allow a suitable commuter aircraft a non-standard departure or approach pattern, thereby keeping the aircraft clear of the larger aircraft.

Typical of such an operation is the use of a STOL aeroplane, such as the de Havilland Canada Dash 7, using area navigation equipment and MLS (Microwave Landing System) final approach aid for stub runway operations following a separate access landing system.

This technique could be used more widely than at present, but there is often an element of opposition to be overcome. Similarly, true STOL aeroplanes can use specially designed strips constructed close to city centres. This type of operation is currently being developed in London's dockland.

All of this activity in the United States is constantly being monitored, often with some trepidation, in Europe. Deregulation on the American pattern is unlikely because of the national interests involved. Instead Europe looks towards a more cautious method of liberalisation. Critics of the current system

Known to Eastern Airlines as the Whisperliner, the wide-bodied Lockheed L1011 (*above*) uses three Rolls-Royce RB 211 turbofans. The new Swedish-American Saab SF 340 commuter airliner uses General Electric CT7 turboprops; it is seen here in the colours of Crossair of Switzerland, the launch customer (*right*). Four Pratt & Whitney PT6 turboprops power the de Havilland Canada Dash 7 (*below*), pictured with the earlier but still highly popular DHC-6 Twin Otter in the background at Plymouth Roborough airport, from where Brymon flies to London and other British destinations. De Havilland Canada has been successful in North America sales, as well as in Europe and Asia.

Commuter airliners

It is arguable that in the early days of air transport all aircraft could be termed commuter airliners. Then as technology advanced the air transport system evolved as two general classes. Long haul was concerned with operating scheduled services on designated routes while short-haul operators were concerned with local services often operating from a fixed base.

The large, long-haul carriers did not concern themselves with the needs of small communities and the United States, so often the vanguard of progress in air transport, began to certificate local service carriers.

Since that time the demands of a domestic market and the operation of often conflicting market forces have led to new regulations and qualifications which have considerably clouded the issue. In an attempt to define a 'commuter' the American Federal Aviation Authority (FAA) defined such an operation as a Part 135 operator that makes at least five scheduled round trips per week between two or more points or carries mail.

Unfortunately due to past custom and experience many minds are tied to the overriding impression that commuter aircraft are always small aircraft and this has been reinforced by past regulations which have variously used 5,670 kg (12,500 lb); 60 seats or less and payload not exceeding 3,175 kg (7,000 lb) as commuter limitations.

Whatever the classification, the regional airline industry has shown considerable growth in the last decade and a half and particularly since the Airline Deregulation Act of 1978, which established a free market for US airlines. The effects of that law have literally been felt worldwide and have created a demand for commuter services in many countries other than the United States.

Inevitably the appetite for unrestrained competition has not been accepted as eagerly in all other markets. Entrenched

Top right: the Embraer Brasilia has been successfully marketed to the North American commuter operators and entered service with ASA in 1985.
Lower right: the Shorts 360 (seen here at Edinburgh) has been equally successful in the relaxed UK airline route arrangements where independent operators are competing with British Airways, the national carrier (Boeing 757 Super Shuttle in background). The Shorts aircraft is manufactured at the company's plant at Belfast.

interests are often loath to face change, especially when a comfortable life style is at stake. Nevertheless, the example of the United States suggests that customers all over the world can be tempted with greater choice of service and lower fares: a heady combination that does not always emerge in the realities of other less competitive markets.

At least it has opened up the whole concept of commuter operations and it is no longer automatically associated with small aircraft. Instead the word has become increasingly accepted as a name for a type of operation which is primarily dedicated to local air services at high frequency with no frills. Consequently, the type of aeroplane is of secondary importance and when this notion is accepted it can be equally easily understood that a Boeing 747 operating a high-frequency, short-haul operation can be justly termed a commuter aircraft. Indeed such a service exists between Tokyo and Osaka.

Nevertheless the old ideas sometimes die hard and commuter aircraft had gained a reputation for austere conditions and less than contemporary technology – in other words an element of the pioneering spirit remained. While this was acceptable to romantics it no longer had a place in the new open marketplace and manufacturers were quick to recognise a growing gap in the future market.

New technology in airframe manufacture and improved-efficiency engines offered the possibility of a new generation of commuter aircraft that could provide the passenger with a degree of

comfort normally only associated with the larger aircraft, while still returning a profit for the owners. Already many of the independent operations spawned in the United States in the post-1978 period were entering into closer association with major operators and it was becoming increasingly important in market terms to attempt to retain an overall product standard throughout a passenger's journey from local airport to international hub airport and on to his overseas destination. The American Eagle set-up is a classic example.

Due to the short stage lengths and associated performance constraints the new aircraft all feature turboprop power while the seating capacity is indicating an inexorable rise from the early 19-seaters to 30-35 and even up to 60 or more as initial designs are stretched. Of course the new high-efficiency designs and engine-matched aircraft have been developed specifically for the requirements and economics of short-sector operations and cannot be matched by earlier jet aircraft such as the Boeing 727 or the McDonnell Douglas DC-9.

New avionics will further assist the flexibility and versatility of the new aircraft by the introduction of flight-management systems and 3-D area navigation, which will allow the aircraft to maintain an optimum flight profile at all stages of the flight. This in turn should allow the more enlightened airport authorities to consider separate access facilities to make rapid departures and arrivals possible. This is not likely to be popular with the major operators, however.

Shorts 360 of Aer Lingus, *c*.1985.

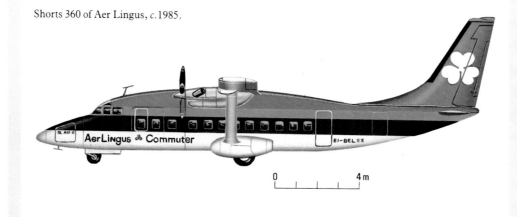

0   4 m

Typical of the new, smaller commuter aircraft is the Beechcraft 1900, a comfortable 19-seater offering pressurised comfort on the power of two Pratt & Whitney PT6A-65B engines. Also at the lower end of the scale is the Dornier 228, which features an element of STOL performance and a ruggedness that will ensure long life in the rather harsher conditions of short-haul operations, often from semi-prepared strips.

These are of course the harbingers of the advanced technology types currently characterised by the Aérospatiale/Aeritalia ATR 42; de Havilland Canada Dash 8; Embraer Brasilia; CASA/Nurtanio CN-235 and the Saab SF340. Meanwhile, the first flight approaches of the British Aerospace ATP, which will offer more than 60 seats and ensure that the continuing market gap at this size will finally be closed. This inevitably means that most or all of the previously mentioned types will undergo a process of stretching with the ATR 42/72 in the vanguard of that process.

Against this background of emerging types there exists a somewhat complex variety of aircraft all contributing to the commuter or regional market. Among this varied equipment some aircraft have established themselves as particular successes and form the backbone of many fleets.

The American-manufactured Metro III is still widely used, while Embraer's Bandeirante is a common sight in the skies of both North America and Europe. In the UK Short Brothers has had a particular success story with the SD3-30 and SD3-60.

Although some observers think that its basic approach lacks the sophistication of the new-generation aircraft, the Shorts

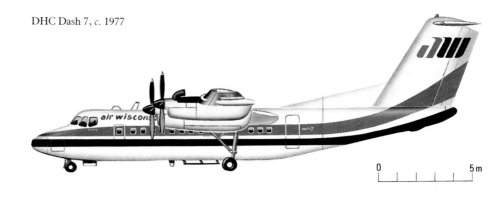

DHC Dash 7, c. 1977

0   5 m

Joining the regional market, perhaps as a replacement for the F28, is Fokker's new 100 airliner which first flew in 1986; it is seen here in the livery of Swissair, the launch customer, and it has also been ordered by US Air.

formula has proved itself to date and it is significant that the company is considering a new design that is virtually a stretched version of the SD3-60.

Fokker's F27 is another worthy type that is being used in the commuter role, often supplementing a major carrier. SAS uses the type for this purpose and it would seem that it will have many more years of service ahead, although the company has felt obliged to enter the future market battleground with the 'new technology' Fokker 50.

British Aerospace is well represented in this class of aeroplane. The original HS 748 continues to attract operators and the latest BAe Super 748 is ensuring that this popularity will continue for some time to come.

Another undoubted BAe success is the Jetstream 31, which has enjoyed excellent sales in the United States. This most attractive of aeroplanes has had something of a chequered past. Originally designed by the Handley Page company, the aircraft had been inhibited by the technology of

The de Havilland Canada Dash 8 (above), powered by two Pratt & Whitney turboprops, carries 36 passengers. Just out of the commuter market and more a regional airliner is the British Aerospace 146 (left) powered by four Avco Lycoming low noise turbofans; a 100 series is pictured in the livery of Dan-Air, the largest UK independent. As well as being ordered by some American operators, the aircraft serves with the Queen's Flight.

the times and then overtaken by the manufacturer's financial problems. Using the latest techniques in materials and manufacture and using the Garrett TPE331 turboprop engine, the BAe Prestwick factory evolved the '31', and its present popularity justifies the faith that the manufacturer had in the concept.

British Aerospace has also developed the BAe 146 family and, in contrast to the usual commuter trend, this aeroplane is powered by four Avco Lycoming ALF 502R turbo fans. Although not a commuter in the classic sense, the 146 certainly appears to be an ideal aeroplane for this type of operation.

Another effective performer in the less sophisticated section of commuter operations is the CASA C-212 Series aeroplane. Its basic rectangular section fuselage allows it to be adapted to a number of roles, military as well as civil, and it has also been built under licence in Indonesia by PT Nurtanio.

As de Havilland Canada's elegant Dash 8 begins to make inroads on the scene the earlier Dash 7 continues to offer excellent service, although its remarkable STOL characteristics are rarely used to full advantage.

Another jet that provides a quiet efficient service to the regional market is the attractive Fokker F28. It often appears at European airports in its Cityhopper guise. Fokker seem set to develop this type of aeroplane by the emergence of the Fokker 100, which as a 100-seater will certainly be extending the role of commuter acceptance.

Perhaps the real key to the design of the commuters of the future will lie in the consolidation of company fleets around key designs. This will improve efficiency and reduce costs by avoiding unnecessary complication in logistics. This trend may be gauged from statistics published in the United States for 1983 which indicate that just 12 aircraft types accounted for nearly 82 per cent of total capacity.

Don Parry.

The vastness of Australia means that the country provides an increasingly large market for passenger aircraft. One of the expanding airlines is Ansett Western Australia, which has equipped itself with the BAe 146.

suggest that it is in total contradiction with the Treaty of Rome, and the European Community is currently much involved in creating a truly European air transport policy.

This will inevitably take time but progress is being made and several agreements have already been made to breach what had previously been considered to be immutable standards of pricing and scheduling.

It all seems a long way from the oil price crisis of 1973, yet events of that time are now having a major effect upon all aspects of the air transport system. The demands of the environmental lobby were already being taken seriously, particularly in respect of noise. Night jet bans were in force and manufacturers were taking steps to improve engine design to reduce noise footprints. The oil crisis gave further emphasis to this work in the search for more fuel-efficient engines as well.

Fuel saving and new technology

Every aspect of the operation was scrutinised in attempts to save fuel. Crew procedures were changed, performance requirements adjusted and computer flight planning also assisted in the search for economy. *En route* techniques were tightened up and crews were careful always to operate at the optimum conditions of height, speed and weight. Meanwhile electronics were coming into greater prominence.

The Boeing 747 was the first airliner to be fitted from scratch with an integrated, automatic inertial navigation system. This offered a

new standard of *en route* precision and allowed air traffic control a degree of flexibility hitherto unknown in the business. This was seen as the basis for a number of innovations that would result in major avionic applications to improve efficiency and reduce crew workloads.

Airbus Industrie was looking ahead to a highly improved version of the A300, while Boeing was developing the 757 and 767, featuring a two-man crew and extensive avionic equipment, including cathode-ray tube presentation of flight data. These technical advances were allied to improved airframe design and fuel-efficient engines with extremely low noise signatures.

Of course this technology is not cheap and at a time when passenger expectation is firmly fixed upon ever-lowering fares, the cost of these new aircraft is rising and they have to be highly efficient. By their very nature airlines are very capital and labour intensive, while user charges and maintenance costs show steady increases.

Airlines are especially prone to recessional cycles as they cannot react quickly to downturns in the market and then often fall behind again when the upturn takes place. In an attempt to alleviate some of the problems of

this type of financial switchback many airlines attempt to use a variety of complex methods of financing new aircraft purchases. Typically many of the world's airliners belong to banks or other consortia and are merely leased to the airline. Similarly the manufacturer has an equally difficult time in achieving sales and has to become embroiled in various lease-back deals or even accepting a competitor's product in part exchange. Recently the use of barter techniques has become popular once more with an order for Boeing 747s being settled by payment in oil.

When in service the new aeroplanes have to be highly productive. This means high load factors, and consequently there are continual fierce marketing campaigns, which again create additional expense. Despite the best of intentions all airlines are faced with a number of variables. Fuel prices have tended to stabilise in recent years and even fallen. Now the trend is being reversed and by the latter part of 1984 these prices were rising quite suddenly in Europe. Oil is purchased in US dollars, so the movement of the dollar against national currency is another problem.

In order to cope with the pressures and introduce new, efficient aeroplanes many airlines have undergone intensive reorganisation and reduction in staff numbers. The search for more cost-effective operations has now resulted in the use of the new 'big twins' such as the extended-range version of the Boeing 767 on the Atlantic.

Singapore Airlines has been a major success story, and one of the carrier's hallmarks has been the introduction of modern, high-technology aircraft like this Boeing 757, used for such services as Singapore to Jakarta (Indonesia).

Airbus Industrie achieved a coup by selling the American airline giant, Pan American Airways, the Airbus, out-bidding its major rival Boeing. At the same time the airline changed its traditional livery to suit the 1980s.

It is entirely understandable that many people feel a degree of apprehension in flying the Atlantic on two engines with just two crew members on the flight deck. No doubt much of this is due to emotional conditioning from the past, although the concept still has to be completely accepted.

There is an obvious advantage in using such an aeroplane in comparison with a three- or four-engined aeroplane; the question remains however – is it a safe operation? Can the desired levels of reliability and operational integrity be maintained? The FAA gave clearance but insisted that the route enabled the aeroplane to remain within two hours' flying time of a suitable diversionary airport.

Already Air Canada and TWA are proving

the point and El Al has also carried out the occasional flight. A powerful commercial message is provided by the fact that using aircraft like the Boeing 767 or A310-300 in this way provides an ideal solution for the long, 'thin' routes, where the capacity of a Boeing 747 is too great. It also allows future operations of a greater frequency, which seems to be the general direction for the future of the industry.

There is already a perceptible move away from the established international gateways to allow a greater freedom of choice and this fits in well with the booming commuter market in the United States. Now analysts are suggesting that the future trend will be towards smaller aircraft and higher frequencies.

Much credence may be given to this suggestion by reviewing the current state of airline orders. As the recession slowly fades the second-hand aircraft market has already been cleared of suitable aircraft and now orders are coming in for new aeroplanes. In particular orders for 150-seaters such as the

A320 and Boeing 737-300 seem set to take off in a big way.

Boeing and Airbus Industrie are locked in a tough commercial battle that has seen the European company win some very important orders. Boeing accepts that its assumption of product leadership is under considerable stress and some analysts suggest that Boeing's recent attempt to purchase the de Havilland Aircraft Company of Canada has much to do with a very real desire to ensure as wide a product range as possible.

New technology continues to evolve and the international aero engine consortium of Rolls-Royce, Pratt & Whitney and Japanese manufacturers has seen its innovative V-2500 engine become established as the lead engine for the A320.

Now the battle for new orders is moving into the area of the very long-range aeroplane. Boeing has started work on the Boeing 747-400; McDonnell Douglas is pushing its DM-11 and Airbus Industrie is projecting a new aeroplane, the TA-11. First blood has fallen to Boeing with the announcement of an order for 20 of the -400s by Northwest Airlines.

The new Boeing has the same fuselage dimensions as the 747-300 but with an increased range of up to 12,900 km (8,000 mi); a fully digital two-man-crew flight deck and improved fuel economy. Advanced engines can be either the Pratt & Whitney PW4000 or the General Electric CF6-80C2, rated at some 56,000 lb of thrust each. It is claimed that the

With overflight rights denied in black Africa, South African Airways acquired the Boeing 747 Special Performance for non-stop services between the Republic and Europe. It also flies to Israel and the Far East.

Avionics for airliners

Thanks to the micro-electronic revolution the key to efficient and economic airliner operation now depends very much on the suite of avionic equipment. Instead of creating an ancillary system known as 'electrics', the contemporary design philosophy emphasises the ability of electronic techniques and computational power systems to maximise the inherent efficiencies of modern airframe and engine designs. By the mid 1980s the avionics suite of an airliner represented some 15 per cent of its selling price and is likely to increase even more in the later versions of contemporary designs or new-generation equipment.

The term avionics now covers a wide range of equipment and can include everything from communications to auto pilots, flight-management systems and electronic instrumentation. Even passengers now have access to air-to-ground communications and there is the promise of much more to come.

Digital techniques revolutionised the look of the flight deck and now new methods of integration and packaging are being introduced to create ever more compact equipment of lower weight, reduced size, increased reliability and considerably enhanced capability.

Already the sight of cathode-ray tube displays is becoming commonplace. Instead of a mass of crowded and confusing electro-mechanical dials the crew now have immediate access to a mass of information which can be called up at will or when it is really needed.

Relevant flight data is presented in an unambiguous fashion and is considerably enhanced by the use of colour. Raw data can be overlaid with weather information.

Initially there was a great deal of resistance to the use of this type of display. Past experience had suggested that CRTs were heavy and relatively unreliable. Perhaps it is sufficient testimony to the rapid advance in technology that while this statement was true a few years ago the entry into service of this equipment in the Boeing 757 and 767 and Airbus A310 has been remarkably trouble free. They have been enthusiastically accepted by the crews and now many older aircraft are being fitted with these devices during overhaul or refurbishment.

In addition to altering the look of the flight deck, avionics have had a direct influence in reducing its size as well. Compact instrumentation allows better and more economical presentation of data and the use of automated navigation and flight-management systems has led to the departure of the specialist flight navigator and flight engineer.

The flight-management system can combine many of the navigation and flight-engineering tasks to ensure that the aircraft is operated as efficiently as possible. In general, this means being at the optimum speed and height and following the most precise route over the ground.

To achieve this the memory must be capable of storing a vast amount of flight-plan data, including departure and arrival information. While in flight the system can accept air data and navigation inputs to correct and refine the required programme and provide a series of inputs to the automatic flight system.

There is considerable promise in the concept of automated flight management techniques and their evolution will see the increasing adoption of all communication management and specialised techniques, including specialist noise-abatement profiles.

Farther along the development road will be true 4-D navigation created by the addition of time to the navigation function and the ability to interface with all types of inputs, including MLS and satellite navigation.

This steady trend towards increasing automation can now be seen in what is being called the flight-control revolution and will be exemplified by the advent of the A320. This aircraft will see the introduction of the sidestick controller in civil aviation, and massive on-board computational power will ensure that the precise flightpath holding is the responsibility of the electronic systems. The computers will be programmed with all of the relevant aerodynamic data such as speeds and attitudes to ensure that whatever the pilot may try to do the computers will always prevent inadvertent overspeeding, stalling or overstressing the airframe.

Such techniques will have a considerable impact on future designs. Currently all aircraft are over-engineered to ensure that an element of ham-fisted handling will not overstrain and damage the airframe. In other words the aeroplane is built more strongly than is theoretically necessary to allow for the human error. This in turn means that the airframe is heavier than it needs to be with a consequent loss in efficiency.

The new flight-control techniques that use cable or fibre optic links to carry the signals are now lighter and more effective and allow the designers to reduce the airframe weight and achieve more efficient aerodynamic results.

There is considerable benefit inherent in this approach: accidental stalling will be impossible; the problems of windshear will be overcome; and the aeroplane will always be in optimum trim. Pilot workload will be reduced considerably and the way will be open to the development of a truly totally automated aeroplane, requiring minimal human interference.

Don Parry

Modern airliners are leading the way in the use of high-technology avionics in civil aviation; both Boeing and Airbus are competing to give the best cockpit environment for aircrew. The use of fibre optics and cathode-ray tube screens is widespread.

new aeroplane will consume 10 to 12 per cent less fuel than a 747-300, depending upon type of engine, with an improvement of up to 22 per cent to 24 per cent in efficiency over the earlier 747-200.

It is interesting to contrast the perceived requirements now being canvassed by the airlines. The big twins are expected to cater for the long, 'thin' routes, while a new class of very long-range aeroplanes is now being actively pursued. Flights are already being undertaken from London to Singapore and Hong Kong and the new aircraft will certainly tend to offer their major advantages in the Far East and the Pacific.

As the industry moves towards the end of the century it has to take notice of the fact that the shape of the aviation world market is changing. The Pacific is becoming what the Atlantic once was. On any one day four out of five large passenger aircraft are in flight over the Pacific area. In the United States a discernible shift is already apparent in business patterns from the east to the west coast. The Pacific Basin is poised to become one of the richest and most dynamic areas of the globe. Little wonder that there is a need for the long-range monster and it looks as if Boeing is back in the lead.

The exalted Blue Riband route between Europe and New York is soon to be eclipsed by non-stop flights between London and Tokyo; Los Angeles and Sydney; Chicago and Seoul.

Several airlines are reviewing the requirement of the new super jet and these significantly include Japan Air Lines, Cathay Pacific, Qantas and Singapore Airlines. This crystallises the grouping that many European airlines see as the most dangerous competition. So deep is this concern that in 1985 a delegation from the European Parliament's Transport Committee was invited to a fact-finding mission by Singapore Airlines.

The delegation was offered every opportunity of appraising the constraints and opportunities which have enabled this airline to achieve its outstanding performance without recourse to state aid.

High staff productivity was very much the key element and contrary to popular misconception it was noted that salary levels at supervisory and management grades are high-

Lufthansa

The merger between Deutsche Aero Lloyd and Junkers Luftverkehr in January 1926 to form Deutsche Luft Hansa marked the beginning of the present-day airline. An innovator from the earliest days of air transport, Lufthansa pioneered long-distance routes using flying boats in the 1930s and in 1938 flew a Focke-Wulf Fw200 Condor non-stop from Berlin to New York – a considerable achievement for a landplane at that time.

Following the resumption of operations after the Second World War, Lufthansa quickly built up a network of services to all parts of the world, benefiting from its position as the national carrier of a leading industrial nation.

Now established as a major international carrier serving over 70 countries, Lufthansa has retained its early pioneering spirit by the launch orders for such significant airliners as the Boeing 737 and Airbus A310.

As well as operating an extensive worldwide system of passenger services, the airline has long been active in the air cargo business, maintaining an important transit operation in several major cities. Lufthansa established German Cargo Services as a wholly owned subsidiary to operate non-scheduled cargo services mainly to Africa and the Far East.

Condor is another subsidiary formed by Lufthansa to operate non-scheduled services, concentrating on charter and inclusive-tour operations for the German holiday market.

At one time the largest (non-American) operator of Boeing airliners, Lufthansa has in recent years selected Airbus A300, A310 and A320 aircraft to boost its short- and medium-range capacity. Most long-haul services are operated by Boeing 747s and DC-10s.

Lufthansa, the Federal Germany flag-carrying airline, operates such aircraft as the boeing 737, for which it was the launch customer. This picture shows the general scene at an airport turn-around, made in less than an hour on European routes.

er than those for equivalent jobs in the UK, Hong Kong, Australia and New Zealand. Although it may be that wage rates for other grades are not so attractive, the company insists that the secret of success lies in the more efficient use of aircraft, human resources and

Below: based in Hong Kong, Britain's Asian airline, Cathay Pacific, has established a reputation for high standard of service and its non-stop route from London Gatwick to Kai Tak International.

superior cabin service, this latter point being emphasised by the carriage of four more cabin staff than are usually used on the Boeing 747. This is a point that is carefully and skilfully put across in TV advertising.

Cargo operations

Although most emphasis is inevitably placed upon the carriage of passengers it must also be remembered that air cargo has always had a role to play. The business has never quite lived up to expectations and cargo airlines seem to be particularly vulnerable, yet it has grown into a major international operation. It is an essential part of the industry and is supported by a complex and comprehensive freight-forwarding sector in all parts of the world.

The freighter aeroplane has often been a conversion of a normal airliner, though in some instances specialised aircraft have been developed. The advent of the wide-bodied aircraft was a boon for the industry as it

British Airways

The ancestry of British Airways can be traced back to 1919 – a year which saw the formation of several European airlines. Air Transport and Travel along with Handley Page Transport began services to the Continent, but although they were later joined by other airlines, competition between them meant that none could be profitable.

The first of many government-appointed committees looked into the problem and concluded that the small airlines should merge into a single carrier. Thus Imperial Airways was established in 1924 with the promise of a government subsidy of £1 million to be used over a 10-year period.

In common with KLM, Sabena and Air France, Imperial Airways concerned itself with the establishment of a network of routes to serve the far corners of Empire. As a result, little effort was devoted to the development of Continental services.

A large fleet of Short C-Class flying boats was flown by Imperial Airways in support of the Empire Air Mail scheme which linked colonies in Africa and the Far East with the mother country. The comparative neglect of European routes led to the establishment of the first British Airways, which was set up in the late 1930s.

Another government committee proposed a division of routes between the two major British airlines, but the onset of the Second World War put paid to such developments and the two airlines were nationalised in 1939 to form the British Overseas Airways Corporation.

After the war, BOAC maintained long-distance services, while British European Airways looked after shorter routes.

Both airlines earned a reputation for pioneering new technology; BEA in the shape of the Vickers Viscount turboprop airliner and BOAC with the Comet and, later, Concorde.

Brought together to form British Airways in 1972, the airline is today one of the largest in the world, with an extensive network of scheduled and charter services. The fleet includes the Boeing 737, 747 and 757, TriStar and Concorde.

One of the strangest aircraft ever developed is the Super Guppy, at one time the largest aircraft in the world. It has been used for carrying spacecraft and other aircraft, especially from factory to factory during manufacture.

Concorde is the flagship of British Airways, and has been one of the most controversial aircraft of recent years. The company claims great success for it on the all-important London-New York route.

Flying Tiger has long held the air cargo banner in the United States, while Europe has tended towards the specialist market with Heavy Lift using converted military freighters (Short Belfast) to create a useful market ranging from the carriage of helicopters to satellites. European manufacturing collaboration has also created the need for the Guppy conversions which carry parts of the Airbus around Europe.

Perhaps the most exciting aspect of the business is the development of the small package express service, which began with the advent of Federal Express Corporation. Since

Cargo and air freight are big business for today's commercial airlines, with many regular passenger services carrying cargo, especially in wide-bodied aircraft. Specialist freight carriers are also used by several airlines.

allowed a realistic integration of cargo and passenger loads to achieve optimum operational efficiency. The trend has been taken a step further by the use of combi aircraft, in which part of the main deck is also used for cargo.

At one time cargo handling tended to be a rather slow process and freighter schedules were often sadly awry. New handling equipment and palletised loads altered that state of affairs and in many cases cargo changes can be made more quickly than loading passengers.

Yet the full benefit of air cargo has probably yet to be achieved. Much of the problem is created by the need for the 'return load'. Often a proposed route has excellent traffic in one direction while the reciprocal service finds it hard to raise any traffic at all.

that time the service has widened and several other operators have entered the fray, including the US Postal Service. In Europe the all-red aircraft of the British Post Office's Datapost are an increasingly familiar sight.

Perhaps the most encouraging thing of all is the public acceptance of such a service. Initially it was seen as something of an emergency measure, but now it is being used more and more as a normal overnight service. The emergency requirement is now seen as the same-day delivery.

Typical of the cyclical nature of this business is the fact that operators are now considering the conversion of aircraft like the Boeing 757 to the role of a package freighter. Other operators fear that suitable aircraft for cargo are becoming increasingly scarce and there are at least two projects (1985) for converting the Lockheed 1011 to a pure freight role.

Air cargo has generated an industry all its own. Extensive ground facilities need to be provided and a complex pattern of interrelated operations and procedures has evolved to ensure the efficient transit of freight over its ground links as well as the airborne segment.

Most of the major airports now have a 'cargo village', where most of the activity seems to take place at night. Today the role of the air freight forwarder is crucial to smooth operation and is likely to grow even more as the diversity of freight services continues to grow.

The future

As the world moves towards the final decade of the century air transport would seem to be poised for another step forward in technology. In the past these have usually meant increases in speed or carrying capacity. This time it would seem that the technology will be aimed at increased efficiency and lower sound levels. New materials are being used to reduce weight and improve reliability. Avionics are taking over many of the flight deck tasks and engines are becoming more efficient in terms of fuel burned.

There is also a discernible trend to a higher frequency philosophy, which could lead to a requirement for a very efficient aeroplane in the 120-150-seat bracket. This development is likely to help in the problem of airport congestion by using more airports, though the major gateways for international operations

seem set to be overcrowded for the foreseeable future.

Massive building programmes continue at many major airports and each new development is fraught with problems of placating the local environmental lobby. Further noise reduction will help this situation and much of the new technology will be applied to the engines.

Perhaps one of the most interesting aspects of this process is the apparent return of the propeller. In order to fly through the air the aeroplane must be able to generate thrust. This is usually done by propelling rearwards a column of air. A propeller-driven aeroplane displaces a large diameter, relatively slow moving column of air, while a jet displaces a small column at high speed. This simplistic situation undergoes several variations with different types of engines, but the overall effect

The future will see the development of quieter, more cost-effective power plants such as the McDonnell Douglas MD 80 Propfan idea. Experiments are quite widespread to meet new economy requirements.

Qantas

Founded in 1920 as Queensland and Northern Territory Aerial Services, Qantas no longer operates services on Australian domestic routes but has become the nation's international carrier.

To link up with Imperial Airways' Empire Air Mail services, Qantas acquired a fleet of C-Class flying boats and the airline's Far East services continued throughout the Second World War. Catalina flying boats and Liberator bombers converted for transport duties flew the eastern leg of the 'horseshoe' route which BOAC operated from Durban during the war.

Although converted bombers in the shape of the Lancastrian continued to be operated after the war, Qantas soon acquired more modern equipment. Super Constellations and Electras were operated before the airline entered the jet age in 1959 by starting Boeing 707 services.

The airline built up a reputation for excellent standards, including an admirable safety record. As immigration built up from a growing number of countries, so the demand for scheduled services grew in parallel, and today Qantas operates a large fleet of Boeing 747s supplemented by some Boeing 767s.

As is so often the case with an airline which rises above the average, Qantas was for long guided by one of the characters in air transport – Sir Hudson Fysh.

Australia's international airline flies the Boeing 747 on overseas routes to North America, Asia and Europe. The aircraft is being supplemented by the Boeing 767. There is now a one-stop route to London from Sydney.

is the same.

As aeroplanes were built to fly faster the propeller became a problem and tended to lose efficiency in excess of around 725 km/h (450 mph). Then along came the jet and speed was again in the ascendant. Of course the pure jet tends to be a voracious fuel-eater and developments like the by-pass and fan jet came along to improve the fuel consumption.

As aerodynamic knowledge grew, it became apparent that various types of propellers could be successfully integrated with turbine engines and it eventually transpired that a form of propeller, now called prop fan, could lead to significant fuel savings. Now several companies are working on demonstrator engines, though they all seem loath to mention the word 'propeller' and have settled instead for a form of 'unducted' nomenclature.

Currently three demonstrator aircraft are being readied – Gulfstream II, Boeing 727 and an MD 80. The excitement can be gauged by noting that this type of technology can result in fuel savings of between 15 and 30 per cent.

Inevitably the concept is still at an early stage and argument is rife over the merits of a geared or ungeared fan and number of blades and stages. Certainly there is a great deal of work to be done and it is necessary to define certain areas such as blade integrity, failure containment, noise and vibration and gearbox configuration. This is quite difficult in terms of engine development but there can be little doubt that this is the next fashion.

There are also indications that designers are looking at the SST once more and much of the interest has been inspired, paradoxically, by recent developments in subsonic aircraft construction. New materials, manufacturing processes and computer-aided design techniques have created an entirely new dimension in aircraft design. If engineers were to build the Concorde from scratch today these techniques could result in it being just half the present weight and using almost 50 per cent less fuel.

Similar advances have been made in engine design for very high-speed aircraft. These include the variable-cycle engine and a range of improvements in materials, cooling and fabrication. Noise would still be a cause for concern but recent research has suggested that coannular nozzles, mechanical suppressors and various types of acoustic shielding offer a range of options not previously available.

As in other areas the major influence in the operation of these new engines will be electronic systems that will ensure optimum operational results at all times. The precision and delicacy of these computers will ensure that the engine is never overstressed either mechanically or thermally, thereby ensuring increased reliability and reduced maintenance.

Increasing use of automated electronic systems to provide easier and more efficient operation will diminish the role of the human crew. Satellites now seem set to play a major role in airline communication and navigation. Rather surprisingly the merchant marine has been more innovative than the airlines in the use of this type of technology. The advent of the Boeing 747 was an excuse for much discussion on satellite communications and provision was made in the airframe for suitable antennas.

With hindsight it would seem that the proposed rentals and operating costs of such equipment were more than the airlines would be prepared to consider. Instead, long-range communications continued to depend upon HF techniques and crews became philosophical about the limitations of the system.

Now the airlines are being offered a service based upon the satellites of the International Maritime Satellite Organisation (INMARSAT). Three areas are likely to be exploited to provide air traffic control communication, operations communication links with airline offices and an airborne public communications service for the passengers.

Similarly satellite navigation techniques can offer the facility of providing a high degree of automation and precision for all types of commercial aircraft and has the ability of offering the potential for safe let-down procedures at almost any airfield in the world. This latter potential could be of major significance to many of the small operators, especially in the Third World.

Lockheed's proposal for an advanced supersonic transport has been superseded by President Reagan's 'Orient Express' sub-orbital transport idea and its rival development, Britain's HOTOL. These are scheduled to fly by the next century.

It all seems very exciting and for those who still have a strong aeronautical imagination there is the ultimate prospect of a large, very high-speed, sub-orbital machine capable of flying from Europe to Australia in minutes rather than hours. New engines which can be loosely described as 'air-breathing rockets' would be the base technology for such a device.

Extensive use of electronics would create an unprecedented ability for automated control of all functions and creates the acceptability of the emergence of an airliner that did not require a human crew.

Such suggestions now seem unrealistic and it is worth remembering that eminent minds in the 1930s condemned supersonic flight as 'impossible'. At the 1984 Farnborough Air Show a small model on the British Aerospace stand was perhaps a portent of the future. Called HOTOL, which stands for Horizontal Take Off and Landing, the model suggests a configuration of aerial vehicle that could use a new engine now being developed by Rolls-Royce.

This engine breathes air, as with a conven-tional jet, during the initial stages of flight through the atmosphere. Consequently it does not require to carry as much fuel as a standard rocket. This results in a lighter, smaller and cheaper machine that can switch over to conventional rocket fuels as it rises to the edge of the usable atmosphere.

Yet, whatever the exciting prospects of the future of technology the airlines are still firmly wedded to the economics of the present. Cheap air fares are here to stay for the foreseeable future. Airlines like People Ex-press in the United States and Virgin Atlantic in Europe have proved that there is a large market for the no-frills service.

The aftermath of deregulation which has opened up the business to free market forces has also created uncertainty and confusion, but this is just one aspect of the market revolution currently affecting most of the world's airlines.

For a long time the international air routes were dominated by American and European airlines. This has now passed and their market share is slipping while the airlines of the Far Eastern region are showing dramatic gains.

Asia is the area of the world tipped for further growth in civil air transport, particularly the People's Republic of China with its large population, vast area and rapidly developing economy. This DHC-6 Twin Otter operates there.

Commercial pressures and economic reality are forcing European governments to revise protectionist policies to allow greater access to their areas by the dynamic Asian newcomers.

The years of being cushioned against open competition on many international routes are now coming to an end. Rationalisation is at hand. The fight to maintain load factors is tough and the product has to be constantly reviewed. Sadly in contrast with many Euro-pean operators the Asian airlines have proved too often to be more innovative in marketing and to offer a more attractive in-flight service.

As the industry moves towards the last decade of the century the older-established airlines face tough, uncompromising decisions merely to stay in being, while those of the East face exciting prospects of expansion in an area that seems set to be the only part of the world expecting to achieve any real, significant economic growth.

5

HELICOPTERS

The helicopter is the child of the post-1945 period, gaining a baptism of fire during the Korean war, and coming of age during the Vietnam war. On the civil front, the exploration for offshore oil and the helicopter's ability onshore to take people and supplies into remote locations has gained it credibility. Everyone is familiar with the helicopter, too, as a lifesaver, especially during winter storms when military crews risk their lives lifting stranded sailors and people lost on mountainsides. In many countries there is also a major emergency medical service to lift casualties from road accidents to hospital in the critical one hour after injury.

The helicopter has often been called the unique flying machine, with its special ability to take off and land vertically, to fly forwards, backwards and sideways, and to remain hovering in the same position. Yet it was not until after the Second World War that the concept of a rotary-wing aircraft began to catch on with the interested parties who had the money to invest in its development – the military.

It has been recorded that simple rotary devices, toys more than anything else, were made in China but not developed further. It was not, in fact, until the times of Leonardo da Vinci that the first ideas for making a practical vehicle were put down on paper and studied by contemporary Italian scientists. It was not until 1842 that interest was again shown in a rotary-wing flying machine and a steam-powered device was built for manned flight. Two important inventions at the beginning of the 20th century are still important today: Charles Renard's articulated rotor blades helped to give inventors a practical way of altering the angle of incidence of each blade when attached to a central shaft, and in Italian scientific circles, G A Crocco caused a major stir with his patented cyclic pitch control. This enables the angle of incidence of the blades to be altered to allow forward and upward flight (when the angle allows air to be deflected downwards), downward and backward flight (when the angle allows air to be pushed up) and hovering flight (no incidence).

With the major technical problems solved in respect of rotor blades attached to a central shaft, it was left to the French Breguet Brothers to design a flyable helicopter. In 1907, their Gyroplane No 1 took to the air powered by a simple 50 hp Antoinette engine

and given lift by four double rotors at the ends of arms of tubular steel. The first flight was not free flying: as there were no flying controls other than an engine throttle, the machine was man-tethered to the ground. But later that same year, another Frenchman, Cornu, using a 24 hp Antoinette engine, powered twin rotors into the air.

The helicopter's development then went into the doldrums, but some work was continued, especially in Denmark, Spain and Argentina, where cyclic pitch control was first installed and flown in a helicopter by the Marquis de Pescara. In the inter-war years, a Russian emigré to the United States, Igor Sikorsky, continued to develop helicopters. His company is now the largest manufacturer of helicopters in the world.

On the other side of the Atlantic, the Second World War had prevented France's Breguet Company from continuing its work of developing the true helicopter from the auto-gyro. The latter has free-rotating blades not linked to the powerplant, lift being generated by the forward motion on the aircraft. In Nazi Germany, the helicopter was one of the advanced aeronautical ideas pushed by the Hitler régime and carried through by Profes-sor Henrick Focke, who later went on to develop world-beating fighter aircraft. The Focke design was the Fa 61, which first flew in 1938, but it did not see series production. Another German had the honour of producing the world's first production helicopter, this

Despite its supposed use by James Bond and military experiments in the 1930s, the autogyro has not been a great success. Today it is used for sport. This one was designed, built and flown by the British world record holder, Ken Wallis.

Juan de la Cierva (d. 1936)

Although Spanish by birth, Cierva went to the United Kingdom to develop his ideas for autogyros. 1920/22: began development work on autogyros at Madrid. 1925: arrived in UK and formed Cierva Autogyro Company Ltd to produce and market the C4 autogyro which had first flown in 1923. The type was built in the UK, France, Germany, Japan and the United States, with unlicensed copies built in the Soviet Union. 1930s: the design in widespread service. 1936: Cierva killed in flying accident. 1950: Cierva company merged with Saunders-Roe Helicopters.

In 1940, Igor Sikorsky, the Russian-born helicopter pioneer, first flew his VS300 in free flight. This introduced the helicopter to the US military and demonstrated its potential. This machine was the first practical helicopter.

During the Second World War, the German military had a world lead in the development of helicopters, such as these Fl 282, the first helicopter to enter service, in 1942. It was powered by a single BMW radial piston engine.

time for the German Navy. This was the Fl 265 from the Flettner Company, which made its maiden flight in May 1939 and was followed by the first operational helicopter, the Fl 282.

The Fl 282 served with the German Navy throughout the Second World War. Sadly there are few records left in existence, but it is widely known that these helicopters were embarked in armed merchant cruisers and larger warships for operation evaluation.

The German Navy's U-Boat arm had an interest in helicopters and autogyros which led to over 200 Focke Achgelis Fa 330 one-man scout autogyros being produced. Their role was to be carried by U-Boats during operational patrols, to be made up on the casing and towed behind the U-Boat to search 'over the horizon' for enemy surface craft, whether as targets or to give warning of possible attack. A self-powered version was not proceeded with because of the Allied advance into Germany in 1945, but the ideas and notes of the leading German designers were used by both the United States and the Soviet Union in the immediate post-war years.

Before this, however, the Sikorsky design team constructed the VS-300 of tubular steel and on 14 September 1939, Igor Sikorsky himself – complete with homburg hat – flew the aircraft at his factory's sports field at Stratford, Connecticut. The design was developed and the helicopter made its first free flight in 1940 – followed by several record-breaking events, including endurance and

Focke-Achgelis Fa 223 of the German Luftwaffe, c.1942.

0   5m

Anton Flettner (1885-1961)

German by origin, Flettner's name has become synonymous with the development of rotary-wing aircraft in Germany. 1905: developed control systems for airships (Zeppelins). 1920s: worked on marine propulsion systems using vertical rotating cylinders. 1922: began helicopter development. 1933: tethered trials. 1936: first flight of Flettner Fl 185, his only single main rotor design and later abandoned. 1939: first flight of twin-intermeshing rotor Fl 265. 1940: Fl 265 tested by German Navy. 1942: Flettner's Fl 282 introduced into service but did not achieve production status. 1947: Flettner invited to emigrate to the United States and work for US Navy.

Igor Sikorsky (1889-1972)

Born in Kiev, Russia, 25 May 1889; father physician and professor of psychology at University of Kiev, mother also qualified physician, although not in practice. His mother's deep interest in the art, life and work of Leonardo da Vinci had influence on Sikorsky's early life – he built a rubber-powered model helicopter at age 12. 1903: Entered Imperial Naval Academy at St Petersburg. 1906: Resigned in order to pursue a career in engineering. He entered Kiev Polytechnic Institute after a period of study in Paris, but left his studies after one year, concluding that the sciences as then taught bore little relationship to the solving of practical problems. 1908: Embarking on a journey through Europe, he was influenced by the European inventors, who were then trying to emulate the Wright Brothers' achievements. On his return to Kiev, Sikorsky concluded that the only way to fly was 'straight up', and, with financial help from his sister Olga, he began construction of a helicopter in May 1909, using an engine which he had purchased in Paris. Both this first attempt, and a second version with a larger engine failed to fly, and Sikorsky realised that the state of the art, and the materials and expertise available, were not advanced enough yet. 1910: Turned his attention to fixed-wing designs, and flew his S-1 biplane. A version with a larger engine,

the S-2, was followed in quick succession by the S-3, S-4 and S-5, the latter, by the summer of 1911, powered by a 50-hp engine and capable of flying for over an hour at reasonable heights. He was awarded International Pilot's Licence Number 64 at this time. The S-6 series led to orders from the Russian Army, and was followed, in what was to become characteristic style, by the first-ever four-engined aircraft, the 'Le Grand' of 1913, which Sikorsky first flew in that year. This

design had many advanced features, including a completely enclosed cockpit and cabin. 1919: Following the Russian Revolution, Sikorsky decided to leave his homeland and settle in the United States. Following several years as a teacher and lecturer, Sikorsky, together with some associates, formed the Sikorsky Aero Engineering Corporation in an old barn on Long Island. The company expanded to become part of United Aircraft Corporation, and by 1929 occupied a large factory at Bridgeport, Connecticut. 1931: The first S-40 flying boat, successor to the S-38 amphibian, entered service with Pan American World Airways; by 1937 the four-engined S-42 was flying transatlantic and other long-distance routes in commercial service. 1939: Began design work on the VS-300 helicopter, and with the designer at the controls, this helicopter flew for the first time on 14 September of that year. In an improved version of the same design, Sikorsky established an international endurance record of 1 hr 32.4 secs on 6 May 1941. The Sikorsky company went on to pioneer military and commercial helicopter designs and operations, with Igor Sikorsky himself retiring as engineering manager in 1957, but remaining active as a consultant until his death on 26 October 1972 at Easton, Connecticut.

Bob Downey

speed. Without doubt, this was one of the most important milestones in the development of the helicopter, especially in the military field. Yet the Sikorsky design is reported not to have been as advanced as the German designs already flying.

Development continued until 1943, when the prototype was posted to a museum, but not before the American military had backed the idea for the development of the military helicopter in the guise of the Sikorsky R-4 Hoverfly – the first American helicopter to go into mass production and to be used oper-

Sikorsky's R-4 Hoverfly 1 was the first helicopter designed from the outset for military use. It entered service with the American, British and Canadian military during the Second World War, starting in 1943.

ationally. It flew from merchant ships and US Coast Guard cutters and was used in rescue operations in Alaska and Burma – anywhere that it was impossible for a conventional fixed-wing aircraft to visit. The Sikorsky Hoverfly entered service with the Royal Navy in 1945 and later with the Royal Air Force (RAF) was part of the King's Flight.

Frank Piasecki (b. 1919)

Began his rotary-wing career as an engineer with the Kellett Autogyro company and Platt Le Page in the 1930s and 1940s. 1943: formed the P V Engineering Forum, later known as Piasecki Helicopter Corporation (1947) and designed and built the USA's second helicopter type, the XHRP-1. Late 1940s: continued to design tandem rotor helicopters for the US armed forces before resigning as chairman of the company in 1955. 1956: formed the Piasecki Aircraft

The second helicopter to be flown in the United States was the Piasecki PV-2, which made its first public flight in September 1943 and led eventually to the first tandem-rotor helicopter. It was known as the Piasecki HRP-1 'Flying Banana' and was ordered for the Navy in February 1944. This type was the forerunner of several tandem-rotor logistics helicopter designs, including the Boeing-Vertol CH-47 Chinook, which is in widespread defence force service and lately in commercial use. The HRP-1 was one of the first aircraft to be engaged in anti-submarine warfare (ASW) in the hunter role, when trials

Corporation (after the helicopter company had been acquired by Boeing) and began work on experimental concepts of vertical flight. 1965: first flight of the Pathfinder with shrouded tail rotor and stub wings, and with retractable undercarriage. 1980s: designed the Heli-Stat, using balloon envelope and helicopter-outriggers for lift and direction; possible applications for heavy-lift and overwater reconnaissance.

Nikolai Ilich Kamov (1902-1973)

Born in Siberia, Kamov was one of the greatest Soviet rotary-wing designers. 1926: graduated from Tomsk Institute in railway engineering. 1928: qualified as pilot and worked on flying boat and seaplane designs. 1929: working with others, produced the first autogyro in the Soviet Union, based on the Cierva C 8 and known as the KaSkr-1. 1931: with the TsAGI design bureau, designed the A-7 and worked on other projects during the Second World War. 1945: set up his own design bureau and produced co-axial rotored helicopter, Ka-8, from which were developed the Kamov series of helicopters, including the Ka-25 (known as Hormone in the West). 1973: died near Moscow and his bureau continued with helicopter development in his name.

Mikhail Leontyevich Mil (1903-1970)

Born in Siberia and destined to be the best known of the Soviet helicopter designers. 1926: graduated from Tomsk Institute. 1929: involved in work on the KaSkr-1, with Kamov and others. 1930s: involved in the design of autogyros and early helicopter experiments. 1936: became Kamov's deputy, and undertook war service with autogyros. 1947: given consent to found his own design bureau and developed the first Soviet helicopter, the Mi-1 (known as Hare in the West). Amongst the designs of the bureau have been the Mi-4 (Hound), Mi-10 (Harke), Mi-24 (Hind) and Mi-26 (Halo). 1970: died near Moscow.

ican aerospace giant. Its major project for the 1980s is the Apache advanced attack helicopter, and there are high hopes of receiving a contract for the new US Army Light Helicopter Experimental or LHX.

In the Soviet Union, the helicopter developed in the gyrocopter and autogyro sectors until the Kamov Design Bureau was established in 1945, to be followed by the more successful Mil Bureau in 1947. The former concentrated on contra-rotating systems, whilst the latter followed the traditional main rotor 'schools' around the world. To a certain extent, the Russian aircraft industry has always tended to copy the Western designs and adapt them for its own uses in the more extreme climates of the Soviet Union.

Helicopter programmes began again in France after 1945, and after several reorganisations of the state-run industry, Aerospatiale

were carried out with dipping sonar transceivers, which were lowered through the fuselage into the sea whilst the helicopter was hovering. The Piasecki tandem designs progressed through the Retriever, Transporter and Pathfinder, before the designer joined forces with the Vertol Aircraft Corporation, which subsequently became Boeing-Vertol and is based in Philadelphia.

The other American design company which was carrying out pioneering work in the 1940s was the New York-based Bell Aircraft Corporation, which specialised in smaller helicopters, including the Bell 47 Sioux which has been in widespread military and commercial use since 1946. In fact, the Bell 47 was the first helicopter to be given a commercial certificate of airworthiness, allowing it to fly fare-paying passengers. Eventually Bell moved to Fort Worth, Texas, and has become part of the multi-national Textron group.

The great American design genius and industrialist Howard Hughes did not come into the helicopter business until the 1950s, when his company built the then largest helicopter in the world, the Hughes XH-17 Flying Crane. It was constructed to a design by Wallace Kellett, whose small design house and manufacturing plant had been bought out by Hughes in 1949. In 1985, the company became the McDonnell Douglas Helicopter Company, following acquisition by the Amer-

Above right: Kamov Ka-15 Hen of Soviet Navy, *c.* 1955. It was the first Kamov design to enter mass production, between 1953-63 and was armed with depth bombs.

President Eisenhower was the first American leader to use helicopters, and a Bell 47 is seen here in the grounds of the Washington White House. Later presidential helicopters have included the Sikorsky HH-3, and soon the VH-60 Black Hawk will enter service. Presidential helicopters are traditionally flown by the US Marine Corps.

AGUSTA A129 MANGUSTA

Country of origin: Italy.
Role: Light attack and armed reconnaissance.
Rotor diameter: 11.9 m (39.04 ft).
Length: 12.27 m (40.27 ft).
Max weight: 3,700 kg (8,157 lb).
Engines: 2 × RR Gem 2-2 Mk1004 turboshafts, 952 shp.
Max speed: 278 km/h (150 kt).
Range: 629 km (340 nm).
Weapons: 8 × TOW or 6 × Hellfire, guns and rockets.

was formed in January 1970. Meanwhile in Italy, the Agusta Company, which was founded in 1907 but now also is state-owned, began with licensed production of American designs before beginning its own design and production facilities. These culminated in the first really successful designs – the Agusta 109 and 129 Mongoose and EH 101 (a joint programme with Britain's Westland) – in the early 1980s. Post-war German designs have stemmed from the MBB (Messerschmitt-Bolkow-Blohm) company, which was formed in 1969 and today has strong links with India, Japan and Indonesia, where licensed production of helicopters has been under way for some years.

In the United Kingdom, the gyroplane was developed for military and sports use during the inter-war period, but it was not until 1949 that the first proper helicopter was given a

Among the export customers for the Bristol Sycamore helicopter were the Federal German armed forces (illustrated) and the Royal Australian Navy. In British service, the helicopter saw action at Suez (the first helicopter assault), Aden, Kenya and Malaya.

certificate of airworthiness, when the Bristol 171 Sycamore entered the civil field. It was also a highly successful military helicopter, giving the British forces experience in search and rescue (SAR) and troop transportation. Britain's first twin-rotored helicopter was the Bristol 173, which had been designed to meet a Royal Naval requirement for an ASW helicopter, but it proved to be underpowered and was the victim of political reassessments of the Fleet Air Arm. The only manufacturer in the UK is now Westland Helicopters, who amalgamated with Bristol and Saunders Roe in 1960 to make a powerful world force for licensed production and indigenous design. The company has always had a close working relationship with Sikorsky Aircraft and in 1948 began the first of several joint ventures of licensed production with the S-51 Dragonfly for military use. The design was developed by Westland into the commercial Widgeon. Later other civil types were tried, including the unique Fairey/Westland Rotodyne, which made its first flight in November 1957, and the Westland Westminster flying crane. Both of these failed to enter operational service.

It was in Korea, in 1950-53, that the helicopter really began to make an impact on the military front, especially for troop movement and search and rescue (SAR) operations, even into the front-line positions and occasionally into the enemy's hinterland. The first helicopters to be used in this way were the Sikorsky S-51 Dragonflies (some flying from Royal Navy aircraft carriers) and Sikorsky H-19 Chickasaws. Users were predominantly the US Army, Marines and Air Force, with valuable contributions from the US Navy, especially at sea. Many a Commonwealth naval aircrew owes his life to a USN rescue 'whirlybird'.

Sikorsky S-51 Dragonflies were built under licence in the UK (and were copied by Soviet designers). The type saved the lives of numerous downed aircrew in Korea when flying from aircraft carriers and later in Malaya, operating from air bases.

Towards the end of the war, helicopters were becoming very common on the battlefield and by May 1953 there were nine squadrons of H-19s operating with the US Marines alone. At the front line, the Mobile Army Surgical Hospitals (MASHs) were supplied with wounded by the first Bell H-13 Sioux models, which carried the wounded in panniers on either skid. Their glass bubble silhouette became a common sight to the troops in the mountainous Korean terrain.

Since 1953, however, the world has never been totally at peace and the helicopter has played an important part in all post-Korean conflicts. It flew supplies to the beleaguered defenders of the French outpost at Dien Bien Phu in what is now called Vietnam, then French Indo-China. The French used the Sikorsky Chickasaw for similar operations in their later colonial war in Algeria, whilst the British began assault operations for heli-borne troops with Whirlwinds and Bristol Sycamores during the short-lived Suez 'police operations' in November 1956. The Royal Navy and the British Army had already pioneered jungle trooping operations in Malaya and Borneo during the counter-Communist operations, later also during confrontation with expansionist Indonesia in the late 1950s and early 1960s.

The American presence in South-East Asia was aided more than almost anything else by the helicopter. The piston engine was slowly giving way to the gas turbine with its more economic and powerful performance, better safety and smoother operation. These developments were especially noticeable in the army support helicopters used by the US Special Forces and later by the US Army, Navy, Marines and Air Force units, plus those of her allies, including Vietnam, Korea, New Zealand and Australia.

By far the most widespread helicopter in all Vietnam operations was the Bell UH-1 Huey – the world's most produced helicopter – which first flew in 1956. This theatre of operations

Arguably the most famous helicopter in the world, the Bell UH-1 Iroquois, or 'Huey', has seen service with many nations and in several wars since Vietnam. The leader of the Royal Australian Air Force pair is fitted for special operations tasks.

AEROSPATIALE SA 319B ALOUETTE III

Country of origin: France.
Role: Observation, scout and naval.
Rotor diameter: 11.02 m (36.15 ft).
Length: 10.17 m (33.38 ft).
Max weight: 2,250 kg (4,960 lb).
Engine: 1 × Turbomeca Astazou XIVB, 593 shp.
Max speed: 220 km/h (118 kt).
Range: 630 km (340 nm).
Weapons: 4 × AS 11 or 2 × AS 12 missiles; 1 × Mk46 ASW torpedo.

against Soviet-made Syrian tanks during the invasion of Lebanon (1982). In the South Atlantic, the Royal Navy, Royal Marines, Army Air Corps and Royal Air Force contributed to a very great degree in the eventual recapture of the Falklands and South Georgia in 1982. Despite losing about 25 helicopters, including at least six Wessex HU 5s in the ill-fated *Atlantic Conveyor* alone, the Fleet Air Arm flew 21,050 hours 10 minutes in assault, casevac, support and anti-submarine roles. One detachment of Sea King HC 4s lifted

also saw the development of the helicopter gunship in the shape of the Bell AH-1 Huey Cobra and of the heavy-lift helicopter in the Boeing Vertol CH-47 Chinook and the Sikorsky CH-54 Skycrane. The US Marines favoured the Sikorsky CH-53 for heavy-lift operations and for the combat SAR missions deep into Viet Cong-held territory or even North Vietnam. The 'Jolly Green Giants' were successful in saving many a stranded aircrew or fighting man from potential death.

The wars in the Middle East and in Africa have seen the extensive use of the helicopter. The Israeli Defence Forces have pioneered assault operations in modern times and have been very successful in deep-penetration raids on enemy strongholds, like, for example, the raid which captured an Egyptian Soviet-made radar station *complete* between the 1967 Six Days War and 1973 Yom Kippur War. During the Rhodesian bush war, the government's troops were very successful in adapting the Aerospatiale Alouette III as a gunship-assault craft and flew many successful missions against the guerrillas in the bush, as well as 'out-of-country' sorties in Hueys.

In recent years, attention has been focused on small anti-tank helicopters such as the Hughes 500MD Defender in Israeli operations

Bell AH-1S/TOW Cobra of Israeli Air Force, c.1982.

0   3 m

Sikorsky's CH-54 Tarhe (*above middle*) is known to many as the Skycrane. It served in Vietnam as a retriever of downed aircraft, and is still used by the US Army Reserve and several civil operators. The helicopter does not have a fuselage but is fitted with

various detachable pods, such as a hospital, command post or emergency accommodation, as required, or can be used without pods for load-lifting of stores and weapons.

413,683 kg (912,000 lb) and 520 troops during the conflict.

The civil helicopter

The civil helicopter took longer to develop fully than its military counterpart, although the world's first scheduled helicopter service was commenced in October 1947 in the Los Angeles area, where Los Angeles Airways (LAA) flew the US Mail in the Sikorsky S-51. The first civil airworthiness certificate had been awarded to the Bell 47 as early as 8 March 1946, but development for military uses took precedence, especially when its potential was seen during the Korean War.

In Britain, helicopter development, both in terms of production and of usage, followed closely behind the Americans. The first commercial operations were flown in East Anglia by British European Airways (BEA), based at Peterborough, using the same S-51 Dragonflies as LAA, but passengers were not carried commercially until 1950, when BEA experimented with a service from Cardiff to Liverpool, again using the S-51, but the service terminated in 1951.

Interestingly enough, it was from Belgium that the first international helicopter flights were made. In September 1953, four years after its first flight, Sabena, the Belgian State Airline, used the Sikorsky S-55 to link Brussels, Rotterdam and Maastricht (the Netherlands) with Lille (France). In New York, the city centre (downtown) was linked for many years to the outlying airports by a helicopter service using initially the Boeing Vertol 107, a twin-rotor design flying from the Pan American building's rooftop in Manhattan. Later, New York Airways also operated the same service, ferrying first-class transatlantic passengers to the city using the Sikorsky S-61. But the service folded in 1979 following two accidents that caused much undeserved public criticism of the helicopter. In 1981, however, a local helicopter specialist

Providing first-class passengers with rapid transfer from New York's airports to the Manhattan downtown area, this Aérospatiale Dauphin is operated by New York Helicopters on a scheduled service.

One form of scheduled service is the use of helicopters to support offshore oil and gas platforms around the world. Here in the Gulf of Mexico, the service is provided by a Petromex SA 330J Puma which can carry 21 passengers and two crew.

saw an opportunity to use a smaller, more cost-effective type and commenced a service with the Aérospatiale SA 360C Dauphin, flying the same routes as the now defunct New York Airways. It is currently a great success, and has been joined by Pan Am's service with the Westland 30. The first British civil helicopter for almost a generation, the Westland 30 is also operating in the short to medium distances in the southern North Sea, flying out of East Anglian fields to the offshore gas platforms.

In Britain, too, the scheduled helicopter service has begun again between London's Gatwick and Heathrow airports, using the Sikorsky S-61N, operated by British Airways and British Caledonian Helicopters, on behalf of the British Airports Authority, called the 'Airlink'. The most regular services in Britain, though, are the shuttles between the Scottish and East Anglian coasts and the North Sea oil platforms. All around the North Sea, but particularly in Britain, the Netherlands and Norway, there is intense day-and-night all-

weather helicopter flying, initially to support the exploration of the North Sea's gas fields, later the oilfields and now to resupply the production platforms with men and materials. Several large companies operate in the North Sea area – British Airways Helicopters and Bristow (UK), Helikopter Service A/S (Norway) and KLM Helicopters (the Netherlands).

Another European company to take on the Americans in their own backyard is Aérospatiale. The company is now the world's greatest helicopter exporter and many of their machines, from Alouette III to Super Puma, are operated in the United States by airlines, offshore companies and corporate concerns. The Alouette III and the Twin Star (or Twin Squirrel) are also used extensively in the Emergency Medical Service (EMS) role. American freeway accidents are often attended by helicopters carrying medical staff who can begin emergency life-saving measures on a victim before and during helicopter transfer to hospital. It is interesting to reflect, however, that the first service was in Switzerland and that today it is the Federal Republic of Germany – with the surface area of just one American state (Oregon) – which has the largest helicopter EMS fleet in the world. Simply by dialling 110 a helicopter of the Katastrophenschutz, usually an MBB BO 105 or Bk 117, may come to the scene instead of an ambulance road vehicle. In the Bavarian Alps, there is considerable co-operation between the authorities of Germany, Austria and Switzerland, especially during the winter sports season.

The helicopter is widely used as man's servant and in many cases its work is essential – nowhere more so than in agricultural operations. Even in the Soviet Union, where the national airline, Aeroflot, is the sole operator, flying everything from the single-engined Mi-1 to the mighty Mi-26 that has the cargo capacity of the C-130 Hercules, the

Increasingly throughout Europe and North America, road accident victims are taken by helicopters to hospital – in this case by an Agusta A109A Mk II with a special stretcher kit and provision for nurse attendants.

Company helicopters

Air travel is increasingly important for many large businesses worldwide, so much so that many national, international and multinational companies have found it advantageous to buy and operate their own aircraft. Many companies have seen the inherent flexibility of the helicopter as an added benefit to the advantages of rapid air travel.

Most of the companies that operate an aeroplane or helicopter do so mainly for senior management, although not exclusively for this purpose, as there are a number of firms involved in high technology which use their corporate aircraft as test and demonstration platforms in addition to the transport uses. But undoubtedly the primary use of a corporate aircraft is as efficient transportation for senior staff between places that are important to the concern's business. In many cases, the speed and efficiency of such a means of travel can prove more cost-effective than using scheduled airlines. The helicopter offers the additional advantage of being able to operate from almost any location, while still carrying a useful number of passengers in a cabin sufficiently quiet to permit a useful meeting to take place

SIKORSKY S-76 Mk II

Country of origin: USA.
Role: Business.
Rotor diameter: 13.4 m (44 ft).
Length: 13.2 m (43.4 ft).
Max weight: 4,672 kg (10,300 lb).
Engines: 2 × Allison 250-C30 turboshafts, 650 shp.
Max speed: 287 km/h (155 kt).
Range: 889 km (480 nm).
Passengers: Up to 12, with 2 pilots.

agricultural helicopter is widely used. The British Helicopter Advisory Board reports that there are over 3,000 such helicopters in use behind the Iron Curtain, as against only 1,000 in the USA and 60 in the United Kingdom. The civil helicopter has an important place in the future development of our world.

The battlefield helicopter

Although some of the original pioneering work in military development was carried out against a maritime background, the helicopter found its first niche in the battlefield of the post-World War Two era.

Until a few years ago, it was still virtually a fair-weather weapon, but modern developments in the field of avionics – radar, radio, altimeter, laser guidance and night-vision goggles – have transformed the helicopter into

AEROSPATIALE AS 355F2 Ecureuil 2

Country of origin: France.
Role: Business or ambulance.
Rotor diameter: 10.7 m (35.1 ft).
Length: 10.9 m (35.9 ft).
Max weight: 2,540 kg (5,600 lb).
Engines: 2 × Allison 250-C2 0F, 420 shp.
Max speed: 278 km/h (150 kt).
Range: 705 km (305 nm).
Passengers: Up to 5, with 1 pilot.

during the course of the journey.

There are two main ways in which a company can make use of the helicopter for its corporate transport needs. Firstly, it can lease or buy the helicopter outright, and appoint pilots as salaried employees. The necessary maintenance work will, as a result of the specialised equipment and facilities needed, usually be conducted by an agent who is appointed or approved by the manufacturer of the helicopter. Secondly, the company can decide to operate a helicopter on a regular basis by chartering it from one of the many independent operators, who provide a wide variety of types for hire at competitive rates, complete with crews.

the all-weather, all-terrain weapon that can deliver weapons to a precise point as selected by the battlefield commander and return, even under enemy fire. It can also lift loads from the supply areas to the front line, or blunt an armoured advance with laser-guided 'smart'

missiles, or gather intelligence about enemy movements without being observed using ground cover and the newly developed mast-mounted sight atop the main rotor head. Soon, the use of satellite positioning equipment will mean the helicopter's navigation will be greatly enhanced, giving the ability to deliver troops, equipment or prepare an ambush position without the need for complex directions or maps, to an accuracy of less than 10 m (33 ft).

Such is the effectiveness and flexibility of the helicopter, there are still a number of first-generation, single-engined helicopters flying, especially with the smaller nations of the world. The simply operated and maintained training types, such as the Bell H-13 Sioux, Hiller UH-12 Raven and Hughes TH-55 Osage, still provide many Third World countries with their only battlefield helicopters. Others have progressed to the simple scout and/or attack helicopters, such as the Aérospatiale Alouette II/Lama, Alouette III, Bell UH-1 Huey, Mil Mi-2 'Hoplite', Mi-4 'Hound' and Westland Scout.

These first-generation helicopters carry such weapons as the general-purpose machine gun (GPMG), 70 mm (2.75 in) rockets and wire-guided anti-tank/ship missiles. They are very much limited by the weather and

The Aérospatiale (Sud) Alouette II is one of the simplest helicopters for civil and military applications. It first flew in 1955 and 1,300 were built up to 1975. Many are still in service for a variety of tasks, including pilot training.

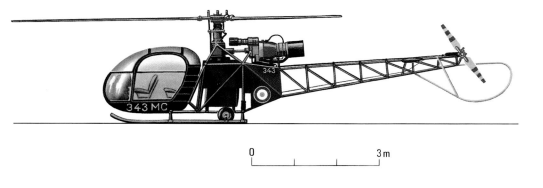

Battlefield Helicopter Tactics

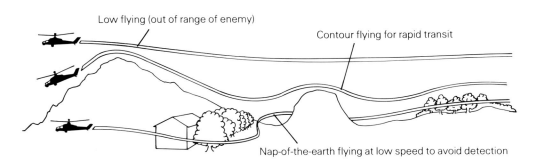

Low flying (out of range of enemy)

Contour flying for rapid transit

Nap-of-the-earth flying at low speed to avoid detection

BELL AH-1S COBRA

Country of origin: USA.
Role: Attack and anti-tank.
Rotor diameter: 13.41 m (44 ft).
Length: 13.59 m (44.6 ft).
Max weight: 4,535 kg (10,000 lb).
Engine: 1 × Avco Lycoming T53-L-703, 1,485 shp.
Max speed: 315 km/h (170 kt).
Range: 587 km (317 nm).
Weapons: 8 × TOW, cannon, machine guns and rockets.

Described as a terror weapon by Afghan guerrillas fighting the Soviet invaders, the Mi-24 Hind assault helicopter is also respected by NATO commanders who are concerned by the type's firepower and ability to land troops behind the lines.

operational conditions with primitive radio and navigation aids, but still do a fine job.

In most people's eyes, the Vietnam War was synonymous with the development of the helicopter gunship. Initially, the Bell Huey was deployed with various machine gun, cannon and rocket armaments to suppress the enemy during landing operations in hostile territory. The US Army and Bell saw the requirement for a specially dedicated attack helicopter gunship and so the HueyCobra was born. The first prototype made its maiden flight in September 1965 and the type entered service in 1967; both single-engined versions and those with twin units are flying today, showing the parallel of the design with the development of the UH-1 Huey into the twin Huey and 214 SuperTransport.

The Cobra is now in service with several different, widely ranged nations, both ashore in the battlefield and afloat with marine and naval units – USA, Israel, Japan, Jordan, Greece, Korea, Iran and Spain. For many countries such a design is too costly and a more versatile helicopter has been ordered and operated.

The Hughes 500 Defender is one such helicopter, being capable of effective anti-tank operations with TOW missiles or rocket pods, or for light observation tasks, which could even result in it being used for QRF (quick reaction force) sorties with small parties of special troops. The British attempted to penetrate the market dominated by the Americans with the Westland Lynx-3, but there is more interest in the highly effective, battle-proven naval version. In Germany, the MBB combine designed and built the Federal German Army's first anti-tank helicopter, the BO 105 (PAH-1), which is also in service with Dutch and Spanish forces in various roles. The highly successful Aérospatiale Gazelle, also built under licence in a number of countries, including Yugoslavia, is a battlefield light observation helicopter which can also be equipped with anti-tank missiles, rocket and cannon to give it an enemy force suppression role.

The heavy-lift and troop transport helicopter has also been pioneered in the Soviet Union, by the Mil design bureau. The first of 5,000 Mi-8 Hip helicopters took to the air in 1962 and production of this 24-place,

12,000 kg (26,455 lb) machine continued until 1979. Later versions were interim gunships, soon to be replaced in front-line service from 1972 by the powerful Mi-24 Hind, which is now in its fifth variant, Hind-E, equipped with air-to-air capable anti-tank missiles and cannon. According to current Western thinking, a number of Hinds would be used in wartime to drive a wedge into NATO anti-tank defences, particularly helicopter attack teams. This has led to the development of self-defence systems for helicopters, including fire-and-forget missiles, such as the General Dynamics Stinger II, Shorts Starstreak and BAe Thunderbolt, plus chaff and metal smoke launchers to screen the launch helicopter. Other Soviet Battlefield helicopters include the Mi-10 Harke flying crane, and the latest development, the powerful 80-place, 56,000 kg (123,200 lb) Mi-26 Halo, which is thought to have entered service in 1983 in Afghanistan; there are now some 2,500 helicopters operating in that country.

Perhaps the most interesting development

MIL MI-24 HIND D

Country of origin: USSR.
Role: Attack and assault.
Rotor diameter: 17 m (55.77 ft).
Length: 17 m (55.77 ft).
Max weight: 10,000 kg (22,045 lb).
Engines: 2 × Isotov TV-3-117 turboshafts, 2,200 shp.
Max speed: 366 km/h (198 kt).
Range: 600 km (324 nm).
Weapons: 12.7 mm cannon, 4 × Swatter and/or rockets; can carry up to ten troops.

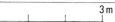

Agusta A129 Mongoose of Italian Army Aviation, c.1986.

Developed from the successful Lynx series, the Westland Lynx-3 has found that the battlefield helicopter market is difficult to penetrate even with a proven system. Lynx-3 can carry the latest weapons, including TOW and Hellfire, as well as cannon and machine guns in the attack role. Designated a multi-role helicopter, it can also carry anti-tank teams and small assault parties and be used for casualty evacuation. The helicopter is fitted with special blades to decrease noise.

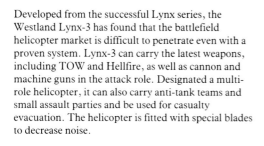

Vietnam – a new advance in military helicopter techniques

Many major conflicts since the Second World War have been notable for an increase in the role played by the helicopter – Vietnam especially.

In the 1950s, the Korean War saw a significant increase in the use of helicopters to move troops tactically about the battlefield, but a decade later, in Vietnam, the concept was developed to a level never seen before. It is to the 1st Cavalry Division (Airmobile) of the US Army that most of the credit must go.

The 1954 agreement to partition the country into North and South Vietnam brought only a brief respite in what had begun as an anti-colonial war against the French. American involvement grew, but by the early 1960s the Viet Cong guerrillas were strong enough to capture a provincial capital near Saigon.

Clearly a radical new approach was necessary if the North Vietnamese forces – soon to be backed by regular troops – were to be defeated. Helicopters operating in support of the French and American forces had already played an imporant part in the war, but the deployment of the 1st Cavalry Division led to new developments which have influenced subsequent conflicts.

Although a standard division of 16,000 men, it had nearly five times the number of helicopters operated by an infantry division. Of its total of about 450 helicopters, the majority were Hueys – some fitted as gunships to escort other unarmed helicopters – but the fleet included many Chinook transports and even four Skycrane heavy-lift helicopters.

To qualify for service with the division, ground vehicles had to be air portable; even large vehicles were of a type which could be quickly disassembled for transport by helicopter and the division dispensed with artillery which could not be easily lifted.

The concept developed by the 1st Cavalry Division relied on the use of helicopters to carry out reconnaissance in force from heavily fortified bases such as Da Nang and Chui Lai. The purpose was

Lockheed AH-65 Cheyenne development helicopter for the US Army's advanced aerial fire support system programme.

often concerned with 'sweeping' an area that had been occupied by the Viet Cong, although in its first year in Vietnam – 1965 – the 1st Cavalry fought a bloody battle with a division of the North Vietnam Army at Plei Me. Both sides suffered heavy casualties and the result of the battle was inconclusive.

However, the 1st Cavalry continued to refine its techniques and it quickly became evident that a special helicopter gunship would be necessary; simply fitting machine gun mounts in the cabin of a Huey, together with a pair of rocket pods, was not a suitable answer to the requirement for an armed escort. The heavily laden armed Hueys could not keep up with those carrying troops, so the Huey Cobra was quickly developed by Bell and introduced into service in 1967.

Helicopters like the small Hughes OH-6A Cayuse were highly successful in supporting ground troops in jungle fighting. With its characteristic 'egg' shaped cabin, the Cayuse was regarded as good in surviving crashes.

The Bell design had in fact been the loser in an earlier competition with Lockheed, but the winner proved to be too complex and too expensive, so the Cobra was quickly added to the 1st Cavalry armoury. Capable of carrying 52 rockets in addition to a machine gun and grenade launcher, the Cobra could 'hose down' an area before the arrival of an air cavalry assault unit.

The 1st Cavalry's helicopters were to prove useful not only on attack missions – they also brought succour to the beseiged Marines at Khe Sanh by flying in supplies and taking casualties out. Finally the 77-day seige of the combat base was lifted when 30,000 US and ARVN (Army of the Republic of Vietnam) troops were flown in by transport helicopters. This success in 1968 could not affect the outcome of the war, which led to America's withdrawal in 1972. However, the 1st Cavalry had had a decisive influence on the use of helicopters in modern warfare.

Brian Walters

BOEING VERTOL CH-47C CHINOOK

Country of origin: USA.
Role: Medium support and lift.
Rotor diameter: 18.29 m (60 ft).
Length: 15.54 m (51 ft).
Max weight: 22,680 kg (50,000 lb).
Engines: 2 × Avco Lycoming T55-L-712, 3,750 shp.
Max speed: 297 km/h (160 kt).
Range: 424 km (229 nm).
Weapons: None, can carry at least 44 troops.

---

in battlefield helicopters is the advanced attack helicopter, capable of penetrating enemy defences to inflict severe damage on the rear echelon forces. The US Army selected the Hughes (now McDonnell Douglas) AH-64 Apache for this new role and the helicopter entered service at Fort Hood, Texas, in the spring of 1986. The first operational unit becomes active in the Central Region in 1987.

Other advanced anti-tank helicopter designs include the Eurocopter PAH-2/HAP/ HAC-3C, a Franco-German approach for

anti-tank and anti-helicopter operations beyond the year 2000. Meanwhile, the Italians, with support from the British and Dutch military, have developed the A129 Mongoose for the light attack helicopter role. Modern attack helicopters must be reliable, maintainable and survivable on the modern battlefield, using modern weapon systems, such as third-generation missiles, which will use laser designation. The US Army has brought the Rockwell/Martin Marietta Hellfire into service on the Apache and later it will be carried by

the Black Hawk transport in a secondary role; this missile is an interim measure between wire-guided systems and the modern third-generation types.

To support the advanced attack helicopter of the US Army, Bell Helicopter has totally rebuilt the OH-58A Kiowa to AHIP (Army Helicopter Improvement Program) standards to cope with the demands of the modern battlefield, which will be a high-threat environment. This type brings the mast-mounted sight into service to give the ability

In Grenada, the Sikorsky UH-60A Black Hawk won its spurs as a tactical transport helicopter in the hands of the 82nd Airborne Battalion carrying US Rangers to assault the barracks of the Cuban-led rebel government forces.

for the observer to detect targets over the cover of natural or man-made obstructions; from this position the laser designator in the sight can be used to identify targets for the Apache. In the short- to mid-term, the Cobra will remain as a substantial part of the attack helicopter force.

To support ground troops armies provide light support and medium support helicopters for mobility. Most Western forces have used the Bell Huey for many years, although the British and French forces use the Aérospatiale Puma and the US Army is now bringing the Black Hawk into service. Westland Helicopters had hoped to interest the UK RAF in the Rolls-Royce-powered Westland 30. Heavier

lift work in the RAF and US Army is performed by the Boeing Vertol Chinook, which has been ordered in quantity by Japan, and the US Marine Corps have the Sikorsky Super Stallion coming into full-scale service. These craft are capable of lifting 12.1 and 18.3 tons respectively.

For the future, the US Army and US Marine Corps have funded the development of the Bell-Boeing JVX (Joint Service Vertical Take-Off Experiment) Osprey, which is destined to fly in 1987 and enter USMC service in 1995, replacing the CH-46 Sea Knight.

McDONNELL DOUGLAS AH-64A APACHE

Country of origin: USA.
Role: Advanced attack and anti-tank.
Rotor diameter: 15.74 m (51.64 ft).
Length: 9.75 m (32 ft).
Max weight: 7,500 kg (16,500 lb).
Engines: 2 × GE T700-GE-701, 1,690 shp.
Max speed: 378 km/h (197 kt).
Range: 611 km (330 nm).
Weapons: 16 × Hellfire and 30 mm chain gun, or rockets.

For medium transport roles, the Boeing CH-47 Chinook is highly respected, especially with its triple hook arrangement. This is a UK Royal Air Force Chinook HC 1 carrying 105 mm field guns and crews into a firing position for exercises.

The naval helicopter

Today, the helicopter has made its home at sea. It performs many roles far quicker than warships and often with greater effect. It also has such an important position in modern naval warfare that there are not many surface combat ships over the size of corvette which do not have the facilities – flight deck and hangar – to operate a helicopter.

Although the first helicopters went to sea during 1944-45 in the battle of the Atlantic, it was not until the endurance and load-carrying performance had been developed that the machine has become an indispensable weapon system. The Royal Navy put its first ASW helicopter to sea in August 1957 when the Westland-built Whirlwind was declared operational; this was a development of the Sikorsky HO4S with primitive ASW systems, and although the later Whirlwind HAS 7 was equipped with dipping sonar, the anti-submarine torpedoes were carried in a second helicopter, making it a task force weapon rather than an individual ship's system. This was to come later with the development of the Westland Wasp HAS 1, the world's first small ship helicopter, which, although no more than a weapon carrier without sensors, will remain in service until 1990. The Wasp was the first British naval helicopter to use missiles in anger, being partly responsible for the disabling of the Argentine submarine *Santa Fé* off South Georgia in 1982, the first action in the successful liberation of the island.

The Whirlwind was replaced by the faster, better-endurance turbine-powered Wessex, which was an improved version of the H-34, and fulfilled anti-submarine and assault roles. It was widely used in Malaya, Aden and the Falklands conflicts, being used for various secondary roles, but is scheduled to leave the service in the late 1990s. In turn the Wessex began to be replaced for front-line anti-

The helicopter has proved invaluable at sea. The first successful small ships' helicopter was the Westland Wasp, seen here at Portland, Dorset. The type has also been exported to New Zealand, Brazil, South Africa and Indonesia for naval tasks.

WESTLAND NAVY LYNX

Country of origin: UK.
Role: Naval and shipborne.
Rotor diameter: 12.8 m (42 ft).
Length: 10.61 m (34.83 ft).
Max weight: 4,876 kg (10,750 lb).
Engines: 2 × RR Gem 41/43 turboshafts, 1,000 shp.
Max speed: 278 km/h (150 kt).
Range: 593 km (320 nm).
Weapons: 4 × Sea Skua or 2 × Stringray, guns, depth charges and other missiles; can carry up to 9 people.

By far the best medium ASW helicopter in the world, the Westland version of the SH-3 Sea King has sold well. It is seen here in Royal Australian Navy colours, operating from HMS *Invincible* in 1983 during a worldwide RN deployment.

submarine duties by the Sea King in 1969. This is again American derived. It is designed by Westland, but in its Advanced Sea King form has been procured by the Indian Navy to carry the Sea Eagle missile supported by the latest MEL Super Searcher radar.

The naval helicopter now operates in a wide variety of roles: anti-submarine warfare (ASW), anti-surface vessel warfare (ASVW), as well as troop and logistical lift, airborne mine countermeasures (AMCM), search and rescue (SAR) and airborne early-warning (AEW) roles. Most modern warships with helicopters use them for hunter-killer ASW operations, with twin-engined safety and all-weather capability, the limits being the ability to launch and recover from the heaving deck. Examples of the newer types include the Agusta-Bell 212, which has been developed

Supporting the US Marine Corps and US Navy worldwide, the Sikorsky CH-53E Super Stallion is the largest and most powerful Western helicopter in production. One version has been designed for aerial mine sweeping, another for assault tasks.

from the American Huey and is now at sea with the Italian, Turkish, Spanish, Peruvian and Venezuelan navies, and the NATO standard small ship helicopter, the Westland Lynx, which equips the frigate and destroyer-type warships of France, Federal Germany, the Netherlands, Denmark, Norway, Brazil and the UK.

It took the US Navy longer to adapt the helicopter to small ship operations than for the Europeans, partly because the Americans went down the remotely-piloted helicopter

SIKORSKY CH-53E SUPER STALLION

Country of origin: USA.
Role: Heavy lift and assault.
Rotor diameter: 24.08 m (79 ft).
Length: 22.35 m (73.32 ft).
Max weight: 33,339 kg (73,500 lb).
Engines: 3 × GE T64-GE-416 turboshafts, 4,380 shp.
Max speed: 315 km/h (170 kt).
Range: 2,075 km (1,120 nm).
Weapons: Machine guns for combat rescue, otherwise none; can carry up to 55 troops.

road first. Having found this to be a dead end, Kaman was asked to develop its successful utility and SAR helicopter into the air vehicle for the Light Airborne Multi-Purpose System (LAMPS) programme to provide ASW support for the surface combat ships of the famed 600-ship US Navy. The 6,124 kg (13,500 lb) SH-2F Seasprite is equipped with the standard Mk46 anti-submarine torpedo but not with dipping sonar, which would allow the helicopter to localise the submarine threat. The newer destroyers of the US Navy have begun receiving a highly complex but highly ship-dependent replacement, the Sikorsky SH-60B Seahawk, which has also been ordered by Australia, Spain and Japan. The Seahawk is the air vehicle for the LAMPS-3 programme, which uses the helicopter purely as a tool without the ability to operate independently.

Larger ASW helicopter assets are the Sea Kings carried aboard the US Fleet Carriers and which are due to be replaced in the mid- to late-1990s, perhaps with the SH-60F Seahawk, specially modified with dipping sonar or even with the EH 101 helicopter. This machine, which flies in late 1986, has been jointly developed by the Anglo-Italian EH Industries consortium to fulfil the replacement needs of the Royal and Italian navies, both of which still operate the Sea King and will do so for several years to come. The advantage of the EH 101 lies in its three engines, which give better shipboard performance and safety margins. In addition, its size and dimensions, whilst close to the Sea King, will allow it to be carried on the flight decks of the newer and smaller escorts, such as the new British Type 23 'Duke' Class frigates.

On the other side of the Iron Curtain, the Russians seem to have taken up the development of naval helicopters rather later than the West, and in doing so initially copied a successful American design, the Sikorsky

France's anti-submarine forces include the Super Frelon. Equipped with dipping sonar and torpedoes, the SA 321 has a range of 820 km (508 mi). It has also been bought by the People's Republic of China, Libya and South Africa.

S-55. From this basic design, not only was a most successful piston-engined tactic transport helicopter born, but also the first Soviet operational ASW helicopter, the Mi-4 Hound; it entered service in 1953 as a transport. Now reduced to second-line duties, the helicopter nevertheless played an important role in the development of ASW techniques for the Red Banner Fleet and its Warsaw Pact allies. Not operated at sea, the Mi-4 has been replaced from 1975 by the Mi-8-derived Mi-14 Haze for on-shore, coastal ASW operations.

Afloat, the Soviets have used the Kamov design bureau's contra-rotating, no tail rotor designs, such as the 7,300 kg (16,093 lb) Ka-25 Hormone, which is embarked in carriers and escorts. The prototype was the small Ka-20 Harp, which was first seen by Western observers in 1961 and later developed into the Hormone, which has been identified in three variants: Hormone A (ASW); Hormone B (OTHT – over-the-horizon targetting); Hormone C (SAR, vertrep – vertical replenishment, HDS – helicopter delivery service, COD – carrier onboard delivery). The Hormone first entered service in 1967, but has been replaced in the newer carrier and surface combat ship designs by a developed version, the Ka-27 Helix, powered by two Glushenkov GTD-3F turboshafts (900 shp each), first identified in November 1981. Even the Ka-27 is not thought to be equipped with active dipping sonar, relying instead on passive sonobuoys, which are tactically better at the detection of nuclear-powered submarines.

The French, still the greatest exporters of

Top: one of the two types of land-based helicopter used by the Soviet Navy for anti-submarine patrol is the Mil Mi-4 Hound, developed from a battlefield transport helicopter design. The other is the Mi-14 Haze.

military and civil helicopters, have been surprisingly slow in producing sound naval helicopter designs, although the Alouette III is in service with several navies, including France's Aéronavale and the Indian Navy. The role has been mainly supportive, carrying out SAR, HDS and limited weapons work. The larger, 13,000 kg (28,660 lb) Super Frelon, now outdated, is still the current French naval medium ASW helicopter, being also in service with Libya and China, as well as carrying the distinction of the largest operational helicopter built in Western Europe. The Super Frelon, powered by three Turbomeca Turmo IIIC turboshafts (1,550 shp each), can carry homing torpedoes or launch the sea-skimming anti-ship missile, Exocet, in which role it is believed to have been used in the Iraq–Iran conflict. The French naval modus operandi is to use one helicopter as the scene-of-action commander (using radar and

Above: an Aérospatiale SA 365F Dauphin II carrying AS 15TT anti-shipping missiles. This helicopter has recently entered service with the Saudi Navy and may have been sold to Iraq. China is making the type under licence for its army.

dipping sonar), whilst one or more ASW torpedo-armed variants deliver their weapons to the appropriate spot. On board ship, the 19.4 m (63.7 ft) long fuselage can be folded to allow ease of stowage below decks.

As a replacement for the Super Frelon, Aérospatiale has developed the Super Puma, and to replace the Alouette III, there is the 3,900 kg (8,600 lb) SA 365F Dauphin 2, which has been ordered by the newly formed Saudi Arabian Naval Air Arm. The Dauphin/AS 15TT combination is in direct competition with the Lynx/Sea Skua package operated by the Royal Navy; the French Navy has a substantial number of AS 12 missile-armed Lynx afloat as well. Larger helicopters are now being fitted to carry anti-ship missiles, extending the strike arm of the surface warship well over the horizon, say 200 km (108 nm), and these include the Sea King (Sea Skua for Federal Germany, Exocet for Pakistan and Sea

Falkland rescues

In a period of less than three months during 1982, the aviation elements of the Royal Navy, Royal Marines, and the British Army were engaged in a war which was totally unforeseen, and which did not comply with any of the long-term strategic scenarios favoured by defence planners. That war was in the Falkland Islands, and it was one in which the helicopter came of age for the British armed forces, not only as a battlefield system, but also as a vital element in the saving of a great number of human lives.

Helicopters were used in a number of distinct ways for rescue purposes. Firstly, the Royal Navy's Sea Kings – both the anti-submarine (ASW) and commando support versions – were used extensively in all of the actions in which ships were hit. Helicopters were on the scene within minutes to assist the crew of HMS *Sheffield* after the attack which resulted in the loss of that ship, the Sea Kings bringing in medical supplies, personnel and rescue aids such as breathing apparatus, and evacuating the more seriously injured survivors. The Sea Kings were also used in a number of SAR operations, and flew standard combat search and rescue sorties in support of the Sea Harrier and RAF Harrier pilots flying from the carriers of the Task Force. Sea Kings were also first on the scene when HMS *Coventry* was attacked and sunk, and according to one Royal Navy source, the helicopters were the sole factor in the aversion of a major tragedy, and the minimal loss of life.

The most spectacular contribution made by helicopters to rescue operations during the Falklands conflict was undoubtedly the heroic action at Bluff Cove, when Argentine A-4 Skyhawks attacked and inflicted severe damage and casualties on the Welsh Guards embarked in the landing ships, *Sir Tristram* and *Sir Galahad*. The entire sequence of events was seen later all over the world, having been recorded by the cameras of the British news teams reporting the conflict. It has been reported that, when the Argentine pilots who had carried out the attack saw the film themselves, they spontaneously cheered the extreme bravery of the Sea King aircrews, many of whom had flown well into the flames and thick, acrid smoke in their efforts to use the rotor downwash to drive dinghies and liferafts away from the burning ships. The action was an object lesson in the

capability of the helicopter and the true heroism of the crews.

The helicopters of the Army Air Corps and the Royal Marines – mainly Westland Scouts and Gazelle light observation helicopters – played an equally important part in the land battle. It was the bravery of their pilots, operating very close to the ground, over hostile terrain and often under direct enemy fire, which kept many a forward position resupplied with ammunition, and brought many wounded men back to the main field hospital for urgent medical attention. Without the dedication of these pilots, many survivors would have died, or suffered far more from their injuries. In this respect, the British Army helicopters were used exactly as planned, with the Scout – already obsolete by 1982 standards – bearing the brunt of these medical evacuation and resupply flights, with casualties being carried on internal or external stretchers. The Army Air Corps also participated in SAR duties in support of downed Royal Navy and Royal Air Force Sea Harrier and Harrier pilots, again often behind enemy lines and under severe fire. A number of light helicopters were lost to ground fire while carrying out such tasks.

Helicopters were used for rescue purposes on the Argentine side, too, especially in SAR flights in support of their attack aircrews and resupply flights from the mainland, and, to a lesser degree, the evacuation of casualties from forward areas, although the increasing air superiority of the British curtailed much flying later in the campaign.

Bob Downey

Troops and sailors are landed aboard a British fleet auxiliary after evacuation from the San Carlos bridgehead during the Anglo-Argentine conflict in 1982. The helicopter is a Sea King HAS 2 of 826 Squadron.

Equipped with the Northrop Seahawk forward-looking infra-red for night search and rescue, this US Coast Guard Sikorsky HH-52A Seaguard has been replaced by the similarly equipped Aérospatiale HH-65A Dolphin, imported from France.

Eagle for India) and the Seahawk (Penguin for the US Navy).

Although SAR missions are not exclusively flown at sea, the majority of those who find themselves in difficulties are associated with the sea – boating accidents, ditched aircraft, cliff rescues, oil rig disasters. The main purpose of the extensive SAR service around the coasts of Europe is however directly related to the combat rescue of downed aircrew in the event of war. Other countries have important civil-aid groups using rescue helicopters, such as the US Coast Guard, which in peacetime is a purely civilian body, but in war comes under the control of the US Navy; this service has just replaced its ageing S-62 Sea Guard helicopters with the SA 366G Dauphin 2 Short Range Rescue helicopter. The USCG is active in the Atlantic, Pacific, Caribbean, Great Lakes and Arctic areas. Early American space-flight recoveries were actually made in the Pacific by the Sikorsky SH-3 Sea Kings of the US Navy, which dropped divers on the ditched craft and collected astronauts using the Pugh Net, still used today for fire-fighting rescues from tall buildings.

In both the military and civil spheres, the helicopter has carved an important and essential niche for itself with its economy and service potential. The future is bright in terms of the high-technology input, which sadly is basically derived from the military programmes of various governments. New equipment being developed will make the battlefield helicopter a better weapon platform and give the aircrew a better chance of survival, at sea the naval helicopter has greater endurance and better safety, while the commercial helicopter will be reaching out to the more distant offshore oil and onshore natural resource locations. The helicopter is here to stay, but its form may not always be too conventional.

6

ROCKETS, MISSILES AND SPACECRAFT

From earliest Chinese experiments with rockets, the thought of putting a man into space has been a dream for each succeeding generation, yet it was mostly due to Hitler's V-2 technology that the first manned missions could become feasible. This also led to the development of guided missiles, now so critical for national defence and carried by warships, warplanes and armoured vehicles around the world. Some of the most advanced devices are the 'smart' missiles of the modern fighter.

From the first Russian Sputnik, through the American moon landings, to the Shuttle flights of today and tomorrow, space flight has come to influence daily life on this planet – communications, specially made drugs, weather reports and intricate non-gravity manufacturing are amongst the civilian benefits of the space race.

In Man's attempts to defend himself against a hostile neighbour, he has striven to produce a weapon which will outwit his opponent. Even with the early battles involving stones, bows and arrows and spears the need was to create a greater firepower at a greater range. As the role of explosives became more established, together with the capability of controlled flight, technological advances in guidance systems and control brought the possibility of altering the path of the munition after its launch to cope with the evasive target.

Within the overall field of rockets and missiles comes a range of unguided as well as guided vehicles, but this chapter deals with guided rockets and missiles which have some form of power at least at the point of launch.

Power for these unmanned vehicles can be either solid fuel, used generally for missiles required to perform over the shorter ranges, jet engines for the longer-range cruise missiles, or liquid rockets for any vehicle which travels into space. Jettisonable boosters can also be used to assist the launch.

Guidance can be by a number of methods, not only by heat seeking as many people believe. Undoubtedly heat seeking is an important and effective means of locating a target, especially as it is passive and can only be decoyed by an alternative heat source. But a shortcoming of this method is that the hottest part of an aircraft is its jet pipe, which means it is most effectively destroyed after it has passed its target, and may already have inflicted damage.

Another important form of guidance, which gives an all-weather capability, is semi-active or active radar. Both methods send out a signal which can be located and possibly jammed by a target. Semi-active guidance requires the target to be illuminated the whole time by the launch platform until impact, the signal being sent out from the launch area, bouncing off the target and then being received by the missile seeker, which is then able to adjust its path accordingly. An active

Demonstrating the importance of guided missiles to the modern air defence fighter, this US Air Force F-15A Eagle carries wing-mounted Sidewinders and fuselage-recessed Sparrow medium-range air-to-air missiles.

homing head is programmed for the target position just prior to launch, and automatically adjusts its path for target movement until impact. This is known as a fire-and-forget missile.

There is still a good case for optical guidance. It keeps the man in the loop, allowing him to identify the target as hostile before launching the missile. This system is of course passive because no signal is transmitted and it cannot be jammed. However, it relies upon clear daylight, although a clear night capability can be introduced using infra-red (IR) sensors.

All these guidance systems are used in the relatively short-range tactical missiles, the guidance systems for the long-range strategic missiles being inertial (not requiring any external guidance information) and capable of preprogramming to approach the target from a variety of flight profiles.

Early missiles were developed as miniature unpiloted aircraft carrying a warhead. They could sometimes be controlled from the ground or from a companion aircraft, or were preprogrammed for a set flight path and range before falling with no guaranteed accuracy on to a target area. They were either surface-launched or air-launched and the targets were on the ground. The object was to bomb areas at random at low cost and risk in an effort to overwhelm the defences by saturation. The earliest effective example was the German V-1 flying bomb.

As the development of electronics and guidance allowed greater reliability it became practical to carry missiles on aircraft for use as stand-off weapons against surface targets or against other hostile aircraft. Equally a need developed to defend ground targets against air attack, the initial method being radar-guided missiles against medium- to high-altitude aircraft, often using proximity fuses setting off large warheads to overcome any guidance inaccuracies.

A guided missile or rocket is now a very complex miniature aircraft which has to undertake an often complicated role without any human input. If any faults occur in development, no-one is there to recover the situation, and the vehicle is usually a total loss. A missile must be developed to work first time, every time in all environments, however hostile. Solid state microelectronics give lighter weight, greater reliability and improved performance. But expense has to be kept to a minimum to make the product affordable. Missiles have to be developed for specific tasks, such as anti-armour, anti-radar or even anti-satellite. Each missile system has its own clearly defined performance and capability and has to be capable of evolution to meet the changing threat.

The early pioneers

The Chinese pioneered the use of rockets in the Sung Dynasty just over 1,000 years ago, but without any guidance after launch. However, they did not maintain this lead, and had to start from scratch again in the early 1960s.

The early guided missiles were in effect relatively slow-flying pilotless aircraft carrying a payload of explosives and directed by radio control.

As early as 1915, in Britain, Professor A M Low had adapted an early form of television as guidance for a pilotless aircraft, a number of small aircraft powered by a short-life 35 hp ABC flat-twin engine being produced. Radio control was used for guidance, but no ATS, as they were known, saw active service. Professor Low also did some preliminary work with a radio-controlled rocket in 1917, but it never progressed beyond the experimental stage.

Almost certainly the first guided missile in the world to see active service was the bomb-carrying target Larynx developed by the Royal Aircraft Establishment (RAE). It was designed to a British Air Ministry requirement issued in 1925 for a surface-to-surface missile capable of carrying a 91 kg (200 lb) warhead 322 km (200 mi) in one hour. Power came from a 220 hp Armstrong Siddeley Lynx engine. The initial test flight down the Bristol Channel was followed by trials on desert ranges in Iraq under the direction of Sir George Gardiner.

In 1940, Miles Aircraft proposed a Gipsy Major-powered pilotless aircraft capable of carrying a 454 kg (1,000 lb) bomb at 483 km/h (300 mph) aimed at enemy cities. This aircraft, known as Hoop-la, could have been produced in large numbers at low cost, but no official interest was shown.

As early as 1916 in the USA, Dr F W Buck designed and built a number of aerial torpedoes with biplane wings, the latter being released at the required distance from the launch point. In 1917, also in the USA, a gifted engineer, Charles Kettering, worked on a series of light pilotless aircraft weighing 136 kg (300 lb) and capable of carrying their own weight of explosives up to 100 km (62 mi) with a target hit accuracy of 91 m (300 ft). The flying bomb was called the Bug and the arrangements being made to set up the first squadron in France in October 1918 were stopped by the Armistice. Also in production at the time of the Armistice was the Modisette Hot Shot, a fabric-covered spruce biplane powered by two Ford auto engines. It featured lateral stabilisation, and a clockwork longitudinal control which put the elevators hard down over the target, releasing three 20.5 kg (65 lb) bombs and disconnecting the wings, so that the missile could follow the bombs.

Early German experiments were more in the air-to-surface tactical class, the SSW of the First World War being developed for the German Navy as early as October 1914 as a

Konstantin E Tsialkovsky (1857-1935)

Tsialkovsky was a Russian schoolteacher who never built a rocket in practice, but developed the theoretical principles. 1883: he established that a rocket would work in the vacuum of space by the recoil effect of its exhaust gases, and not by its action of pushing on the air. He studied advanced concepts of liquid-fuelled spaceships and the mathematics of space travel. 1903: his first rocket design proposed a fuel mix of liquid hydrogen and liquid oxygen, later investigating other fuel combinations. He proposed the gyroscopic effect of a revolving flywheel as stabilisation and developed the theory for rockets to leave and re-enter the Earth's atmosphere. For control in space he suggested a gimballed jet nozzle or jet vanes in the rocket exhaust. He also proposed multi-stage rockets for launching into space and pressure cabins for the crew.

Colonel William Congreve (d.1828)

Congreve was a British military officer who made a detailed study of unguided land-to-land rockets against troop concentrations. Around 1800 his work started at the Royal Laboratory at Woolwich, near London, using as a basis the amusement rocket. By improving the powder mixture of the rocket's fuel, the range was increased to 2.5 km (1.5 mi) and the weight was increased to as much as 19 kg (42 lb). A light launcher was constructed. 1805: the Congreve rockets were used in combat for the first time, fired from ships against Napoleon's forces in Boulogne, starting many fires. The use of these rockets followed at a number of battles in Europe and America, where they were very effective, an instruction book being published by Colonel Congreve in 1814.

Robert Hutchings Goddard (1882-1945)

Goddard was one of the forerunners of modern rocketry. In his early years he flew a successful liquid-fuelled rocket at his farm in Massachusetts. It was a simple device with the combustion chamber and exhaust nozzle above the propellant tanks and held by pipes that carried the fuel and oxidant. It flew for 2.5 seconds and went 56 m (184 ft). He was the first to adapt the gyroscope to guide rockets, the first to install movable deflector vanes in an exhaust nozzle, and the first to use automatically deployed parachutes for instrument recovery. Goddard proposed the use of ablative compounds, which are materials that char and partially burn, removing excess heat. In a controlled re-

America's great rocket pioneer, Robert Goddard, pictured in 1918 experimenting with a solid-fuel rocket; in 1926 he moved into liquid fuelled types. The Houston manned space flight centre is named after him.

entry, a thin covering could be used to ensure the survival of a spacecraft. Goddard also proposed re-entry at a shallow angle, during which the air would brake the vehicle to a point where it would fall vertically to Earth.

Right: in 1929, Goddard launched his first 'weather rocket' and recovered the pressure, temperature and photographic data collection devices by parachute; he later used multi-chambered rockets to improve performance.

remotely controlled glide bomb. Flight testing started in January 1915, and control was by wire guidance. On reaching the target, the two halves of the aircraft separated, releasing a torpedo just above the surface of the water. Trials progressed satisfactorily, with the missiles being launched from Zeppelins by April 1917 and carrying war loads of up to 1,000 kg (2,205 lb). Before they could be used in combat, the Armistice was signed.

Further German experiments during the Second World War on missiles included air-launched tactical glide torpedoes and a glide bomb, and the Mistel composite aircraft, where a manned fighter flew attached to a pilotless, explosives-loaded bomber, releasing it in line with its target. Over 250 conversions were made, but no serious use was made of this weapon. The German V-1 and V-2 flying bombs were more successful guided weapons.

In 1942, the Imperial Japanese Navy began to develop the Funryu series of tactical air-to-surface guided missiles. Starting with the Funryu 1 as an anti-ship missile armed with a 400 kg (882 lb) warhead, guidance was by radio command, but the programme was abandoned before production commenced.

The Japanese Imperial Army Air Headquarters began to develop a series of tactical air-to-surface missiles. The basic research of the I-GO-1-A, the first of the series, was completed in mid 1942, and the design was then assigned to Mitsubishi for completion in late 1943. The missile was a wooden aeroplane configuration powered by a 240 kg (529 lb) thrust rocket engine. It was radio-command

guided and carried an 800 kg (1,764 lb) warhead, set off by an impact fuse. The first unguided test launches were made in mid-1944, with the guided flights soon after. But no production was undertaken. The I-GO-1-B was a smaller missile assigned to Kawasaki, and used the same radio guidance as the 1-A. It was powered by a 150 kg (331 lb) rocket motor and carried a direct-action fused warhead of 300 kg (661 lb). The first aerial launch was towards the end of 1944, testing building up to a rate of some 20 missile launches per week by the end of the year. Approximately 180

preproduction test missiles were built, but the missile did not enter production. The third and final missile was the I-GO-1-C, an air-to-ship missile still in the early stages of development at the time of Japan's surrender.

During the First World War the Italian inventor A Crocco commenced the development of a glide bomb. An evaluation batch of over a dozen Telebombs was produced after the war and tested between 1920 and 1922. Guidance was by a primitive autopilot, with the gyro and servos fed from an air bottle. It was claimed that it could carry a bomb of 80 kg

Hermann Julius Oberth (b.1894)

In 1927, Oberth was one of the founder members of the German Society for Space Travel (VfR). As a spare-time interest during the First World War he studied the concept of space flight and proposed a ballistic missile powered by liquid-propellant rockets, which could carry an effective warhead over long ranges. The ideas were turned down by the German War Ministry. Despite official indifference Oberth continued his studies, corresponding with Goddard and

Tsialkovsky, gaining some recognition in the mid-1920s. 1930: the theoretical deliberations were replaced by practical experiments in Germany using an old ammunition dump near Berlin as a proving ground. Many of the early static tests ended in the rocket exploding before launches could be attempted, but the first successful flight was on 14 May 1931. With a shortage of funds military sponsorship was required, the German Army taking over what was to be developed into the V-2 flying bomb.

The German V-1 and V-2

One of the most infamous guided missiles was the German Fieseler Fi 103, more commonly known as the V-1 or 'Doodlebug'. It was a pulse-jet powered early cruise missile launched from ground ramps and aimed indiscriminately at whatever target came within its range. Go-ahead for the project was given in June 1942 and guidance was by a preset compass heading, an aneroid for height and an air-log propeller which depressed the elevators at a preset range to put the bomb in a dive. The first V-1 hit London on 13 June 1944 and although over 22,000 missiles were produced, only some 13,000 were launched, nearly 2,500 falling on Greater London. It was possible to shoot the V-1s down either from the ground or the air – another method avoiding debris from the explosion in a tail chase being to topple the gyro by tipping the Doodlebug wing with the fighter wingtip. The straight-wing pilotless airframe could carry 850 kg (1,870 lb) of explosives.

A more advanced weapon was the ballistic A-4 missile, known as the V-2. This vertically launched rocket-powered missile was developed by Dr Wernher von Braun from the late 1940s, the first unsuccessful test launch being on 13 June 1942. The second on 16 August exceeded the speed of sound and the third on 3 October was a complete success. The V-2 could be launched from mobile platforms and guidance was by a system of pendulums, gyros and accelerometer. The missile was launched in line with the great circle route to the target and climbed to about 96 km (60 mi) above the earth before falling without warning on its target. About 10,000 were built, the first firing being poor shots at Paris on 6 September 1944. This was followed by a barrage of over 1,000 aimed at London; the 975 kg (2,150 lb) warheads caused similar damage to the V-1s. Over 4,000 were fired in combat up to March 1945 with a further 600 fired in troop training.

Philip Birtles.

The first of the terror weapons used against the United Kingdom during the Second World War was the V-1 (or Fieseler Fi 103) 'cruise missile' (*top right*), which was replaced by the advanced V-2 (*right*), seen here with its launch equipment when the British, Canadian and US Armies overran the launch sites in north-west Europe. The V-2 rocket, against which there was no defence, formed the basis of both American and Soviet missile technology immediately after 1945.

(176lb), but development was abandoned.

In the summer of 1944, the United States Army Air Force (USAAF) converted a small number of battle-weary B-17 and B-24 bombers to pilotless drone missiles, known respectively as BQ-7 and BQ-8. Stripped of all their unnecessary equipment the aircraft were filled with 9,072kg (20,000lb) of Torpex, a highly powerful explosive, and then taken off by a crew of two. Once set on course, the warhead was armed and the crew handed over to radio control, before abandoning by parachute. The B-17 conversions, known as Aphrodite, were assigned to the 562nd Bomb Squadron operating from Fersfield in Norfolk. At least eight combat missions were flown, but with poor results.

Early dog-fight missiles

As electronic guidance systems became lighter and more compact for missiles, it became more practical to consider launching from aircraft, first of all against ground targets. With the greater sophistication of seeker heads, initially using infra-red (IR) to home on the heat of the target, it became possible to consider replacing guns with missiles in air-to-air combat. In this case, lighter weight and high-hit probability were paramount, otherwise a defending aircraft would quickly lose its weapons and be less effective than with gun armament. Few aircraft are available just to defend against other air-to-air targets, since with the high costs of combat aircraft in the modern age they have to carry a variety of weapons for a range of targets, and usually the self-defence dog-fight missiles are hung on as an afterthought.

Britain commenced dog-fighter missile development in 1949 with the Fairey Fireflash, seen here carried by a Supermarine Swift F7. The intention was that the missile could engage fighters at stand-off range.

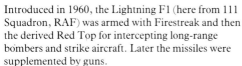

Introduced in 1960, the Lightning F1 (here from 111 Squadron, RAF) was armed with Firestreak and then the derived Red Top for intercepting long-range bombers and strike aircraft. Later the missiles were supplemented by guns.

Once again the Germans pioneered the guided air-to-air missile, but despite completing their development, none reached combat status. The first air-to-air missile in the world was the HS 298, which had slightly swept-back wings mounted on an oval section fuselage. The 25kg (55lb) warhead could be triggered by either a stand-off impact fuse or a proximity fuse, and power came from a rocket motor. Over 300 missiles were fired on test, but development was abandoned in December 1944.

Another important German air-to-air missile was the wire-guided X-4 developed under the guidance of Dr Max Kramer. The cruciform swept-back wings were mounted on the centre of gravity and directional control was by cruciform rudders. The pilot guided the missile by using a small joy-stick and the 20 kg (44lb) warhead was to have an acoustic proximity fuse so that it would fire when the sound of the target was detected. By late 1944, about 1,300 missiles had been produced, the first air firing being from an FW 190 on 11 August 1944. Some missiles were fired in anger, but none were delivered to combat units before the end of the Second World War.

In Britain, dog-fight missile development commenced with the Fairey Fireflash in 1949. The unpowered missile was accelerated to Mach 2 after launch by a pair of jettisonable solid-fuel motors and its guidance was by radar beam riding, requiring the target to be illuminated by the launch aircraft until impact. The first unguided firing was in 1954 and subsequently 300 missiles were issued for service trials and training with No 1 Guided Weapons Development Squadron at RAF Valley, equipped with specially modified Swift F Mk 7 aircraft.

Development of the first production British air-to-air missile began with de Havillands in 1951, leading to the first of many totally successful guided firings in 1954. With IR homing on to the heat of the hostile aircraft exhaust, this was a tail-chase missile, initially code named 'Blue Jay', but later to become Firestreak, using valve technology. It entered service on a trials basis with the Sea Venoms of 893 Squadron RN in 1958, and later was the primary armament for the Javelin, Sea Vixen and Lightning, and is still in service in the mid-1980s.

The world's first air-to-air guided missile, the Hughes Falcon, is pictured immediately after launch from an F-106 Delta Dart fighter, used to defend North American (both US and Canadian) airspace during the 1960s and 1970s.

The logical development of the Firestreak was the de Havilland Red Top, using more advanced solid-state electronics giving greater reliability. The range was increased from 8 to 12 km (5 to 7.5 mi) and a larger warhead gave greater lethality with overall performance improved by a new motor and seeker head, with the capability of attacking from any direction. These Mach 3 missiles were issued to the RN Sea Vixen Mk 2 units and RAF Lightning units from 1964, remaining in service with the latter.

Future developments involve collaboration with Federal Germany and the USA on the Advanced Short Range Air-to-Air Missile (ASRAAM) for the NATO nations.

The world's first operational guided missile was the American Hughes-developed Falcon family of short-range missiles, some with IR guidance, but the remainder with semi-active radar. Production began in 1954, the first missiles arming the F-89H Scorpions, and later the F-102. The missile was declared operational in mid-1956 and progressive development increased the range from 8 to 11 km (5 to 7 mi) and the speed from Mach 2.8 to Mach 4. Nearly 50,000 of the early series missiles were produced, which was over three-quarters of the overall total, later improvements including greater range, improved guidance and a more powerful warhead.

The market leader in the short-range dog-fight missiles is the Raytheon Ford Aerospace AIM-9 Sidewinder family of missiles with derivatives forming the basis of development in other countries. With the correct external pylon connections and simple electronics, the Sidewinder can be fitted to a wide variety of aircraft for self-defence in the minimum of time. The missile first entered service in 1956 and tens of thousands have been sold to over 30 countries. Apart from one of the early versions, the Sidewinder has an IR

FORD AEROSPACE SIDEWINDER AIM-9L

Country of origin: USA.
Role: Air-to-air short-range dog-fight missile.
Length: 2.85 m (9.35 ft).
Payload: Classified.
Launch weight: 89.3 kg (196 lb).
Propulsion: Solid fuel rocket motor.
Maximum speed: Mach 2.5.
Range: 17.7 km (11 mi).
Military load: Annular blast fragmentation warhead sheathed in a skin of preformed rods triggered by a proximity fuse.

sensor and a proximity-fused conventional warhead. The early versions were pursuit missiles, but improvements included increased range to over 6.5 km (4 mi) and a more sensitive sensor.

BRITISH AEROSPACE SKY FLASH

Country of origin: Great Britain.
Role: Medium-range air-to-air missile.
Length: 3.68 m (12.08 ft).
Payload: 40 kg (88 lb).
Launch weight: 193 kg (425 lb).
Propulsion: Hercules/Aerojet high-impulse solid motor.
Range: 50 km (31 mi).
Military load: Continuous rod + warhead triggered by active radar fuse.

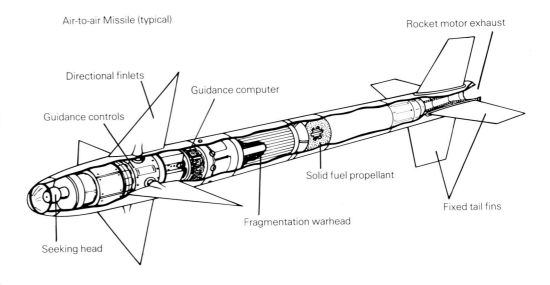

Air-to-air Missile (typical)

Directional finlets

Guidance computer

Guidance controls

Seeking head

Fragmentation warhead

Solid fuel propellant

Rocket motor exhaust

Fixed tail fins

A Japanese development of the Sidewinder was the AAM-1, and this is being followed by the much improved AAM-2, capable of collision-course interception. The Shafrir is an Israeli development of the Sidewinder started in 1961 using an early AIM-9 as a basis. By the mid-1960s Israeli improvements had largely overtaken the American development, with an increased-diameter body, housing a larger warhead, increasing lethality. A further development was the Python 3.

France is another major missile producer, one of the main manufacturers being SA Matra, who developed the R550 Magic dog-fight missile. It has an IR seeker and travels at Mach 3 over ranges of up to 9.5 km (6 mi). It provides strong competition to the Sidewinder with sales to 15 countries, and to make it more acceptable it has interchangeable attachment points with the American missile. Company-funded development commenced in 1968 with official adoption the following year. It can be launched within a 140° forward hemisphere at an altitude up to 18,000 m (59,000 ft) at speeds in excess of 1,300 km/h (800 mph) while pulling 6 g. The first guided firing was made in January 1972 with service entry in 1975.

South Africa manufactured its own dog-fight missiles known as the Whiplash and the V3 Kukri, the latter entering production in 1975. They have IR seekers and the Kukri reportedly has a range of 4 km (2.5 mi). Brazil commenced development of the IR-seeking

Piranha in 1977, but the production decision is yet to be made.

Longer-range hostile aircraft are dealt with by using the medium-range larger missiles. Here the market leader is the Raytheon SIM-7 Sparrow, together with its derivations such as the British Skyflash and the Italian Aspide, using semi-active radar homing. Meanwhile, SA Matra developed the R511, later to be replaced by the Super 530.

First-generation land-launched missiles

Tactical land-launched missiles are used for anti-aircraft defence or anti-armour warfare.

During the Second World War the British Army undertook development of one surface-to-air missile (SAM) system, known as Brakemine. Funding was practically non-existent, resulting in slow and ineffective development. Work started on the system in April 1943, the proposed guidance being radar beam riding. Design commenced in February 1944 of a missile with monoplane wings and six boost motors. The first unsuccessful launch was at Britain's Walton-on-the-Naze AA range in September 1944, trials continuing for three years. The first ten firings were unguided, but radio guidance was used subsequently. The airframe was improved from No 20 onwards, launch being from an inclined ramp. Despite encouraging progress, the low funding dried up after the end of the war and the programme was abandoned.

Early practical SAM systems were designed to counter the medium- to high-altitude bomber threat, the hope being to replace interceptors. The removal of the pilot from the loop was not achieved, maintaining the flexibility that a missile system cannot achieve.

Still protecting British air defence installations today, the Bristol Bloodhound entered service in 1957. An uprated version (illustrated) with improved guidance and countermeasures was introduced in 1964 and also exported to Switzerland and Sweden.

In 1949, English Electric were awarded a contract to develop Thunderbird, originally known during testing as Red Shoes. Power came from four jettisonable solid-fuel boost rockets for take-off and a liquid propellant sustainer motor. Semi-active radar guidance was used and a continuous rod warhead was fitted. Following successful trials it was redesigned as a (British) Army Weapons System consisting of a number of mobile units. Each Army unit had a Battery Command Post, Tactical Control Radar and Height Finder Radar. These controlled up to six firing troops each with a Launch Control Post, Target Illuminating Radar and three launchers. The missile range was up to 75 km (47 mi) and the first regiment was operational in 1959. The improved performance Thunderbird 2 entered service in 1963, remaining in operation until 1976.

The Royal Air Force (RAF) medium- to high-altitude defence was provided by the Bristol Bloodhound, under the code name Red Duster. The semi-active radar guidance operated the missile's wings, the tailplane remaining fixed, and launch was from a 45° elevation ramp on a fixed-base turntable. The zero-length launch was assisted by four solid-fuel boosters, and the missile cruised on upper and lower ramjets at over Mach 2. First deliveries were made in 1957 and the first deployments to protect the RAF V-Bomber bases were 1958-61. In 1958, development of Bloodhound 2 commenced, resulting in improved missile performance, including better lethality against low-level targets. The new

A Mirage F1 carries a Matra Magic dog-fight missile on the outboard pylon and two Matra Durandel runway cratering types inboard, underwing. Magic has been operational in the Middle East and the South Atlantic.

version entered RAF service in 1964 and export sales were made to Sweden and Switzerland. Bloodhound 2 is still used to defend RAF airfields in the eastern counties of England.

The American Nike Ajax became the first SAM system in the world to enter operational service, with the initial deployment near Washington DC in December 1953. It was a large and complex fixed-base installation with much of the support services located underground. Guidance was radar controlled, developed from Second World War gun radar,

the targets initially being acquired by a surveillance radar feeding the information into computers. The missile was a canard configuration with a large stage 1 boost motor giving 26,762 kg (59,000 lb) thrust for 2.5 seconds. The sustainer motor was rated at 1,179 kg (2,600 lb), giving a range of 40 km (25 mi) and maximum speed of Mach 2.3. By early 1958 around 16,000 rounds had been delivered to 40 US Army battalions, and deliveries of the much-improved Nike Hercules had also begun. This used basically the same installation facility to optimise on the

necessary investment.

The world's first active homing SAM system was the Boeing Bomarc, a pilotless interceptor for area defence, similar in concept to the Bristol Bloodhound, but scaled up for America's territorial needs. A development contract was placed in 1951 initially with the designation XF-99, finally becoming CIM-10A. The missile was designed to cruise at Mach 3, with peaks to Mach 4 over a range of up to 708 km (440 mi). The first of these fixed-base area defence weapons were delivered in 1957, the last unit being deactivated in 1972.

During the mid- to late-1950s, some development work was undertaken in France on SAM systems, but when the American Hawk system was selected for area defence, the national systems were shelved.

Anti-armour systems have well defined heavy, metal, slow-moving targets, requiring a simple but effective missile and guidance. When armour became too thick to be penetrated by special shells, the hollow-shaped-charge warhead was developed to focus a burning charge through the metal. The majority of the anti-tank missiles are subsonic and wire guidance is used, which is also immune to countermeasures.

In 1944 the Germans began development of X-7, known as Red Riding Hood, to defend against the Soviet armour. During ten months of development it proved effective against the heaviest armour, and evaluation batches reached combat troops in small numbers, although no missiles were produced in quantity.

With the end of the war anti-armour weapon development came to a halt despite the availability of the German research. In 1948 studies began in France on anti-tank weapons and early development work commenced in 1952 by a group who were later to become Nord Aviation. The result was the small and cost-effective SS.10 with a 5 kg (11 lb) warhead capable of penetrating 406 mm (16 in) of armour. Development was completed in 1955 and it remained in production until early 1962, when nearly 30,000 missiles had been delivered.

In Britain, Vickers developed the Vigilant in 1956 as a private venture to keep their guided weapons team together after many projects were cancelled. First firing was in 1958 and not only was it ordered by the British Army, but also gained some export success. The 6 kg (13.2 lb) warhead could penetrate

A standard NATO air defence weapon is the Hawk, which first entered service in 1954 and is only now being replaced in Europe. The standard, mobile, three-missile launcher is used to form defensive belts.

To give the soldier on the ground protection against main battle tanks, many nations have developed portable battlefield anti-tank systems, like the Vickers Vigilant (*left*). To give medium strike power armies use tactical missiles, sometimes nuclear-armed, like Honest John (*below*) and for air defence, the transportable Rapier low-level air defence system (*bottom*) is in widespread service. Rapier was particularly active in the Falklands and has since been sold to the US Air Force to guard its British bases, and to Indonesia for national defence purposes. A tracked version is in service with the British Army of the Rhine, having been developed originally for the Shah of Iran's army but not delivered.

more than 559mm (22in) of armour.

Early anti-armour missile development in the USA started with the large and cumbersome Dart in late 1951. It had large cruciform wings and tail set at 45 degrees to each other and was truck-launched. Work was terminated in 1958.

Current land-launched weapons

As with any weapons system, the threat changes, and either the defensive systems have to be capable of evolution to meet this threat or new systems must be developed. Sometimes the threat changes to avoid the performance envelope (the methods and capabilities) of a class of defences, requiring not only new systems, but the retention of the old systems in order to provide a comprehensive and effective cover.

An example of this change in strategy is for attacking aircraft to approach the target area at the lowest possible altitude, to avoid detection by radars until it is too late. Vital military targets such as airfields, docks, supply dumps and radars have therefore to be protected by quick-reaction low-level air-defence systems capable of destroying a hostile aircraft before it releases its war load.

The British system in this category is the British Aerospace Rapier family of weapons systems. When deployed optically, Towed Rapier gives a passive clear daylight capability, which with IR enhancement can give a clear-day and night defence. For all-weather coverage the Blindfire radar can be added. Tracked Rapier provides a low-level air defence for armoured formations in the battlefield and Rapier Laserfire provides forward troops with a defence against air attack.

Deliveries of Rapier commenced to the British Army and RAF Regiment in mid 1967,

BRITISH AEROSPACE RAPIER

Country of origin: Great Britain.
Role: Low-level air defence system.
Length: 2.24 m (7.35 ft).
Launch weight: 42.5 kg (94 lb).
Payload: Approx 1 kg (2 lb).
Propulsion: Solid fuel.
Maximum speed: Mach 2.
Range: 10 km (6 mi).
Military load: Small semi-armour-piercing warhead with impact fuse.

Developed jointly by France and Federal Germany, HOT is an anti-tank missile which can be launched by helicopter (French Army Gazelle illustrated) or from ground vehicles. It uses a wire system along which guide instructions are passed.

and the missile, which flies at Mach 2 powered by a solid-fuel motor, actually hits its target. During its service it has clearly demonstrated its accuracy and was particularly successful in the Falklands campaign without the Blindfire radar. The Rapier system is also well capable of evolutionary development to meet the changing threat well into the next century. It has been widely exported, including amongst its overseas customers Indonesia, Switzerland, Australia and the USA.

Amongst the closest competitors to Rapier is the Franco-German Euromissile Roland, which is complex with all the system mounted either in a wheeled container or on a tracked chassis. Its proximity-fused missile is loaded automatically. Development commenced in 1964, but it was protracted due to high cost and complication. The US Government

Developed to protect South African installations, the French-designed Cactus system has sold well around the world, as the Crotale missile, for air defence purposes; there is also a naval version in service.

shelved development of their version for the same reason. The first system entered service in 1977. The missile flies at Mach 1.6 and has a range of up to 6.5 km (4 mi).

France also produced Crotale as their national low-level air-defence system, initially at the request of the South African government. Like the other systems it is designed to defend against supersonic attack aircraft flying from just above the ground to 3,000 m (10,000 ft). Work commenced on its development in 1964 and production commenced in 1968. A derivative for Saudi Arabia is known as Shahine.

The Ford Chaparral has been developed in the USA starting in 1965 and firing the Sidewinder missile from a tracked chassis. First deliveries of this fairly cumbersome system were made to the US Army in July 1978.

Amongst other low-level air-defence missile systems are the Swiss Skyguard and Italian Spada systems, which fire either the Sparrow or Aspide air-defence missiles, and the Swiss/American ADATS with an anti-armour capability as well. None of these have yet been adopted for service.

Current anti-armour systems such as Swingfire, HOT and TOW are designed to penetrate modern armour at ranges of up to 4 km (2.5 mi), and like the early missiles are often wire guided and have hollow-charge warheads. Development of the British Aerospace Swingfire commenced with Fairey in 1959 and after being taken over by the British Aircraft Corporation, it entered service with the British Army in 1969. The Euromissile HOT is a high subsonic, optical, remote-guided, tube-launched anti-armour weapon, development of which started in 1964, with a capability of penetrating 610 mm (24 in) of armour. The major American anti-armour weapon which has been sold worldwide is the Hughes Aircraft-developed TOW. It can be launched from ground vehicles or helicopters and first entered service in 1970, combat status being achieved in Vietnam and the 1973 Middle East War. Replacements for these systems are being developed for the mid 1990s capable of dealing with spaced, active or top armour.

In addition to the long-range anti-armour systems are the medium- to short-range infantry portable systems, the Euromissile Milan being an example. Design was completed in 1963, the 2 km (1.25 mi) range missile

HUGHES TOW

Country of origin: USA.
Role: Long-range anti-armour missile.
Length: 1.162 m (3.8 ft).
Launch weight: 20.9 kg (46.1 lb).
Payload: 3.9 kg (8.6 lb).
Propulsion: Solid fuel sustainer motor with launch booster motor.
Maximum speed: 1,003 km/hr (623 mph).
Range: 3,750 m (12,300 ft).
Military load: Pacatinny Arsenal shaped-charge warhead.

As a result of the Vietnam war, the Americans have developed the TOW wire-guided missile, which like HOT, can be helicopter-launched, as here from a Bell AH-1G Cobra. It can also be used from ground vehicles.

BRITISH AEROSPACE SEAWOLF

Country of origin: Great Britain.
Role: Point defence of ships against air and missile attack.
Length: 1.9 m (6.25 ft).
Payload: 14 kg (31 lb) warhead.
Launch weight: 82 kg (180 lb).
Propulsion: Bristol Aerojet solid fuel rocket motor.
Maximum speed: Mach 2.
Range: 5 km (3 mi).
Military load: Conventional warhead detonated by proximity or contact fuse.

being housed in a launch tube and wire-guided. The hollow-charge warhead will penetrate up to 353 mm (13.9 in) of armour and production runs will exceed 200,000 missiles for the European and Middle East markets.

Naval weapons

Naval weapons are both defensive and offensive, tactical and strategic. In defensive terms ships need to protect themselves and other ships against air attack and ship-launched missiles. The underwater threat from

submarine-launched torpedoes is somewhat harder to deal with. On the surface the ships defend themselves against air attack using point defensive systems, such as the developed British Aerospace (BAe) Seawolf missile, which can knock out hostile aircraft, approaching missiles and even shells from guns. The only way this system can protect other shipping is to be placed between it and the launch platform. The more mature British

Right: for anti-shipping strikes, the French have designed the AS 15TT (*tous temps*/all weather), to be carried by the SA 365F Dauphin II helicopter. It has entered service with the Saudi Navy.

Below: an outstanding naval air-defence missile is the Shorts Sea Cat used by all types of warships, like this 'Leander' Class frigate, as protection against low-flying aircraft and missiles. It was used successfully in the Falklands.

McDONNELL DOUGLAS HARPOON UGM-84 A/C

Country of origin: USA.
Role: Ship- and air-launched anti-ship missile.
Length: 4.58 m (15.02 ft).
Payload: 227.6 kg (500 lb).
Launch weight: 667 kg (1,470 lb).
Propulsion: Aerojet solid boost motor and Teledyne CAE J402 turbojet.
Maximum speed: Mach 0.85.
Range: 109 km (68 mi).
Military load: Conventional warhead fused to explode after impact.

Aerospace Sea Dart system, however, provides a high-level area defence against attacking aircraft, and also has a capability against surface shipping. The development of Sea Dart began in 1962, with the first development firing in 1965, and service entry in 1973. It has a range of up to 30 km (18 mi) and can reach an altitude of at least 15,250 m (50,000 ft). With improvements it has been offered as a new land-based area defence weapon.

An early British-developed anti-aircraft missile was the Seaslug. The first launch of this beam-riding missile was in 1951, and it is not due for retirement until the late 1990s. Short Brothers also developed the short-range Seacat, starting work in the late 1950s with first firing trials in 1962.

American anti-air defence missiles for ships include Terrier, fitted to destroyers over 4,000 tons, which was introduced in the late 1950s and is still effective. The medium-range equivalent to Terrier is Tartar, used to defend destroyers and cruisers. The larger long-range Talos has been deployed on board USN cruisers. The Standard missile family is now replacing the three earlier systems.

Anti-ship missiles can be either air-launched or ship-launched, and are often of the sea-skimming type which penetrate the hull of the target and detonate inside, immobilising the operating capability of the ship. The shorter-range anti-ship missiles are represented by the Aerospatiale AS 15TT and the BAe Sea Skua, both launched from relatively light combat helicopters such as the Aero-

spatiale Dauphin and Westland Navy Lynx. They both use semi-active radar guidance and have a range in the order of 16 km (10 mi). Examples of larger anti-ship missiles are the Aerospatiale Exocet and the BAe Sea Eagle. Exocet can be ship-, helicopter- or fast-jet-launched, as can Sea Eagle, and it travels at Mach 0.93 over ranges in excess of 64 km (40 mi). The newer Sea Eagle is powered by a microturbo jet engine, giving it a range in excess of 80.5 km (50 mi) at Mach 0.9, both missiles being guided by active radar seekers,

BRITISH AEROSPACE SEA EAGLE

Country of origin: Great Britain.
Role: Long-range sea-skimming anti-ship missile.
Length: 4.2 m (13.78 ft).
Payload: Classified.
Launch weight: 600 kg (1,323 lb).
Propulsion: Microturbo jet engine TR1-60.
Maximum Speed: High subsonic.
Range: 100 km+ (62 mi+).
Military load: HE warhead, detonates after penetrating ship's hull.

Although it was in its jet-launched version that it attracted attention during the Falklands conflict, the Exocet can also be launched by helicopter (as it was in the opening stages of the Iraq-Iran War) and be carried by warships.

and in the case of Sea Eagle being capable of approaching a target from a different direction to that in which it was launched.

In the early 1960s, the British nuclear deterrent was transferred from the RAF V-Bomber Force to the Royal Navy's nuclear submarines. The basis for this strategic nuclear capability was the American Polaris, the first of which was launched in 1960. In its developed form it carried a single nuclear warhead over a range of 2,880 km (1,790 mi). With progressive modernisation of Polaris, known in Britain as the 'Chevaline' Project,

Blue Steel was the British nuclear deterrent in the years before the submarine-launched Polaris became available. It was carried by the RAF's V-Bomber force of Avro Vulcans and Handley Page Victors based in England.

which cost about £1,000 million in the late 1970s, 16 missiles with multiple warheads could be carried by each submarine, maintaining a credible deterrent until the introduction of the Trident system in the 1990s. The decision for Britain to adopt the American Trident 2D5 missiles aboard newly built submarines was announced in July 1981. Meanwhile the American Government bridged the gap between Polaris and Trident with the two-stage Poseidon system.

With submarines not only providing a major threat to surface ships, but also carrying the nuclear deterrent, weapons had to be developed to attack these targets. The classic American anti-submarine rocket system is SUBROC, launched by one submarine against another. A rocket powers the nuclear depth bomb above the surface until the target area is reached guided by inertial navigation. The rocket separates, allowing the bomb to seek out its target. Surface shipping can use

LOCKHEED POLARIS A3 – UGM-27C

Country of origin: USA.
Role: Submarine-launched nuclear strategic missile.
Length: 9.85 m (32.29 ft).
Payload: Classified.
Launch weight: 15,876 kg (35,000 lb).
Propulsion: Solid fuel.
Maximum speed: Classified.
Range: 4,635 km (2,880 mi).
Military load: Three 200 kT nuclear multiple re-entry vehicles.

Below For many years, American and British nuclear deterrence has been in the hands of missiles like Polaris (still in British service and illustrated here), Poseidon and later Trident, now in US service and awaited by Britain.

the American ASROC rocket system or an Australian-developed system known as Ikara, where a torpedo-carrying rocket is launched from a surface ship, and the lightweight torpedo is dropped to seek out its target submarine.

Future development of ship-launched naval weapons includes vertical launch to cover hostile missiles and aircraft approaching from any direction, containerisation for easier installation on a wider range of ships, and lighter weight to allow more to be carried.

At the centre of political controversy in the 1980s, cruise missiles have a feature of forward military planning since the end of the Second World War. It can be argued that the V-1 terror weapon was the first cruise and since then there have been a number of developments, presently culminating in the ground-launched systems deployed in Europe, the air-launched cruise missiles of the USAF's Strategic Air Command and submarine-launched systems carried around the oceans of the world by the Soviet Red Banner Fleet.

Dr Wernher von Braun (1912-1977)

Von Braun joined the German Society for Space Travel in Berlin in 1930 as an enthusiast in rocket propulsion. He prepared his doctoral thesis at Kummersdorf near Berlin in 1932, sowing the seeds for advanced rocket development. 1933: von Braun was put in charge of German rocket development at Kummersdorf, later moving to Peenemünde, the major rocket establishment. From the experiments the A-4 or V-2 was conceived as initially a mobile battlefield weapon capable of carrying a warhead of one metric tonne up to a range of 275 km (171 mi), but seen by the engineers as a logical step to space travel. With the end of the Second World War many of the German rocket team were given the opportunity to work in the USA and von Braun became Project Director of Guided Missile Development at Fort Bliss, Texas, the first US evaluation A-4 launch being from the White Sands proving ground in New Mexico on 16 April 1946.

With the failure of the American Vanguard rocket, von Braun was asked to develop the Redstone battlefield missile for the launch of satellites into space, which became the Jupiter-C, and achieved three successful test launches before carrying a payload. To progress from satellite launching to moon travel von Braun's group at Huntsville conceived in 1958 the Saturn IB, the first of the development

Intercontinental Ballistic Missiles

The Second World War German rocket developments, led by Dr Wernher von Braun, prepared the foundations for the American and Soviet intercontinental ballistic missiles and vehicles for space travel. Both nations obtained access to the V-2 rocket and the engineers were split between the two countries, von Braun becoming project director in the USA.

The first intermediate-range ballistic missile was Jupiter, developed in the early 1950s by von Braun and his team at Redstone Arsenal. Much of the body of the vertically launched missile contained the fuel to power the 68,040 kg (150,000 lb) thrust rocket engine. The inertial guidance section was above the fuel cells and above that in the nose of the missile was a one-megaton nuclear warhead protected against kinetic heating by an ablative re-entry shield – one with a controlled burn-up. The first representative firing took place in September 1956 when an altitude of 1,097 km (682 mi) was reached and production

batch of 10 being launched in October 1961. In October 1968, a Saturn IB put a manned Apollo 7 into earth orbit, leading to the Saturn V developed by the National Aeronautical and Space Administration (NASA) for putting men on the moon. In March 1970, Dr von Braun resigned as director of the Marshall Space Flight Center where he had worked for 10 years, to join NASA in Washington. On 1 July 1972, he joined Fairchild Industries as VP for engineering and development. Wernher von Braun died in a Virginia hospital in June 1977.

Wernher von Braun

Blue Streak was to have been a British intermediate-range ballistic missile, but the ability of newer American missiles to target the USSR from North America rendered it obsolete and it was cancelled in 1960.

commenced in November 1957 at the rate of four per month from the Chrysler Michigan factory. The first full-range mission to 3,180 km (1,976 mi) was made in July 1958, the first units having been equipped in January of the same year. In addition to US deployment, other units were equipped in Italy and Turkey as part of NATO defences, but due to lack of USAF interest Jupiter was withdrawn by 1965.

With Jupiter developed for the US Army, the USAF decided to develop the Thor IRBM (Intermediate Range Ballistic Missile) in November 1955. Douglas Aircraft were awarded the prime contract, making the first delivery in October 1956. Power, range and guidance principles were the same as for Jupiter, a copper heat sink re-entry vehicle being used to protect the warhead. The major difference in concept was that while Jupiter was mobile, Thor operated from fixed bases. The development programme started with four failed firings, but the system was declared operational for service in Great Britain in 1959, and the first missile became operational in December. Thor was withdrawn in 1965, having equipped three squadrons.

having equipped three squadrons.

Using the abandoned Navajo technology, Atlas was the first Allied Intercontinental Ballistic Missile (ICBM). As one of the biggest strategic missile programmes, the USAF feasibility studies commenced in 1954. In 1955 the Convair Division of General Dynamics were awarded the prime contract to produce the missile in a vast new factory near San Diego.

The first launch of an Atlas test vehicle, with only the boosters fitted, was from Cape Canaveral in June 1957, but this had to be aborted due to a fault in one booster. Following a number of early tests the first fully representative successful flight was made in August 1958 over a distance of 4,023 km (2,500 mi). Initially guidance was radio-inertial, which was accurate, but did not allow salvo firings. Later, the Titan inertial navigation system was fitted. The first operational launch was from Vandenberg AFB in September 1959, initially from ground-level launching pads. Concealment was later improved by launchers recessed into the ground at dispersed sites, and with the increased Soviet threat the silos were hardened to withstand missile attack. The major shortcoming was the 30-minute reaction time caused by the need to elevate the missile for launch. The whole Atlas

Below: Titan was for many years the most powerful ICBM in the world. Sited in deep silos in the heart of the United States, from which it could be launched, it was thought to be safe from a 'first strike' by enemy missiles.

Above: Blasting into the American sky is a Minuteman ballistic missile, which supplemented the Titan II programme but which is due to be replaced by MX and Midgetman in due course, provided that Congress approves the huge cost.

force was deactivated by 1967.

As an insurance against the failure of Atlas, the Martin Titan was developed from 1955. This was a two-stage missile, with the upper stage igniting in space. The robust structure allowed greater launch accelerations and a new factory was built at Denver, Colorado, to build this ICBM. The original inertial guidance system was passed to Atlas in exchange for the latter's radio-inertial system, and the development programme was protracted, a first-stage-only powered launch being made in February 1959. A year later a Titan flew over 3,540 km (2,200 mi), limited from its maximum of 12,875 km (8,000 mi) by fuel loading. Deployment was in hardened silos, giving a reaction time of 20 minutes, and first operational capability was in April 1962. The missile carried a 4 MT warhead, but all were withdrawn by 1966 due to obsolescence. The

MARTIN TITAN II

Country of origin: USA.
Role: Intercontinental Ballistic Missile.
Length: 31.4 m (103 ft).
Launch weight: 149,688 kg (330,000 lb).
Payload: 18 MT warhead.
Propulsion: Aerozine 50 and N_2O_4 rocket fuelled Aerojet-General engines.
Maximum Speed: Mach 5.
Range: 15,000 km (9,325 mi).
Military load: GE Mk 18 MT nuclear warhead.

missiles were replaced by the instant silo-launched Titan II, very little being in common with the earlier missile. A new inertial guidance system was fitted and an 18 MT warhead could be flown over a maximum range of 15,000 km (9,325 mi). Titan II entered service in 1963, and by progressive update has remained effective for two decades, but is now being gradually retired.

By the mid-1950s, there was a need to dispense with large vulnerable ICBMs and produce a lower-cost smaller and simpler weapon powered by solid fuel to give an almost instant reaction time and use dramatically less manpower. The answer was to upgrade the Minuteman in 1957, the prime contractor being Boeing. The first launch was in February 1961, when the missile flew 7,400 km (4,600 mi) completely successfully. The normal range of this three-stage missile system was from 10,000 km (6,214 mi) to 13,000 km (8,078 mi) depending on the variant, and a 1.3 MT warhead was carried. Minuteman was silo-launched, but did not require the exhaust ducting of Titan, and two men could control each flight of 10 missiles. Development of Minuteman continues, including an increase in power, greater accuracy and the installation of three independent warheads.

Because of the vulnerability to attack of the easily targeted fixed missile installations the MX system, or Peacekeeper, began development in the mid-1970s. Air-launching, a rail-road scheme and a network of underground tunnels were all studied, but in the end a fixed installation was chosen, mainly on the grounds of cost. Forty missiles have been approved, each with 10 warheads, and deployment should be complete by 1989.

Currently under development is the Midgetman ICBM, which can be vehicle-launched or placed in super-hard silos. The initial operational capability is expected in 1992 and MX technology will be used wherever appropriate.

Soviet missile systems

As is to be expected, the Soviets have the full range of missile capability, having started along the same lines as the USA, with the V-2 research by the German scientists. With little information on performance and capability coming directly from behind the Iron Curtain, much of the data on these systems must be obtained by inspired guesswork. Many of the early ground-to-air and ground-to-ground missiles were much larger than their Allied equivalents, but they were produced in significant numbers, and in a number of different types. At least fourteen different types of air defence missile systems are produced, ranging from the fixed-base Guild, Guideline, Goa and Gammon to the manportable SA-7 Grail, the latter deployed widely and suitable for terrorist use against airliners. A truck-mounted version of Grail is known as SA-9 Gaskin, guidance being by IR. The current mobile anti-aircraft systems are Ganef, Gainful and Gecko, all of which can be used for low-level air defence. Future systems with even better performance, including ranges of up to 1,000 km (620 mi) and an altitude of 30,500 m (100,000 ft), are expected in service over the next decade.

Missiles such as the ABM-1B Galosh have been developed to counter American ICBMs, and this system is soon to be replaced by the long-range SH-4 with loiter capability and the SH-8 hypersonic anti-missile missile.

The tactical battlefield surface-to-surface weapons commenced with the plentiful truck-mounted Frog-7, but these are being replaced by the SS-21. The SS-23 is a new tactical ballistic missile, and the nuclear SS-22s are now being deployed in Eastern Europe.

Soviet strategic ICBMs started with the outdated Sandal and the Skean, but the

Below: the Soviet Union countered American ICBMs by developing the anti-ballistic missile known as Galosh, used primarily to defend Moscow but later also deployed to defend against possible attack from the People's Republic of China.

Bottom: Soviet aircraft-launched missile development has included many systems which appear to be copies of Western developments, including the Atoll, used by MiG-21 and other Russian fighters for air-to-air combat.

replacement is the mobile SS-20, of which at least 400 are deployed in the Warsaw Pact nations. Each missile can deliver three warheads over a range of 4,828 km (3,000 mi) and the deployments continue to grow. Long-range ICBMs launched from silos have been in service since the late 1950s, with ranges varying from 8,050 km (5,000 mi) to 12,070 km (7,500 mi) depending on the type. The three early ICBMs were replaced by three new systems during the mid 1970s, large numbers now being deployed.

A formidable range of anti-armour weapons has also been produced, not just for Soviet use, but exported widely to friendly nations throughout the world.

In the air-to-air category the earliest Soviet dog-fight missile was the AA-1 Alkali, development of which started in the 1950s. The missile was a little larger than the American Sidewinder. The second generation AA-2 Atoll is much more comparable with the Western missiles, with further developments under way. The longer-range air-to-air missiles such as AA-3 Arab are much larger than the Western equivalent, but later missiles are reducing in size.

The air-launched weapons aimed at ground targets are generally very large and have ranges up to 645 km (400 mi). These large missiles are either carried on underwing pylons or partly recessed in the bomb-bay of Backfire bombers. The long-range missiles are normally fitted with nuclear warheads and can fly at speeds up

Defending Russian, Warsaw Pact and client state armies and terrorist groups against vehicular attack is the AT-3 Sagger, which is also mounted on battlefield helicopters, like the Mil Mi-8 Hip-E and Mi-24 Hind-D; it has been used widely.

The Earth from space, with the continent of Africa in the centre. Much of today's multi-national effort is directed towards earth resources for the benefit of all nations. Work in remote sensing includes crop and forest protection.

to Mach 3.5.

To cover anti-armour requirements the land-launched AT-2 Swatter and AT-3 Sagger can be helicopter-launched, and the tube-launched AT-6 Spiral was first identified in 1977 as an air-launch-only system.

The Soviet Navy has not neglected its missile armoury, with weapons in all the categories of surface-to-air, surface-to-surface, anti-submarine and strategic, submarine-launched ballistic missiles (SLBM).

Initial trials commenced in 1955 with Sand missiles launched from converted diesel-powered submarines, leading to the short-range SS-N-4 Sark, and the SS-N-5 Serb SLBM. Trials continued on improved systems, including the SS-N-6 Sawfly, the SS-N-8, which is slightly larger than the US Trident, and the latest SS-N-20, the majority of the missiles having liquid fuel motors, causing storage difficulties.

Anti-ship missiles can be long-range cruise types such as the SS-N-3 Shaddock or the SS-N-7 Siren, both of which can be submarine-launched, the latter from a submerged position. The larger surface ships, including the Kiev class of aircraft carriers, have eight launchers for the SS-N-12 Sandbox anti-ship missiles with a range in excess of 500 km (310 mi). The SS-N-2 Styx is small enough to be carried by fast patrol boats, but lethal enough to destroy a capital ship. For anti-submarine defence the SS-N-15 rocket was developed for the early 1970s and is now being replaced by the nuclear warhead SS-N-16. There are also at least seven SAM systems operational with the Soviet Navy, the earlier ones based on their land-launched equivalents, and newer systems being developed for the larger warships now coming into service.

Spacecraft

Space technology is moving into a new era away from political- and prestige-driven projects towards commercial and humanitarian efforts. For the first time, there is the possibility that money may be made in space and wide-scale benefits could be reaped by all mankind. Investigations now being made on Spacelab, the European-built manned space laboratory which flies on Space Shuttle, could lead to the commercial production of materials and pharmaceuticals in space, where the absence of gravity can increase yields tenfold and allow the formation of materials not produceable on Earth.

The view from above

Ever since man took to the air, he has realised that the view from above is both beautiful and useful. The early astronauts took pictures of the earth with hand-held cameras and it was quickly realised that such images could provide a wealth of data about weather, crops, irrigation, pollution, fisheries and a host of other things. Today, dedicated satellites are in orbit providing continuous data about the earth, and more are planned. The gathering of such data from orbit is known as remote sensing. The USA has the Landsat system, which has been in existence for some years and so far has been a government-run project, although it could be sold to a commercial operator. France is about to establish the SPOT system, which will be commercial from the start, although there has been government assistance.

One of the most attractive aspects of remote sensing is that it can be of enormous value to developing countries, which would otherwise have little to gain from space technology. There are problems, however, in gathering and distributing the data. Those that need it most may be least able to pay for it or may not have the equipment to receive it. There are questions about who should see the data first. Remote-sensing satellites may discover vast wealth in a developing country in the form of mineral deposits, for example. That country may be kept in ignorance about the scale of the find until an outsider has negotiated access.

TV via space

Problems in radio and TV transmission on earth have been solved by communication satellites (comsats). Only a small section of the entire range of non-visible wavelengths (known as the electromagnetic spectrum) can be used for radio because other parts are swamped by natural interference or can be used for line-of-sight communications only. A limited number of stations can be fitted into the HF, VHF and UHF wavebands and higher frequencies tend to become restricted to line-of-sight. Relay stations can be used, but this detracts seriously from the usefulness of radio communications, which depends in part on the mobility of the receiver.

The solution is to put the relay in space on board a satellite, where it can always be 'seen' by both transmitter and receiver. Of course, to remain in orbit the satellite has to travel around the earth at high speed. Due to a fortunate coincidence, it is possible to place the satellite in an orbit so that it travels round the earth at the same speed as the earth rotates under it. This is called geostationary or geosynchronous orbit. It is at an altitude of 36,200 km (22,500 mi) and satellites in it appear to hover over a point on the equator.

The first comsat was Early Bird, built by Hughes Aircraft of the USA and launched on 6 April 1965. Since then there has been a proliferation of comsats for TV, telephone, telex and other communications. The latest use is in direct broadcast television satellites (DBS). Existing TV networks require the establishment of large transmitting stations at frequent intervals. Many people are familiar with large TV transmitting aerials near their homes, mainly because TV transmissions are short-range over the surface of the earth. Major projects are now underway to put TV repeaters in geostationery orbit, where each

Despite the setback of the Challenger loss, the American Shuttle programme is destined to continue into the next decade. The re-useable spacecraft carries both military and commercial payloads, having the ability to launch satellites.

One of the immediate benefits of space technology for mankind has been the development of worldwide television communications, such as this French DBT satellite: Satellite television is becoming widespread, particularly in America.

station would require only one or even part of one satellite. The uplink would be transmitted at one frequency and the downlink would come back at another.

But there are problems: allocation of frequencies and slots in geostationary orbit requires international agreement, and confining transmissions to the area desired is difficult. India was the first nation to try an experimental DBS system bringing TV to people, many of whom had never seen television before. A mixture of educational and entertainment programmes was transmitted for several hours each day using a radio-location device called a transponder on a NASA satellite. The experiment was successful, but the signals could be received as far away as France. This could cause several problems for future DBS systems. Eastern Bloc governments may not be particularly happy at their subjects receiving Western programmes. DBS companies will have difficulty charging for their services, but one solution is to encode the signals and charge rental on decoding equipment.

The Moon's legacy

Much of today's space technology has come from what people remember as the race for the Moon. In fact, it was very much a one-horse race, with the Soviet Union pulling out very early on in the proceedings. It is difficult to say whether it was a good thing or not but many of the early space programmes were driven by

Yuri Gagarin (1934-1968)

Yuri Alexeyevich Gagarin was born on 9 March 1934 in Klushino, near Smolenski, his father was a carpenter. 1955: Started flight training and parachuting and then joined the Soviet Air Force, flying jets from Orenburg on the River Ural. 1960: Became one of the first cosmonaut trainees. 12 April 1961: Became the first man in space aboard Vostok I, which was launched from the Tyuratam Cosmodrome on top of an AI rocket. 'The sky looks very, very dark and the Earth is bluish,' he said. Gagarin was promoted to Major during the flight. He went on to spend many years as a fine ambassador for the Soviet space programme and he was on the verge of a second career as an astronaut when he was sadly killed in an air crash on 27 March 1968. The Soviet government

paid him the supreme posthumous tribute by placing his ashes in the Kremlin Wall on Moscow's Red Square.

political and prestige forces rather than serious scientific investigation. Being first, biggest and best is often more important to governments than doing a job correctly or by the most efficient means. So it was hardly surprising that the USA was spurred into action when, on 4 October 1957 the USSR became the first

nation to put a satellite in orbit. What's more, Sputnik 1 weighed 83.5 kg (184lb), considerably more than the USA's first satellite whose launch vehicle, in any case, exploded on the launch pad on 6 December 1957, which was to make the USA even more determined.

Sergei Pavlovich Korolev (1910-66)

Korolev is best known for his work on Sputnik, but he was widely involved in many other parts of the Soviet space programme. He began his work as a member of the 'Group Studying Rocket Propulsion'. He worked with Tupolev, during which time he was recognised as one of the greatest aerospace engineers and had much to do with the creation of a line of aircraft bearing the prefix Tu. 1930s: Spent years in prison on suspicion of treason. He spent four years in Special Prison Number 4 flight testing with a group working on rocket-assisted take-off for aeroplanes, bombardment rockets, an in-flight emergency acceleration rocket for conventional aircraft, and pure rocket aircraft. 1945/6: Worked on the repair of a German V-2 base, interviewing ex-Peenemünde engineers and technicians, some of whom were put to work on various research projects. 1947: Produced the first all-Soviet ballistic missile (RI). 1957: Worked on the R7, which was originally developed as an ICBM, but proved to be an excellent rocket booster. On 3 August the first R7 launch was made

and six weeks later the first Preliminary Satellite or Sputnik was put in orbit. Korolev's later achievements included putting the first man and woman in space in the Vostok capsule. He was also involved in the Proton launcher and in the development of robot lunar probes and landers, communication satellites, navigation and military reconnaissance satellites.

Soviet scientists made space history with the launch of the football-sized Sputnik 1 satellite in 1957, which dented American pride and was the first hurdle in the space race. They have continued since to launch predominantly military space vehicles.

Alan Shepard (b.1923)

Alan Bartlett Shepard was the second man in space although his flight was a suborbital lob. Shepard was born in East Derry, New Hampshire, on 18 November 1923. 1944: Graduated from the US Naval Academy, Annapolis, with a degree in science. 1944/5: Destroyer service in the Second World War. 1947: Gained his wings and became a test pilot developing in-flight refuelling. 1958: He was appointed an air readiness officer for the Commander-in-Chief Atlantic Fleet and his name was automatically put forward as a potential astronaut along with hundreds of others. The successful candidates were called the 'Mercury Seven' and they were introduced to the world in April 1959.

Shepard was among them. 1961: Was chosen as the first American to be sent into space. At 9.32 am local time he was launched from Cape Canaveral in a Mercury spacecraft on top of a Redstone rocket. The flight lasted 15 minutes and the capsule splashed down 478 km (297 mi) down range. 1971: After a long battle to overcome an ear infection which robbed him of flight status, Shepard flew as commander of Apollo 14 and spent over 33 hours on the Moon in February.

Apollo astronauts Edgar Mitchell, Alan Shepard Jnr *(centre)* and Stuart Rossa pose for the camera in front of the Saturn V launch vehicle for Apollo 14 at Cape Kennedy, Florida. This was the third successful moon landing mission in 1971.

First man in space

America's pride was further dented on 12 April 1961, when Yuri Gagarin became the first man in space during a one-orbit flight by Vostok 1. The Soviet Union continued to achieve many other space firsts with the USA always seeming to be some way behind. But the USA had a higher level of technology which would pay off later. America's first attempt to put a unmanned Mercury capsule into orbit for a test flight ended in another explosion when its Atlas launcher failed on 25 April 1961. But on 5 May 1961, Alan Shepard became the first American in space aboard a Mercury spacecraft, although his flight was a suborbital lob which lasted 15 minutes. The first American to reach orbit was John Glenn and it was he who received most of the publicity.

Propaganda in space

By this time both governments had realised the enormous propaganda potential of success in space. The public was fascinated by such exploits and did not seem to heed the cost as it did later. It was in the euphoria of this success that President Kennedy set NASA the task before the decade finished of sending a man to the Moon and returning him safely to Earth. Everyone assumed that the USSR was starting a similar programme, but with hindsight there was little evidence for this. The Soviet Union continued with space firsts such as the first woman in space (Valentina Tereshkova in 1964), the first three-man flight in October 1964 and the first space walk (Alexei Leonov in 1965). These were great achievements, but the Soviet Union did not have the technology to put men on the Moon.

NASA's Mercury flights gave way to the Gemini programme, which constituted a rehearsal for a Moon mission. Edward White made the first US space walk in June 1965, Frank Borman and James Lovell orbited the Earth for 14 days in December 1965 during which they were joined by Walter Schirra and Thomas Stafford in another spacecraft, achieving the first rendezvous. Neil Armstrong performed the first space docking in March 1966, and the Gemini programme culminated in November 1966 with a 2.5-hour space walk by Edwin Aldrin.

NASA had already been developing a large launch vehicle called Saturn when President Kennedy announced the Moon mission. Originally, the launcher was planned to have eight SD3 engines but the Moon mission required more power so the new F1 engine, which could develop 680,400 kg (1,500,000 lb) of thrust, was chosen for the main stage. The number of engines needed depended on how the flight to the Moon was to be made. To take a single spacecraft to the Moon and return it

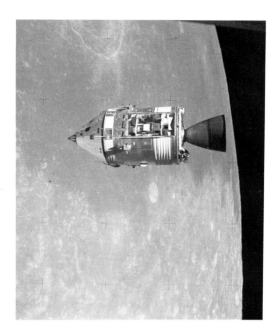

The command and service modules of the Apollo 10, photographed from the lunar module shortly after its detachment from it on its descent to the moon's surface. These modules remained in orbit round the moon during the landing.

complete to the Earth would have required 5.5 million kg (12 million lb) of thrust from eight F1 engines. Less power would be needed if parts of the spacecraft were launched separately or sections left behind en route. The second alternative was chosen, requiring five F1 engines, and so the Saturn 5 was named. The second stage would have five 90,720 kg (200,000 lb) thrust J2 engines 'to be developed' and the third stage would have one J2 engine.

The most demanding part of the Saturn 5 development was the instrument unit (IU). It was the 'brain' of Saturn 5 and performed the pre-launch checkout, transmitted data by telemetry, maintained trajectory as fuel was burned and the atmosphere became thinner with altitude, activated staging procedures, checked systems once in orbit and made the translunar injection. On 1 September 1963, Dr Wernher von Braun, who was director of the Marshall Space Flight Center responsible for Saturn 5 development, reorganised the centre to delegate production of Saturn 5 components ready for full-scale operations.

Flight round the Moon

On 21 December 1968, a Saturn 5 flew men for the first time on the type's third launch. The men were Frank Borman, James Lovell and William Anders in Apollo 8, which orbited the Moon during its six-day flight. Previous Apollos had been launched by the smaller Saturn 1B for Earth orbit rehearsals. Apollo 8 made 10 orbits of the Moon in some 20 hours. Apollo 9 was launched on 3 March 1969 and spent 10 days in Earth orbit performing rendezvous and docking manoeuvres. Thomas Patten and Eugene Cernan of Apollo 10 could have been the first men on the Moon with John Young remaining aloft in the command module if the lunar module had been ready. Problems with the lunar module meant that it could do everything except land on the lunar surface, so Apollo 10 performed a full dress rehearsal for a landing by Apollo 11.

Men on the Moon

On 9 January 1969, NASA announced the crew of Apollo 11, two of whom would walk on the Moon. They were Neil Armstrong (com-

Apollo Moon Landing Sequence

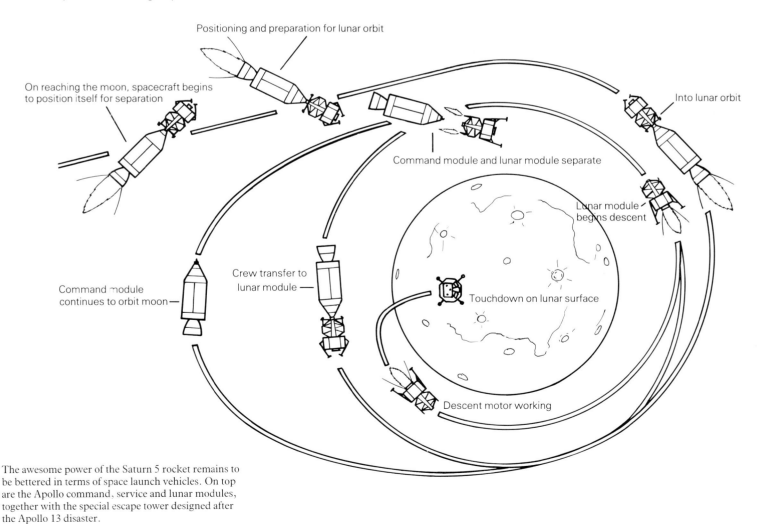

The awesome power of the Saturn 5 rocket remains to be bettered in terms of space launch vehicles. On top are the Apollo command, service and lunar modules, together with the special escape tower designed after the Apollo 13 disaster.

Edwin 'Buzz' Aldrin walks on the surface of the Moon, photographed by Neil Armstrong, the mission commander, during the Apollo 11 mission in 1969. The clear atmosphere makes the infinity of the picture somewhat distorted and unreal.

mander) and Eugene 'Buzz' Aldrin (lunar module pilot). Command module pilot Michael Collins would remain in lunar orbit. The purpose of the Apollo 11 mission was primarily to fulfil the goal set by President Kennedy. Future Moon missions would have a greater scientific content.

The mission began at 9.32am on 16 July 1969, watched by an estimated 600 million television viewers around the world. At T+11min 42sec the third stage shut down leaving Apollo 11 in a parking orbit about 60km (100 mi) up. Towards the end of the second orbit the third stage motor was reignited for 5min 47sec, accelerating the spacecraft to nearly 40,225km/h (25,000mph) and putting it on course for the Moon. En route, the command module Columbia separated from the third stage, turned through 180 degrees, and docked with the lunar module Eagle, extracting it from the third stage.

After a subdued journey the spacecraft passed behind the Moon where, out of contact with Earth, the SPS engine was fired for 357 sec, slowing Apollo 11 into its initial Moon orbit. At T+100hr 12min Eagle and Columbia separated behind the Moon. One revolution later Eagle's descent engine was fired for 30sec, putting the lunar module into an orbit of 92km by 14km (57mi by 9mi). At the lowest point of 14km (9mi) the motor was fired again for 756sec to put Eagle onto the lunar surface.

It was at this point that problems started to arise. At 2 min 11sec into the descent burn, when Eagle was at 14,325m (47,000ft), an alarm light flashed indicating that the onboard computer was being overloaded with com-

mands. Mission control said that they were still clear for a landing, but then another alarm sounded, and finally a third, but mission control continued to tell Eagle that it was 'go' for a landing.

Landing site selected

At 430m (1,400ft) above the Sea of Tranquillity, Armstrong saw that Eagle was approaching a large crater strewn with rocks so he adjusted the flight path to take the lunar module clear of this area. Aldrin called out instrument readings of flight-path angle, altitude, and horizontal and vertical speed. During the final stages of the approach mission control interjected twice, telling Eagle that it had 60, then 30 seconds of fuel left. Eagle landed about 6.5km (4mi) downrange from its intended position, on 20 July 1969.

After some 4 hours on the lunar surface, Armstrong and Aldrin donned the suits and backpacks that would keep them alive in the harsh environment outside the lunar module, where temperatures varied from 240°C (464°F) in sunlight to −173°C (−279°F) in the shade. In addition the suits would protect the Moonwalkers from direct solar radiation and the extremely 'hard' vacuum of space. Armstrong opened the hatch and descended the ladder slowly, making observations as he went. He placed his left foot on the Sea of Tranquillity at 02.56GMT on 21 July 1969, and made that famous statement: 'That's one small step for man. One giant leap for mankind.' The official version of his statement has the word 'a' before 'man', which is probably what Neil Armstrong intended to say. The official version makes sense, whereas what Armstrong said appears to be a contradiction!

Aldrin joined Armstrong on the Moon and after some ceremony the two men continued with detailed observation and description of

the lunar surface. They set up some experiments, and collected 22kg (48.5lb) of Moon rock. Armstrong spent 2hr 14min walking on the Moon, accompanied by Aldrin for 1hr 33min. After 21hr 36min on the surface the lunar module ascent stage left the descent stage behind, taking Armstrong and Aldrin up to rendezvous with Collins in the command module.

Coming home

After both men and samples had transferred to the command module, the ascent stage was jettisoned. Behind the Moon, Columbia fired its SPS motor for 149 sec, putting it into an Earthbound trajectory after orbiting the Moon 30 times in 59 hr 30 min. Apollo 11 splashed down at T+195 hr 18 min in the Pacific Ocean about 338 km (210 mi) from Johnson Island. USS *Hornet*, carrying President Nixon, picked up the capsule from the sea, but the astronauts were transferred directly to a quarantine container, where they spent three weeks to allow the detection of any infection that they might have caught; none was found.

The Apollo 11 mission was a fantastic success that captured the imagination of millions and probably did a great deal to bring unity to an otherwise divided world. What was nearly as spectacular, however, was the speed with which public interest waned during subsequent flights to the Moon.

The Apollo programme was cut, and in December 1972 Apollo 17 made the final flight of the series, reaching the Moon for the sixth time, Apollo 13 having met with near disaster when an oxygen tank exploded due to an electrical fault. The crew of Apollo 13 became dependent on their lunar module as a life boat and it sustained them while they rounded the Moon and came back to Earth. It was detached shortly before re-entry.

The scale of the effort required for the

American orbital and moon missions were recovered in the Pacific Ocean (*below*), courtesy of the US Navy's aircraft carrier groups, whilst Soviet missions have always recovered to the USSR, landing in one of the central Asian republics.

Apollo Moon-landing equipment

The first Moon-landing equipment was designed by the British Interplanetary Society in studies published in 1939 and 1947, suggesting that a manned lunar landing was possible. At a Symposium in 1951 Dr Wernher von Braun and others presented a large-scale plan involving three ships, departing from Earth orbit, landing directly on the lunar surface and returning directly to rendezvous with an Earth Orbital Space Station. In 1959, Rosen and Schwenk presented a two-man lander study at the 10th International Astronautical Congress in London. This involved Direct Ascent from the Earth's surface to a Direct Lunar Landing. This study was called 'Nova', a name that was used for subsequent studies affording lunar landing by Direct Ascent.

By 1962, plans involving Direct Ascent and Return were thought too ambitious and expensive; in addition, President Kennedy's commitment that America would 'land a man on the Moon and bring him back safely' before 1970 did not leave enough time to develop Nova. It was decided to use a single Saturn 5 rocket to achieve the landing, and to reduce the payload delivered to the Moon's surface by Apollo to a bare minimum. This technique involved Lunar Orbit Rendezvous, and required the development of a new lunar landing craft. In 1961, six companies produced feasibility studies; of these,

Grumman Corporation was awarded the contract for the Lunar Module in 1962.

The Lunar Module (LM) was finally delivered two years late, at four times the projected cost; and 50 per cent larger than the initial design size. The spacecraft weighed 16,465 kg (36,300 lb) fully fuelled and 5,262 kg, (11,600 lb) empty; and comprised two stages – the Descent and Ascent stages. The Descent stage incorporated a single engine with 4,536 kg (10,000 lb) thrust, variable between 10 per cent and 65 per cent. The Ascent stage incorporated a single engine of 1,588 kg (3,500 lb) thrust variable. Powered descent proceeded in three phases. The braking phase from 15,000 m to 3,000 m (50,000 to 10,000 ft) and final approach phase from there to 215 m (700 ft) were performed automatically. The two astronauts could view the landing site from the end of the braking phase to touchdown. If the terrain appeared hazardous at this point (3,000 m) new data could be fed into the guidance system to select a more suitable landing site.

The landing phase was usually conducted manually by the Lunar Module pilot using attitude and thrust controls in order to avoid obstacles and land the spacecraft safely. Two minutes of fuel were allowed for this last phase; and wire probes cut the engines automatically on contact with the lunar surface, allowing the LM to fall slowly the last 1.5 m (5 ft) to a soft landing. The astronauts could elect to abort

at any stage, including an unstable surface landing, by separating from the Descent stage and firing the Ascent stage engine. If the landing proved safe, the lunar surface could then be explored. The astronauts carried individual backpacks providing life-support systems. A Lunar Rover was developed by Boeing, and first landed in 1971. It provided a range of 92 km (57 mi). The scientific experiments carried by Apollo were designated the Apollo Lunar Surface Experiments Package (ALSEP) and consisted of a SNAP 27 nuclear generator, a central (communicating) station, solar wind spectrometer, passive seismic experiment, lunar surface magnetometer, and suprathermal ion detector experiment. All ALSEP experiments were finally shut down in 1977.

Following lunar surface exploration, the astronauts initiated a powered ascent from the Moon's surface using the Descent stage as a launch platform. The ascent was synchronised with the approach of the third astronaut, in the command module over the lunar horizon, to ensure a safe docking and return to Earth.

I R Murphy

Below: In 1971, Boeing built a lunar rover for the Apollo programme; in this picture it is seen deployed on the Moon's surface. It was first used by Apollo 15's David Scott and James Irwin in July and August 1971; it was used twice more in 1972.

Star Wars and space weapons

Ever since the first Sputnik reached orbit on 4 October 1957 the attention of military planners has turned towards space. Since the space age began after the full horrors of war were realised there has been much resistance to the militarisation of space, but this resistance is crumbling in the face of enormous pressure from the superpowers not wishing to fall behind in advanced technology. The Soviet Union continually denies that it has a military space programme, but if NATO intelligence is at all reliable this is far from the truth.

The initial military use of space involved reconnaissance satellites used to 'spy' on a potential enemy's territory to gain intelligence data. This has been accepted as a tolerable military use of space, but the placing of nuclear weapons in space is banned by international treaty. This has led to the proliferation of intercontinental ballistic missiles (ICBMs) in the USA and USSR. Until ICBMs were developed foreign attack on the continental USA was virtually impossible. Aggression via land, sea or air could be detected at great distance. But ICBMs gave the USSR and USA the mutual ability to make a nuclear strike in minutes. For many years both nations have sought an effective method of defence against ICBMs and this could be at hand in the USA's Strategic Defence Initiative (SDI) and a similar programme that is undoubtedly being pursued in the Soviet Union.

The initial part of an ICBM's flight takes the weapon just clear of the atmosphere. It is easiest to destroy it during this part of its flight by a number of methods, including lasers, particle beam weapons and ballistic (hypervelocity) weapons. Many people doubt the feasibility of SDI because of the enormous development problems faced. Detecting and dealing with many ICBMs launched at once presents tremendous problems in sensor technology and computing. The power of lasers is being increased continuously, but the most powerful ones today require a plant the size of a small factory to run them. The lasers could be ground based, but the Earth's atmosphere distorts and defocuses the beam. A laser depends on the concentration of energy on a small area for its destructive power. Similar problems are faced by particle beam weapons. If the particles are charged then the beam is bent by the Earth's magnetic field. Research is taking place to develop uncharged particle beam weapons.

The USA has started a US$26,000 million research programme to see if SDI or Star Wars technology is possible. The research is highly classified, but it encompasses many fields such as the use of lasers for very high-speed computing. The speed at which electrical signals travel through a wire is now embarrassingly slow, so attempts are being made to link the electronic components of a computer by laser beams.

Vulnerable satellites

Satellites in orbit are extremely vulnerable. There is nowhere to hide them and detection is easy. It is logical that SDI satellites will become prime targets in a conflict and would have to be made difficult to detect by radar, for example, and the electronics will need to be resistant to the electromagnetic pulses caused by nuclear explosions.

The SDI system is intended to be used for negating ICBMs during any phase of flight. The idea is that ICBMs will thus be made virtually useless and this would lead to a reduction in their numbers or perhaps their elimination, removing the threat of nuclear destruction from most of the world. If the Soviet Union develops a similar system then the balance of power will be maintained. But

Sadly space has already been militarized, and in fact there is little doubt that it always will be, especially with the American Strategic Defense Initiative (SDI) or Star Wars programme, also called the 'High Frontier', and similar Soviet programmes.

the Soviet Union may have the hardware to negate the USA's satellites by putting men into geostationary orbit where most of the satellites will be. Space Shuttle cannot get into geostationary orbit but the USSR is developing a Saturn 5 class launcher that probably can. It is ostensibly for a manned mission to Mars but who in the West really knows?

Ian Parker

Satellites placed in orbit around the Earth are vulnerable to aggressive acts as well as natural damage and malfunction. A future war could start with a battle to destroy satellites and so black out communications and prevent reconnaissance.

Moon landings meant that little thought was given to future projects. After Apollo, NASA flew Skylab, which was developed using leftover Apollo components. The main lab was a converted Saturn 5 third stage, and spare command and service modules were used as ferry vehicles. The Skylab programme began to uncover the promise of manufacturing processes in orbit, but the project ended in 1974 after three flights because there were no more ferry vehicles. After the 1975 Apollo-Soyuz docking in orbit, Americans did not enter space again until 1981.

Shuttle

For many years it had been realised that to use a rocket once only was a large waste of hardware and resources. Both the Americans and Soviets wanted a reusable spacecraft, but it was not until 1972 that Space Shuttle received the go-ahead. By that time, the US space programme had fallen out of favour with many in the US government and there were questions about the need and cost of putting men in space. As a result Shuttle funds were small in comparison with Apollo expenditure and the vehicle has turned out to be very much a cut-price version of what could have been built, albeit a very successful one. The low funding meant that Shuttle could not be totally reusable.

The Space Transportation System (STS), as it is officially called, consists of four major

Shuttle's basic problem is that it has been designed to go into low Earth orbit, and so satellites have been launched on from the cargo bay, but this has not always worked. Shuttle does, however, allow for repair work to be carried out.

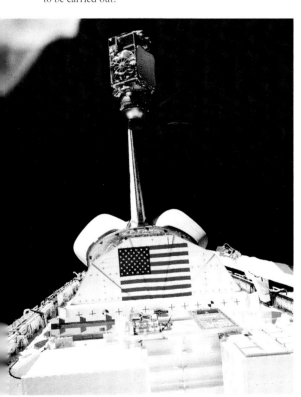

Shuttle's advantage is that it can land conventionally after Earth orbit, albeit at specially prepared airfields like Edwards in California or back at the Cape in Florida. This makes it re-useable and thus more cost-effective.

components: the orbiter, external tank (ET) and two solid fuel rocket boosters (SRBs). On the launch pad Shuttle weighs about 2,041,186 kg (4,500,000 lb) and the engines produce a total thrust in excess of 2,721,582 kg (6,000,000 lb). The orbiter is 36.6 m (120 ft) long and has a wingspan of 23.8 m (78 ft) and can carry up to seven crew members – two pilots, two mission specialists and three payload specialists.

The 46.9 m (154 ft) long ET holds liquid oxygen and liquid hydrogen to fuel the orbiter's three Space Shuttle Main Engines (SSMEs). The 45.4 m (149 ft) long SRBs are strapped on the side of the ET. A computer failure delayed the first flight of Space Shuttle by one day, but on 12 April 1981, 20 years to the day since the first manned spaceflight, STS-1 left the launch pad with John Young and Bob Crippen aboard.

At T+2 min 11 sec the SRBs were discarded and they landed in the Atlantic at T+7 min 13 sec. After SSME shutdown the main tank was detached and it burned up on re-entry as planned. Two days of in-orbit tests followed and after 36 orbits the greatest test of orbiter Columbia began – re-entry. Space Shuttle is protected from the heat of re-entry by thermal tiles; these had given considerable trouble during development, one of the main problems being detachment. Camera observations in orbit showed that a number of tiles had come off from non-critical areas on the upper surface. Photographs of the underside taken from a KH-11 reconnaissance satellite showed the tiles there to be complete.

Re-entry

Some 277 km (172 mi) above the Indian Ocean, Columbia fired its OMS engines to bring the spacecraft down. The orbiter went into communications blackout at T+53 hr 51 min as the searing heat of re-entry blocked radio communications and caused the critical leading edge and nose tile areas to glow orange at 1,482°C (2,700°F). Columbia touched down on a dry lake bed at Edwards Air Force Base in California, landing at 346 km/h (215 mph).

Shuttle flights are now commonplace, performing such tasks as launching satellites, flying Spacelab and engaging in numerous in-orbit experiments. Although Shuttle has been extremely reliable, some of its payloads have not and this has sadly reflected on the STS as a whole. Many of the satellites launched need to reach geostationary orbit out of reach of Shuttle, which cannot go higher than a few hundred miles. To place such satellites in their desired orbit an upper stage has to be used. This consists of one or more rocket motors attached to the satellite. The assembly is deployed from Shuttle, which then moves a safe distance away before the upper stage is ignited.

Unfortunately, there have been a few upper stage failures, some of which have resulted in satellites being lost, but others have ended in remarkable space salvage missions, which have added a new dimension to Shuttle's proven capabilities. Upper stages are usually supplied by the customer and may often be nothing to do with NASA. Two upper stages failed on a flight of Discovery in February 1984, leaving their satellites stranded in useless orbits. The insurance payouts on these two failures were so high that the underwriters started a campaign aimed at NASA to mount a rescue mission for Palapa B2 and Westar 6.

The following November, another Shuttle flight rescued the satellites and returned them to Earth for repair and re-sale. This demonstrated a remarkable Shuttle capability which is likely to be of immense value, allowing satellites to be refuelled, repaired and refurbished either in orbit or on Earth. To exploit this potential fully another spacecraft is needed – the Orbital Transfer Vehicle (OTV). Plans for such a vehicle are now in hand and it will move satellites between their operational orbits and that of Shuttle. However, the programme received a set-back with the catastrophic accident to Challenger in January 1986.

Space Station

Shuttle's apparent incompatibility with satellite launching duties stems from the fact that it was not really designed for such uses. It was built to launch a space station section by section. Had the funding been available it is likely that Space Shuttle and Space Station would have been developed in parallel. But it was not until 1982 that NASA's Space Station planning began in earnest. The unique aspect of Space Station is that it will be permanently manned, unlike the earlier Skylab and the USSR's continuing Salyut series. It will also be an international project. In his January 1984 State of the Union message, President Reagan said: 'We want our friends to help us meet these challenges and share in the benefits. NASA will invite other countries to participate so we can strengthen peace, build prosperity

The next step in space exploration could be the creation of orbiting space stations, as seen in this artist's impression.

and expand freedom for all who share our goals.'

The initial cost of Space Station is put at US$8,000 million with operation starting in 1992. Many questions about the design and configuration of Space Station have yet to be answered, but it is highly likely that it will not look like the traditional cartwheel space station popular with science fiction artists. The spinning cartwheel design would produce artificial gravity and one of the main reasons for going into orbit is to get away from gravity.

Space Station is much more likely to have a scaffolding-like appearance with habitable and experiment modules at one end, supplies and communications equipment in the middle and the power collection and conversion system at the other end. The main uses for Space Station will be microgravity research (zero gravity is never quite achieved in orbit because of the Earth's distorted gravitational field), biological tests, communications, space physics, astronomy and remote sensing. Not all of the Space Station components will be attached to the main structure. Some of the microgravity processes need to be free of the general vibration and disturbance of a manned system. They will probably be housed in free-flying platforms that will be tended by people, requiring an orbital manoeuvring vehicle.

Space Station will be in low Earth orbit, probably at an inclination of 28.5 degrees. This is the easiest orbit to reach from the Kennedy Space Center, but it is not suitable for remote-sensing satellites. In general, they have to be put into polar orbits. Launching to the east near the equator allows the spacecraft to make use of the rotation of the Earth, the surface of which is travelling at about 1,600 km/h (1,000 mph) at the equator. A launch into polar orbit cannot make use of the Earth's rotation and it therefore requires more energy. Polar orbit missions will be flown by Space Shuttle from Vandenberg Air Force Base in California. Flights will head south from the launch pad over the Pacific.

By carefully selecting the height of a polar orbit it is possible for the sensors of a remote-sensing satellite to cover overlapping strips of land as the Earth turns below it.

Future launchers

It had been thought that Space Shuttle would replace all of the existing expendable launch vehicles (ELVs). But Shuttle launch frequency and cargo space availability are well behind plan and ELVs are thriving. Europe's Ariane is the best example, but older US launchers such as Atlas Centaur and Thor Delta are being taken on by private operators. Most of these ventures are looking mainly at the comsat launch market and some new companies are proposing new ELVs for specific

Europe's rival to Shuttle is the so-far unmanned Ariane rocket programme. Of French design, the rocket is launched from South America with commercial payloads. There are plans for a European manned-vehicle, called *Hermes*, by 1995.

duties. One even intends to launch cremation ashes in shiny containers so that loved ones can go out on a clear night to see their relatives in orbit. But ELVs are out-dated and new launcher plans are looking towards full reusability.

NASA is considering a Super Shuttle and the Soviet Union is known to be flying a spaceplane and is thought to have plans for its own shuttle. But the ultimate aim is a single-stage-to-orbit reusable spacecraft. The hardware for such a vehicle does not yet exist (as far as we know), but its attractions are well documented. The best known plans are those announced by British Aerospace and Rolls-Royce for their Horizontal Take-Off and Land (HOTOL) spacecraft. One of the problems facing a single-stage-to-orbit launcher is carrying sufficient fuel. A large proportion of the fuel in a conventional rocket is the oxygen element – the oxidant. In a cryogenic engine, this is in the form of liquid oxygen. Of course, oxygen is freely available in the atmosphere and the HOTOL concept hinges on an airbreathing engine which liquefies oxygen from the atmosphere during the early part of the flight.

HOTOL would take off along a runway, probably leaving behind its undercarriage, and climb to a given height, breathing air from the atmosphere. It would then switch to internal oxygen to achieve orbit. The power-plant is at present highly classified.

HOTOL would probably be unmanned and its re-entry mass could be so low that Shuttle-type thermal protection tiles would not be necessary. Advanced metals would be sufficient. There are some doubts about the feasibility of the project, but the main stumbling block seems to be development costs. The United Kingdom has a history of having good ideas but lacking the confidence to develop them. The European Space Agency (ESA) may be unwilling to take on HOTOL because it is committed to the development of Ariane 5. However, the US Air Force is interested in a transatmospheric vehicle, which is essentially

the same thing. It seems that a single-stage-to-orbit launcher will come sooner or later. The question is, where and when?

Unmanned probes

While much media attention has traditionally focused on launchers and manned missions, unmanned science probes continue extremely valuable work. Some such as Space Telescope remain in orbit while others are sent into deep space, encountering the planets and eventually leaving the Solar System to wander, probably forever, among the stars.

NASA's Galileo Jupiter probe should be the first spacecraft to enter the atmosphere of an outer planet when it encounters the giant in 1988. The probe will be detached from the orbiter about five months before that planet is reached. The two components will follow parallel courses until the probe enters Jupiter's atmosphere at about 185,000 km/hr

(115,000 mph), experiencing a deceleration peaking at 350g. A drogue parachute will deploy 4 minutes after entry, when the probe will be at Mach 0.9. The deceleration module and heat shield will fall away, leaving the probe with its seven experiments.

They will determine the atmosphere's chemical composition, transmitting data back to the probe 200,000 km (124,300 mi) above. The probe is expected to survive for an hour after entry, descending 125 km (78 mi) into Jupiter's crushing atmosphere. The orbiter will then fly close to the natural satellite Io and then begin a 20-month 11-orbit mission phase taking 50,000 high-resolution photographs of Jupiter and its moons.

Voyagers

NASA's Voyagers continue their exploration of the Solar System. Voyager 1 flew past Jupiter in March 1979 and Saturn in November 1980. It is now heading out of the Solar System, observing the solar wind. Voyager 2 passed Jupiter in July 1979, Saturn in August 1981 and Uranus in January 1986 and it should encounter Neptune in August 1989, before following its predecessor to the stars.

Other nations' launchers

The USA and USSR have dominated space for more than 20 years, but other nations are now developing their own space programmes as the full value of such developments becomes clear. Besides the European Space Agency, China, India and Japan are continuing the development of launchers with an eye on the international satellite launch market. China has recently announced its intention to offer the Long March series of launch vehicles, Japan has the N-1, N-2 and H-1 launchers and is working on the H-2. India has the SLV and the GSLV is on the way.

Launcher and satellite reliability is extremely important because of the enormous cost involved. Satellite manufacturers, buyers and operators are therefore very choosy about which vehicle launches their satellites and how their satellites will be handled by the launcher operator. Offering a low launch price is therefore not a sure way of winning a launch contract. The satellite owner has to be sure that its investment is delivered safely into the correct orbit in perfect working order. Assuring owners that this will happen will be one of the main problems faced by countries such as China and the Soviet Union if they go for the commercial market in a big way.

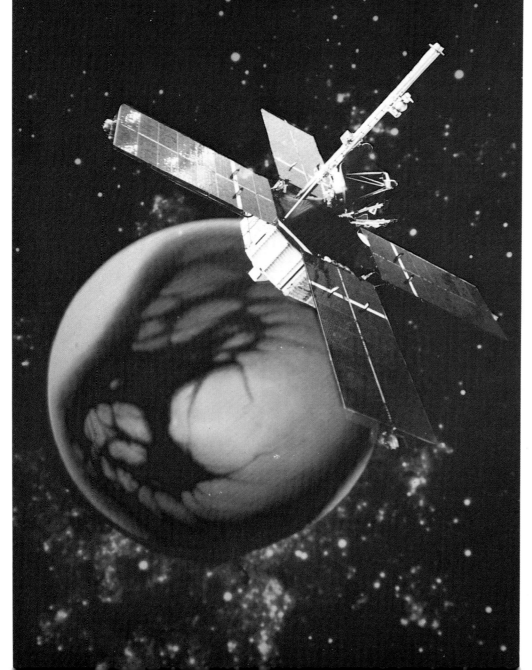

Despite a downgrading of the American space effort after the Apollo programme, the late 1970s saw the Voyager series launches which have flown by outer planets, bringing back the first close-up views of their rings, surfaces and atmospheres.

Launcher specifications

Type	Country	Height	Max Payload	Fuels	Orbit height at max payload
Shuttle	USA	55 m (180 ft)	29,484 kg (65,000 lb)	Solid LOX/LH$_2$	185-1,110 km (115-690 mi)
D-1	USSR	31 m (102 ft)	22,000 kg (48,501 lb)		200 km (124 mi)
Ariane 1	Europe	37.4 m (123 ft)	2,750 kg (6,063 lb)	N$_2$O$_4$-UDMH LOX/LH$_2$	550 km (342 mi)
SLV-3	India	20.2 m (66 ft)	40 kg (88 lb)	Solid	308 - 986 km (191-613 mi)
N-2	Japan	30.2 m (99 ft)	1,110 kg (2,447 lb)	LOX/RJ-1 Solid N$_2$O$_4$/A-50	1,000 km (621 mi)

LOX – liquid oxygen   LH$_2$ – liquid hydrogen   UDMH – unsymmetrical dimethylhydrazine

European Space Agency

In 1975 the European Space Research Organisation (ESRO) and European Launcher Development Organisation (ELDO) merged to form the European Space Agency (ESA), which now has 11 member states. These comprise Belgium, Denmark, Federal Germany, France, Italy, the Netherlands, Spain, Sweden, Switzerland, and the United Kingdom. Canada has an associate membership.

One of ESA's largest and most successful projects has been the development of the Ariane series of expendable launch vehicles (ELVs). This endeavour was started in 1973 and was adopted by ESA on its formation. Ariane was developed at the same time as Space Shuttle and the latter was thought by some to spell the end for ELVs. But to date this has not proved to be the case and Arianespace, the company which operates the launcher, has firm launch contracts on some 32 satellites and reservations on about 16 more. Around 10 of these have already been put in orbit.

Spacelab

A major project transferred from ESRO to ESA is Spacelab, which is a modular space research laboratory flown on Space Shuttle. Combinations of unmanned pallets and manned segments can be assembled in Shuttle's payload bay to conduct research into such things as materials processing, biology, space physics and remote sensing.

In January 1985, ESA's council met in Rome to approve the next series of projects, which will last to the end of the century. A budget increase of 70 to 80 per cent is expected between 1985 and 1990. ESA budgets are calculated in MAU

The European space probe Giotto. Giotto's most remarkable achievement was flying within a few hundred kilometres off Halley's Comet in December 1985 and transmitting live television pictures of the comet's core.

(million accounting units) and an accounting unit is equivalent to about US$0.80. It is set annually on the previous June's exchange rate. The 1985 ESA budget was 950 MAU and the 1990 budget is expected to be 1,700 MAU.

ESA's main projects to the year 2000 will include the development of one or more elements to be attached to NASA's Space Station. This element will be called Columbus and could form the basis for an independent European space station. To service it, ESA will need a manned flight capability and it is likely that the agency will adopt the French space agency's plans for an engineless mini-Shuttle-type craft called Hermes. It would be launched by Ariane 5, now under development by ESA along with the HM60 engine needed to power the new rocket.

Soviet offer

The Soviet Union is a member of the International Maritime Satellite Organisation (Inmarsat) and as such has the right to make bids on Inmarsat satellite launch contracts. Such a bid was made in 1985 and because it was the lowest bid at the time Inmarsat felt under considerable pressure to accept it. The USSR would use its Proton launcher, which has been flying since the 1960s and is therefore well tried and trusted. But the success of such bids will depend in part on a very open attitude towards customers which the Soviet Union is not in the habit of displaying. Owners may want free access to the satellite right up until launch and an assurance that it will not be tampered with.

Amongst the alternative launchers now available are India's SLV-3 (*far left*) and Japan's N series (*left*) of commercial launchers. In addition, the Soviet Union and China (PRC) are trying to tap the lucrative satellite launcher market which so far has been a preserve of the Ariane, Shuttle and some non-NASA American systems. There is no doubt that the future of commercial space exploitation is rosy despite the sad accident to the Challenger in early 1986. The American Shuttle programme alone could not cope with the forecasted demand for placing vehicles into orbit during the next two decades. Commercial space remains an important market although most nations give priority to their military programmes.

7
GENERAL AVIATION

'General aviation' is the term for non-military, non-airline flying, and it covers the microlight, balloon, private aircraft, club aeroplane, glider and business jets. In the 1980s there have been considerable technological developments in this section of the aviation industry, in some areas putting it at the leading edge of research, together with military aircraft.

Learning to fly is an important aspect of general aviation, whether in a powered aircraft or glider, and is becoming more popular around the world through flying clubs.

General aviation (GA) is a term which covers all aircraft other than military and air-transport types. Everything from business jets to hang-gliders is included and the upper size limit is often put at 19 seats. This section of the aviation industry has experienced a severe slump in recent years. In the United States, there is an organisation called the General Aviation Manufacturers Association (GAMA), which, among other things, monitors the performance of its members. About 90 per cent of the world's general aviation is in America, so GAMA figures give a good idea of the state of GA worldwide.

Towards the end of the 1970s, GAMA members were selling some 17,000 aircraft a year, but this turned out to be a peak period and by 1980 the total was down to about 8,800. Since then the decline has continued and GAMA members are now selling about 3,000 aircraft a year. Such a decline in sales has obviously prompted GA manufacturers to look for reasons, but there does not seem to be a single main cause. The poor market has probably been caused by a number of factors.

In addition to the worldwide recession, problems have been caused by the US dollar increasing in value compared with other currencies. As the dollar strengthens it makes US goods more expensive in other countries. This has been a problem for many industries but it is particularly damaging to aircraft sales abroad. Aircraft are expensive items to begin with. Even a light aircraft can cost several

'General aviation' is a wide-ranging term covering most types of non-commercial flying, including flying vintage types like the Percival Gull (*far left*) and businessmen's aircraft, such as the Beech Bonanza, popular in North America (*bottom far left*), with its characteristic V-tail. While enthusiasts still perform 'barnstorming' feats in DH Tiger Moths, like 'wing walking' (*above*), the air taxi and business jet have become a larger and larger part of general aviation. Among the most popular business jets is the HS 125 (*left*), manufactured by British Aerospace to a design by de Havilland Aircraft, and the type has sold widely in the United States, Middle East and even to military customers like the Royal Air Force and the South African Air Force. The business jet of today represents a large investment of high technology and several companies have small fleets of aircraft. Some can now fly from Chicago to Geneva without having to land to refuel.

hundred thousand dollars. If the dollar strengthens against the pound sterling by say 5 per cent, the price of a US-built aircraft in the UK could go up by £10,000 overnight. The customer for such an aircraft may be a flying club or a private operator and a sudden price increase of this magnitude may well prevent a sale being made.

Another possible cause for the falling sales figures is market saturation. It may be that there are simply enough aircraft already made to meet current demands. Many industries go through a boom phase while new customers are being found in great numbers. Later the main sales are made in the replacement market. The car industry does well out of this by building in obsolescence. But aircraft have to be as good in ten years' time as the day they were built. This means that the replacement market is small.

A criticism often made of GA manufacturers is that they are very conservative. The light aircraft being built today are essentially the same as those that were being produced 40 years ago. There has been very little development at the small end of the traditional GA market. But progress has been made at the extremities of the GA spectrum. Microlight and ultralight aircraft have appeared, having grown out of powered hang-gliders, and business aircraft manufacturers are developing types employing advanced materials, such as carbonfibre reinforced plastic (CFRP), and advanced aerodynamic designs such as pusher-engined canards.

High technology reaches GA

For many years it has been realised that metals are not ideal substances from which to make aircraft. Metals corrode, they cannot easily be formed into complex shapes, they are heavy and they need complex and time-consuming fastening systems. Plastics would appear to be far better in all these respects. Why then are there no plastic aircraft? Well, soon there will be.

GA is usually slow to respond to new technology, but materials development in GA is pushing ahead. William Lear was the inventor of the business jet and the company that he founded, Learjet, is still producing designs based on his original concept. But

shortly before his death in 1977 he conceived perhaps the most advanced concept yet seen in GA – Lear Fan. This aircraft is 90 per cent CFRP. Two turbine engines are mounted at the rear and they drive a single rear-mounted propeller via a combiner gearbox. The project has met with considerable financial and technical difficulties and it is uncertain whether the aircraft will ever enter volume production. But it has spurred other GA manufacturers to start similar projects. The US company Gates, which owns Learjet, has teamed with Piaggio of Italy to produce the GP-180 Avanti; Beech is developing the Starship; Avtek is preparing to produce its model 400; and Omac is well advanced with the Omac I. All of these aircraft have taken a leaf out of Lear Fan's book. Some are making use of advanced plastic materials and all have rear-facing pusher engines.

Why plastic?

Early aircraft were made of wood and canvas or some other covering material. This technique of construction is still in use today for some light and homebuilt aircraft and it is hard to beat for small, slow aircraft. But such materials cannot take the stresses felt by high-performance aircraft and so wood gave way to metal as aviation advanced. It is worth remembering, however, that one of the most successful aircraft of the Second World War – the de Havilland Mosquito – was essentially a wood and canvas design.

Aluminium alloy became the most com-

Composite technology for new airframes and structures as well as high performance and reliable engines have led to the development of large, single-engined aircraft like this Cessna Caravan. Several fleet orders were placed in 1985.

Top left: 'general aviation' also includes the simple, rugged and purpose-built agricultural aircraft, so necessary for developing countries. This the Brazilian Ipanema which entered production in 1972; 500 are now in service.

Left: high technology has reached the general aviation scene as demonstrated by the new Beechcraft Starship 1 of futuristic design. This is a scaled-down development airframe; the first full-size aircraft flew in early 1986.

G-BGMW

The Optica is a remarkable design with many of the characteristics of the helicopter, yet it is more cost-effective for certain roles, especially those which require a long endurance but do not require VTOL performance.

Below: early pioneers debated and experimented with control surface designs as in this Short Brothers 'Bird of Paradise' aeroplane design. A constant problem is allowing for centre of gravity constraints, which could lead to a nose-down attitude. Recently, coincidentally with composite technology, changes in conventional design thought have led to the re-introduction of 'tail in front' designs. A new generation of general aviation aircraft, like the Beech Starship 1 and the Avtek, will benefit from this new technology.

mon material from which aircraft were built and even now it is still dominant. It is strong and light and can be machined easily and fastened with rivets, but it does have its limitations. Even aluminium alloy corrodes and it cannot be formed into the more complex shapes that aerodynamic specialists would like. Although light for a metal, it is heavy compared with plastics.

Many people have realised for years that plastics are very attractive materials from which to build aircraft. Plastic can be moulded, it does not corrode, and it can be tailored to have different properties in different directions. For example, a component can be made stiffer in one direction than another.

Aircraft designers have simply been waiting for the right plastic and the right method of manufacture to come along. The two materials now finding increasing use in aviation are carbonfibre reinforced plastic (CFRP) and glassfibre reinforced plastic (GRP). It is difficult to apply mass-production techniques to aircraft made of CFRP or GRP because they literally have to be glued together. But advances have been made in producing very large components to cut down the number of joints. Both CFRP and GRP are composite materials in which fibres are impregnated with a resin. For maximum strength the material is heated in a large oven called an autoclave. The size of components is limited by the size of the autoclave and those manufacturers producing composite business aircraft are investing in

large autoclaves, some big enough to take entire fuselage halves.

Development work is often by hand-layup of composite materials, but when line production starts, automatic techniques will be needed and they are already under development. But the use of composite materials for the primary structure of any aircraft faces one major hurdle – certification.

Any new aircraft has to be tested by the aviation authority in the country of manufacture and possibly by others if the aircraft is to be registered abroad. These certificating authorities examine every stage of development and manufacture of a new aircraft and they enforce very strict standards. The authorities have a great deal of experience in examining new designs in metal but they have little experience in dealing with composite materials and are naturally very wary. They are enforcing more severe tests on composite aircraft and this is causing the designers some problems.

For example, the Lear Fan was subjected to a severe pressurisation test and the fuselage cracked, necessitating a major redesign. The development of these new composite aircraft is very expensive in any case and such redesign work can pose a real threat to the viability of projects.

Another unknown factor is the crash-worthiness of composite aircraft. Metal aircraft crumple during crashes and if the impact is not too great this crumpling can save the occupants. Composite materials tend to shat-

J.T.C MOORE BRAB

The BIRD OF

AERO CLUB of UNITED KINGDOM

ter, so they do not have the same energy absorption qualities. Although composites do not corrode, they do suffer some problems in humid conditions and in bright sunlight, and can deteriorate. But the only way to answer these questions is by extensive testing followed by operational experience.

Tails at the front

In addition to the use of composite materials, the new business turboprops are employing novel configurations, in particular canard layouts. A canard aircraft is one that has the horizontal stabiliser at the front. This configuration is by no means new; some of the earliest aircraft were canards, the term coming from the French word for duck, the duck having a similar shape.

All aircraft need to be stabilised in both the vertical and horizontal plane. The fin and rudder stabilises the tendency to yaw and normally a tailplane stabilises pitching tendencies. To make an aircraft stable in pitch it is necessary to keep the centre of gravity (cg) ahead of the centre of lift (cl). This gives the aircraft the tendency to pitch nose-down, reducing the angle of attack and making the aircraft unwilling to enter an aerodynamic stall. To resist the nose-down moment a tailplane is normally used and most of the time it applies a downward force on the rear of the aircraft. This balances the nose-down tendency, but obviously any flying surface which produces a download works against the wings, which are trying to support the aircraft. If the tailplane is moved to the front to form a canard it then has to produce lift to stabilise the aircraft, resisting the nose-down trim. There are disadvantages to canard designs. For example, the wash from the canard can strike the main plane and the canard can obscure the pilot's view. But more and more designers are looking at canards for general aviation aircraft. The aircraft industry is a conservative one, however, so changes take time to come about.

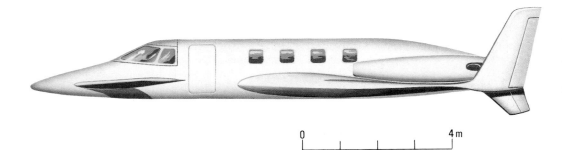

Beechcraft 2000 Starship 1, c.1986 – an example of the new pusher propeller trend.

Below: France's aerospace industry has successfully sold the Falcon series of business jets into the American market on the strength of modern technology with conventional layout. The aircraft, however, faces serious opposition.

BEECH 2000 STARSHIP 1

Country of origin: USA.
Role: Business turboprop.
Wing span: 11.99 m (39.33 ft).
Length: 12.37 m (40.58 ft).
Max weight: 3,566 kg (7,400 lb).
Engines: 2 × PWAC PT6B-35F, 634 kW (850 shp).
Max speed: 667 km/h (414 mph).
Range: 3,224 km (2,003 mi).
Crew: 1 pilot.
Passengers: Up to 9.

Pushers

Most multi-engine propeller-driven aircraft have the engines mounted on the wing with the airscrew at the front. Among the disadvantages that follow from this arrangement are the propeller wash passing over the wing and the propeller tips rotating close to the cabin, making it noisy inside. Several of the new business turboprops have the engines facing backwards with the propellers pushing. This leaves the wing flying in clean air and moves the noise well away from the passengers. These new configurations and materials are producing turboprops which can cruise at 740 km/h (400 kt) and 15,240 m (50,000 ft) – previously the province of jets only.

Bill Lear is credited with inventing the business jet and the Learjet has become probably the best known example. Its original design was based on a military aircraft and since it became available other manufacturers have produced similar aircraft. Cessna builds the Citation series, Dassault produces the Falcon series, and Mitsubishi manufactures the Diamond I and II. Gulfstream

Above right: the Gates Learjet was for many years synonymous with business flying around the world; this is the Lear 55 which has a range of 4,249 km (2,640 mi). The aircraft's performance has been likened to that of a jet fighter in expert hands.

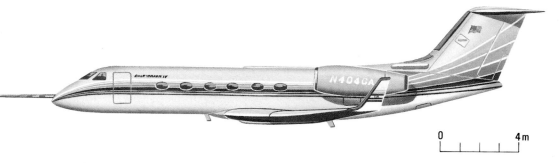

Gulfstream IV first prototype, c.1985; this aircraft is designed to fly business personnel between continents, not just within them.

GATES LEARJET 25

Country of origin: USA.
Role: Business jet.
Wing span: 10.84 m (35.58 ft).
Length: 13.18 m (43.25 ft).
Max weight: 6,124 kg (13,500 lb).
Engines: 2 × GE CJ610-6 turbojets (Model 24C), 1,340 kg (2,950 lb).
Max speed: 877 km/h (545 mph).
Range: 3,250 km (2,020 mi).
Crew: 1 pilot, 1 co-pilot.
Passengers: Up to 6.

builds business jets (the GII, GIII and GIV), as does Canadair (the Challenger). All these aircraft cost millions of dollars and people outside the business aircraft industry often wonder how the cost can be justified.

A business aircraft is often viewed as 'a director's toy'. It is extremely difficult to quantify the value of a business aircraft but many of the world's major companies operate them very successfully, some having aviation departments which not only serve the needs of the employees but fly for other companies as well. These aviation departments often become companies in their own right and may even take on the dealership of a particular manufacturer.

The air-transport industry is not set up to meet the needs of business travellers. Most of its customers are holidaymakers who usually know well in advance when they want to travel and what their destination will be. Business travellers, on the other hand, often have to make journeys at short notice to any destination. In the USA, there are thousands of airports, but only a few hundred are served by the airlines and this number is falling because of deregulation.

Previously, the US government regulated the route structure of the airline industry. Airlines wishing to fly a particular route had to apply for it and non-profitable routes were handed out with the profitable ones. But the system has been deregulated so that any airline can fly any route. This means that the profitable routes are strongly competed for and the unprofitable ones have been dropped

completely. In theory, this should increase the need for business aircraft, but this is only possible if airports remain open. The loss of air transport services to some US airports has meant that there have been substantial closures.

Industry campaigns

The business aircraft industry continues to have to run campaigns to stress its importance. The industry has a tougher time in Europe than in the USA. There is continuing pressure to exclude GA aircraft from major airports to relieve congestion. But one of the most important uses of GA aircraft is to connect with scheduled flights. This will not be possible if they cannot land at the airport.

Smaller GA aircraft tend to be used for training and pleasure rather than business flying, although some people use light single-

CESSNA 150 SERIES

Country of origin: USA.
Role: Trainer and tourer.
Wing span: 9.97 m (32.71 ft).
Length: 7.34 m (24.08 ft).
Max weight: 760 kg (1,675 lb).
Engine: 1 × Avco Lycoming 0-235-N2C, 80 kW (108 hp) Model 152.
Max speed: 202 km/h (125 mph).
Range: 583 km (362 mi).
Crew: 1 pilot.
Passengers: 1.

engine types as self-fly business aircraft. This is more common in the USA than other countries partly because of the size of the country and partly because of the better weather. In Europe, the weather is a limiting factor for many light aircraft pilots.

To fly safely in bad weather the aircraft must carry certain equipment and the pilot must maintain a high standard of flying skill. In particular he must gain an instrument qualification and remain proficient by practising. Most flying schools can offer instrument training.

How an aircraft flies

Whichever aircraft a pilot ends up flying it is likely that he will learn the principles of flight on a GA aircraft. Many armed forces start their pilots off on light single-engine aircraft for basic flying training.

The wing of an aircraft produces lift because the air flowing over the top is travelling faster than the air flowing underneath. This is caused by the shape of the wing's cross-section. There are many different types of wing cross-section and design and development of them is very complicated. But the cross-section of wings on GA aircraft are all quite similar, having nearly flat undersides but cambered upper surfaces. If two molecules of air strike the leading edge together and one goes over the top and the other goes underneath, the particles must meet again at the trailing edge of the wing otherwise the aircraft would leave a continuous area of low pressure

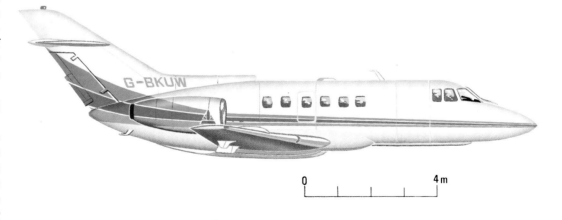

The BAe 125-800 of British Aerospace, c.1985 (*above*), is powered by two Garrett turbofans and is one of the most modern in its class, selling well worldwide. The Piper Apache, c.1960 (*left*), shown in French civil markings, is one of the oldest business aircraft flying, yet in its day it was a world-beater. However, since its design engine technology and airframe design have improved dramatically. Older designs are no longer proving cost-effective.

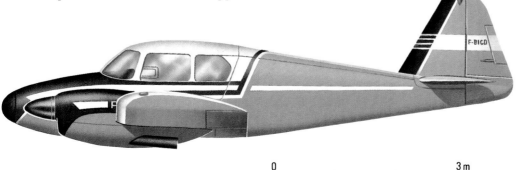

behind the wing. Nature abhors a vacuum, so the molecule going over the top of the wing is forced to travel faster than the one going underneath because it has farther to go. It has to get the energy from somewhere to travel faster. Pressure is a form of stored energy and the molecules travelling over the top of the wing use this energy to increase speed, so the pressure drops. Thus an aircraft wing is sucked into the air.

This process, called dynamic lift, produces most of the lift, but additional lift is provided by the angle of attack (AoA) of the wing. Looking at the wing cross-section, if a line is drawn from the leading edge to the trailing edge, and another line shows the direction of motion, the angle between them is the AoA. It is a variation of the AoA that allows the aircraft to be controlled.

When an aircraft accelerates down a runway the AoA is small or zero. As it approaches its take-off speed it will nearly have enough dynamic lift to leave the ground. The pilot then raises the nose, increasing the AoA. This increases the pressure under the wing, contributing to the total lift and getting the aircraft off the ground. The pilot controls the AoA with the elevators on the tailplane. The pilot has a control stick or yoke that is connected to the elevators and other control surfaces. By pulling the stick back, the pilot raises the elevators. This increases the pressure above the tailplane, pushing the rear of the aircraft down and increasing the AoA.

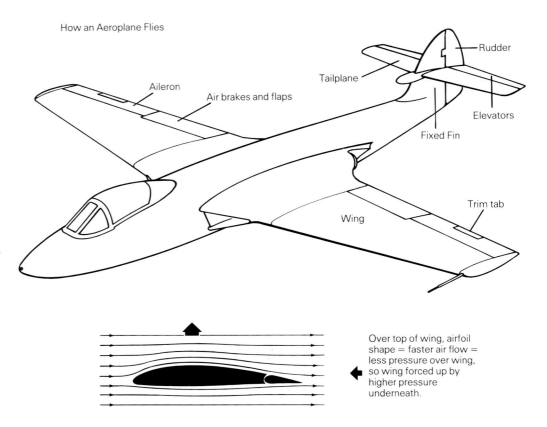

How an Aeroplane Flies

Rudder · Tailplane · Aileron · Air brakes and flaps · Elevators · Fixed Fin · Trim tab · Wing

Over top of wing, airfoil shape = faster air flow = less pressure over wing, so wing forced up by higher pressure underneath.

Roll control

The stick is also connected to control surfaces on the trailing edge of the wing near the tips. These are called ailerons. When the pilot moves the stick to the right, the right aileron moves up and the left aileron moves down. This increases the pressure above the right wing and below the left wing so the aircraft rolls to the right. However, if lift is increased then so is drag. In the above situation the left wing is producing more lift than the right one, so it will be producing more drag. This pulls the aircraft's nose to the left. The pilot is able to cancel this adverse yaw as it is called by applying right rudder. He does this by pressing the right rudder pedal.

Adverse yaw is a secondary effect of the ailerons. The rudder also has a secondary effect. It will roll the aircraft as well as yaw it. If left rudder is applied, the nose of the aircraft will swing to the left. The right wing will move forward and the left wing will move backward with respect to the pilot. As a result, the right wing will produce more lift than the left wing, so the aircraft will roll to the left. The left wing will also experience some blanketing by the fuselage, which will aid the rolling tendency. Rolling can be controlled by spoilers rather than ailerons. Spoilers offer no resistance when closed, but when opened they 'spoil' the airflow decreasing lift and increasing drag.

Most trainer aircraft are powered by piston engines of around 100 hp. The propeller produces thrust in the same way as the wing produces lift. Propeller blades have a cross-section which looks a bit like that of a wing with a flat rear surface and a cambered forward surface. But a propeller blade does not feel a constant airspeed along its length; the tips are travelling much faster than the roots. To cope with this, propeller blades have a section which changes along the length of the blade, and the blade is twisted so that the AoA is greater near the root than it is near the tip.

Amy Johnson 1903-1941

After gaining her pilot's licence in July 1929, Amy Johnson became the first woman in Britain to qualify as a licensed aircraft engineer. She went on to become one of the best-known women aviators of the 1930s. 1930: Became the first woman to fly solo from England to Australia, a flight which she made between 5 May and 24 May in her Gipsy Moth, *Jason*. 1931: De Havilland gave her a Puss Moth, *Jason II*, in which she flew from London to Tokyo and back with C G Humphreys. 1932: Married Australian pioneer aviator James Mollison, the same year in which she made a record solo flight from London to Cape Town in 4 days 6 hours flying Puss Moth, *Desert Cloud*. The return flight took 7 days 7 hours. 1933: With her husband, made a westerly crossing of the Atlantic in 39 hours, flying a de Havilland Dragon called *Seafarer*. The flight started at Pendine Sands, Wales, on 22 July and finished at Bridgeport, Connecticut. 1934: The couple competed in the England-

Amy Johnson.

Australia race with the de Havilland DH88 Comet, *Black Magic*. 1936: set an out-and-return record between London and Cape Town of 7 days 22 hours, 4-12 May, in a Percival Gull. 1939: Became a ferry pilot with the Air Transport Auxiliary but was killed on 5 January 1941 when her aircraft crashed into the Thames Estuary.

Amelia Earhart 1898-1937

One of America's aviation heroines. 1929: Competed as a pilot of a Lockheed Vega in the Women's Air Derby at Santa Monica, California, in August. 1931: Established a women's autogyro altitude record of 5,800 m (19,000 ft) in April. 1932: Became the first woman to pilot an aircraft across the Atlantic Ocean, flying a Lockheed Vega from Newfoundland to Ireland on 20/21 May. In September, she set a women's transcontinental record from Los Angeles, California, to Newark, New Jersey, which she flew in 19 hours 4 minutes. 1935: Made a solo flight from Hawaii to California. 1937: Departed Miami, Florida, in a Lockheed Electra to circumnavigate the Earth at the Equator, with co-pilot Fred Noonan. The couple were last heard of heading for Howard Island in the South Pacific on 2 July and it is widely thought that they crashed-landed on an island held by Imperial Japanese

Amelia Earhart.

forces, who, believing them to be spies, murdered them. This has never been proved.

Variable pitch

Most trainer aircraft have fixed-pitch propellers, but some more complicated touring aircraft have variable-pitch propellers. Commonly, two pitch settings are available – coarse and fine. The effect is similar to having a two-speed gearbox in a car. A fixed-pitch propeller has its pitch set at a value which gives a reasonable performance at both low and high speeds. Improved performance can be obtained by using a low (fine) pitch for take-off and low-speed flight and a high (coarse) pitch for high-speed flight. More complex aircraft have constant-speed propellers in which the pitch is constantly adjusted to keep the propeller at its most efficient rpm.

A rotating propeller affects the aircraft in two ways – by torque and wash. Low-power trainer aircraft do not experience much of a torque effect, but more powerful aircraft do.

Earlier pioneers flew such aeroplanes as the de Havilland 60X Hermes Moth. This beautifully restored example now flies with the Shuttleworth Collection which is kept in Bedfordshire, England. Such designs gave the UK a technological edge.

Learning to fly

A great many people would like to learn to fly but very few do because most think it is very difficult and extremely expensive. In general, anyone who can drive a car can handle a simple light aircraft and while it is not the cheapest of hobbies flying is within the budget of many people if it is approached sensibly. It costs about £50 an hour at 1985 prices to hire a light aircraft and an instructor in the UK. Prices vary from country to country with the USA probably being the cheapest with a typical price of US$40 per hour. When other currencies are strong against the dollar foreigners often learn to fly in the USA on an intensive course which can be completed in a month and often includes full board and lodgings. The total cost can be less than that for a private pilot's licence (PPL) course in other countries and it usually includes travel to and from the USA. Licences gained in the USA can easily be validated in other countries. Weather in the southern states is often very good for flying, so it is almost possible to guarantee a licence.

How much?

Most countries require student pilots to complete about 40 hours of flying training, some of which will be with an instructor and the rest solo. The exact requirement can depend on how quickly the course is completed, whether a school is approved, and of course on the student's ability. But most

The importance of the pre-flight check on the aeroplane is explained to a student pilot about to continue his training in a Cessna 152 at a flying club at Blackbush in southern England. Much flight safety work is in routine and procedures.

people can complete the course in 45 hours, which at £50 an hour gives a cost of £2,250. In addition, there will be charges for books, lectures, and some landing fees, which will bring the price to around £2,500.

In parallel with learning to fly the aircraft, the student will undertake ground study, which will include subjects such as airframes and engines, meteorology, navigation, and aviation law. Examinations in these subjects

will have to be passed and a high pass mark is usually set. Assuming that the student has no experience, the first few hours of airborne instruction cover general handling, usually termed 'effects of controls'. During this period the student will learn to fly the aircraft straight and level, execute turns, make the aircraft climb and descend, and handle the throttle. Once mastered, the student then returns to the home airfield to practise 'circuits', which involves repeated take-offs and landings. This exercise can make the student work quite hard, particularly if there is a crosswind or turbulence. A circuit starts with the take-off run and a climb straight ahead usually to a height of 150 m (500 ft). Circuit direction is usually left, so a left turn of about 90 degrees is started at 150 m. Once the aircraft has reached 300 m (1,000 ft) it is levelled and turned 'downwind' with another 90 degrees turn. The turns will be exactly 90 degrees only if there is no wind and the 'downwind' leg will be exactly downwind only if the wind is blowing straight down the runway. In practice, these two conditions rarely occur and the student will have to allow for wind drift if a perfectly rectangular circuit is to be flown. The second part of the rectangle sees the aircraft descending and landing. Some airfields have circuits different from this because of noise abatement and other requirements.

The de Havilland Tiger Moth (*left*) remains the classic training aeroplane used since the Second World War, but it has been supplemented in many aero clubs by modern designs like the Scottish Aviation (Beagle) Pup monoplane (*below left*).

First solo

However many hours a pilot flies in his or her subsequent career the first solo is the never-to-be-repeated event of a lifetime. Most students are sent solo after about 10 hours and the flight consists of one circuit. Assuming the student can contain himself sufficiently and has enough nervous energy left, further circuits can be flown immediately, but most first-solo-completed students usually repair to the bar for a celebratory drink.

Once practised at circuit flying, the student then departs the airfield again to learn navigation. For the most part this is accomplished by map reading and therefore the aircraft must remain in visual contact with the ground. However, most modern courses include some navigation by radio aids to enable the inexperienced PPL holder to get himself out of trouble when 'temporarily unsure of position' or lost. Some instrument flying is also included, primarily to show how difficult it is. Instrument flying should not be attempted by the PPL holder without an instructor. In the UK, there is a qualification called the instrument meteorological conditions (IMC) rating, which enables private pilots to fly in instrument conditions outside controlled airspace. In other countries a full instrument rating must be gained and this is an expensive and challenging qualification often outside the pocket of a PPL holder.

PPL and after

The PPL course finishes with a general flying test, which is a little like a driving test only far more comprehensive. Once the paperwork has been completed, the pilot is then allowed to carry passengers and exercise the other privileges of the PPL holder. To operate the radio when not under the supervision of an instructor the PPL holder must gain a radio telephony (RT) licence, which is more difficult than might be imagined. The PPL holder can then go on to gain an IMC and instructor's rating and do aerobatics. Some schools offer finance to assist their members to complete courses efficiently. The cost of flying can be brought down after the PPL course has been completed by buying a share in an aircraft or joining a club which operates cheaper single-seat types.

Ian Parker

The engine rotates the propeller in one direction and tries to rotate the aircraft in the opposite direction. When the aircraft is in cruising flight, this can be resisted by trimming the ailerons. When any rotating mass has its axis of rotation swung, it produces a force at right angles to the movement. With tailwheel aircraft having large engines this can have an effect on take-off. As the tail lifts, the nose will try to swing left or right depending on the direction of propeller rotation.

Another cause of swing on take-off is wash. The rotating propeller causes air to corkscrew round the fuselage and strike the tail fin, pushing it either left or right, depending on the direction of propeller rotation. This has to be counteracted by nosewheel steering or differential braking during the initial part of the take-off run and by rudder as speed increases. Whenever the power is changed, the rudder will have to be moved to keep the aircraft flying in balance. Failing to do this is perhaps one of the most common faults among inexperienced pilots.

Engine handling

Engine rpm is controlled by a throttle lever or plunger. This acts like the accelerator of a car, but there are two other engine controls not found in a motor vehicle – the carburettor heat and mixture controls. When air passes through the carburettor it flows through a

The Pilatus Britten Norman Islander is used for passenger and freight services around the world, especially into confined strips using the type's superior short field performance and ease of handling, even when fully loaded.

venturi, which reduces its pressure, sucking in fuel. The pressure drop has two effects – it cools the air and causes water vapour in it to condense. Under certain conditions this can cause ice to build up in the venturi, eventually blocking it. The carburettor heat control diverts hot air to the carburettor, melting any ice that may have formed. It is good practice to apply the carburettor heat once every 20 minutes or so. If the engine rpm falls then there is no ice present. If it rises then ice has been picked up and the carb heat should be left on to clear it.

As altitude increases the air gets thinner, which has the effect of enriching the mixture. To keep the engine running economically, the mixture is progressively weakened as the aircraft climbs.

Flaps

Flaps are attached to the trailing edge of the wing inboard of the ailerons. They increase lift and drag and allow the aircraft to fly in a nose-down attitude, improving visibility over the nose. On a light aircraft the flaps will typically be selectable at three positions – 10°, 20° and 40°. If the runway is short, 10° of flap may be selected for take-off so that the aircraft leaves the ground at a lower speed. Normally, 20° will be selected for the landing, 40° if the runway is short.

The flaps work by effectively changing the shape of the wing cross-section, making the wing into a shape that produces more lift at a given speed or the same lift at a slower speed. The flaps also reduce the stalling speed, allowing the aircraft to fly safely when moving more slowly.

Stalling is caused when the AoA gets too high. If an aircraft is flying straight and level and the throttle is closed, the pilot will have progressively to raise the nose to stop the aircraft descending. This leads to an increasing AoA, but there is a limit to how high it can go. Above a certain AoA the airflow over the top of the wing becomes turbulent and breaks away from the wing surface. Most of the lift is lost and the aircraft descends rapidly. Because the tailplane is still flying, the aircraft pitches nose-down, allowing the airspeed to increase and the stall to be broken. It is important to remember that a wing stalls at a given AoA and not at a given speed. This may become apparent in a tight turn.

An aircraft turns by banking, whereby lift, which always acts perpendicularly to the wings, is partly used to alter the aircraft's direction. If the pilot applied 90° of bank then all of the lift would be turning the aircraft, but there would be none to support its weight and the aircraft would sideslip downwards. Typically 30° of bank may be applied, during which some of the lift will be used to turn the aircraft. To stop the aircraft from descending, the total lift will have to be increased by increasing the AoA.

The pilot increases the AoA by pulling back the stick. If the pilot continues to tighten the turn by applying more bank and pulling back harder, the maximum permissible AoA will be reached. If it is exceeded, the aircraft will stall, even though it may be travelling at high speed.

Richard Evelyn Byrd 1888-1957

Byrd was the first man to fly over both North and South Poles. 1926: With Floyd Bennett, flew from Spitsbergen to the North Pole in a Fokker F.VII-3m. 1927: With Bert Acosta, Bernt Balchen, and George Noville flew the F.VII-3m, *America*, from Roosevelt Field, New York to France in 42 hours between 21 June and 1 July. Rain and fog prevented a landing at Le Bourget and a ditching was made in the sea off Ver-sur-Mer. 1929: With A C McKinley, Balchen, and Harold June flew the Ford Trimotor, *Floyd Bennett*, to the South Pole from Little America on the Bay of Whales on 28/29 November. 1934: Second expedition to the Antarctic made five flights from Little America over the Ross Ice Shelf and Marie Byrd Land.

Landing

Perhaps the most demanding phase of flight in terms of control is landing. The pilot has to line the aircraft up with the runway at a suitable distance so that a gentle descent may be made. This is typically 2.5 m/sec (500 ft/ min) and the aircraft will normally be established on its approach no lower than a height of 150 m (500 ft). A given approach speed should be held accurately with the engine set at a specific rpm. If the air was still this would be a relatively simple procedure. However, the air is rarely still, the wind will hardly ever be straight down the runway and some degree of turbulence may be encountered.

If the crosswind is from right to left, the aircraft's nose will have to be pointed slightly to the right to stop the aircraft drifting left of the runway centreline. It is perhaps surprising that on the approach the speed of the aircraft is controlled with elevators and the rate of descent is controlled with the throttle. If the aircraft's speed falls below the approach speed

Landing on rough strips like at Xaxaba in northern Botswana's Okavango Delta calls for certain bush flying skills and a sturdy aeroplane such as the Cessna Skywagon. Accidents are surprisingly rare, although maintenance costs are often high.

The de Havilland 80A Puss Moth (*right*) was regarded as having special landing abilities in the 1930s, but the modern technology of such aircraft as the Australian Government Aircraft Factories Nomad (*below*) now far exceeds the early designs. Whereas the Puss Moth has been primarily used for fun flying, the Nomad fulfils important transport and paramilitary roles around the world, including a commuter service in the Netherlands and coastal anti-smuggling patrol in Papua New Guinea.

the pilot should lower the nose to speed up. If the aircraft descends below the ideal glideslope the pilot should open the throttle to reduce the rate of descent.

Constant adjustment will be needed if there is turbulence or windshear. Wind near the ground blows slower than the wind high up because of friction with the earth's surface. Sometimes this effect can be very pronounced. As the aircraft descends on approach the

groundspeed may remain constant while the airspeed falls off because the aircraft is flying into slower-moving air. So approach speed has to be monitored very closely.

When the aircraft is about 10 m (30 ft) off the runway the pilot will start to round out to reduce the aircraft's rate of descent. He does this by raising the nose of the aircraft, at which point the throttle is closed. If there is a crosswind and the pilot has approached with

the nose to one side of the runway centreline he will now have to straighten the aircraft up. But if this is done with a conventional turn, the aircraft will immediately start drifting. Instead, rudder is applied to point the aircraft straight down the runway, but bank is applied in the opposite direction so that the aircraft sideslips into the crosswind. With the aircraft held in this attitude, the pilot applies progressively more back pressure to the stick, holding off with the wheels a few feet above the runway. Ideally, the aircraft is stalled down on to the runway but most landings are made with some flying speed in hand.

Microlights and ultralights

Many people would like to learn to fly but find it too expensive at conventional flying clubs. For those who simply want to be airborne in good weather and are not particularly interested in going anywhere, microlights may provide the answer.

Microlights grew out of powered hang-gliders. Early hang-gliders were based on the Rogallo wing, which was developed as a possible parachute/glider for recovering spacecraft during the final stage of re-entry. Early hang-gliders had pure Rogallo wings with highly swept leading edges and a lot of billow in the sail. These very quickly gave way to second- and third-generation hang-gliders, which have much less sweep and billow.

To fly a hang-glider it is necessary to find a hill facing into wind. For training, a hill is chosen which just exceeds the glide angle of the hang-glider so that beginners cannot get very high. The hang-glider is then flown down the hill and carried back up. After some experience has been gained, soaring flight can be attempted, allowing the pilot to land on the top of the hill.

It was not long before hang-glider pilots fitted engines to their aircraft so that they could be flown off level ground. To begin with the engines were strapped under the wing with long drive shafts turning a rear-mounted pusher propeller. This arrangement involved

Howard Hughes 1905-1976

Howard Hughes's name continues to be well known in aerospace, with major Hughes companies building helicopters, arms, and spacecraft, but Howard Hughes was best known in the 1930s. A keen stunt pilot, Hughes suffered a near-fatal crash in his own war film *Hell's Angels* and then trained, under an assumed name, as an airline pilot. 1935: Set a world landplane speed record of 567.21 km/h (352.46 mph) in his specially designed racer *Hughes One*. 1937: Flew the aircraft across America in 7 hours 28 minutes, a record which stood until the days of jet aircraft. 1938: Flew a Lockheed 14 around the world in a record

3 days 19 hours and 14 minutes. The aircraft had a crew of four and they received a heroes' welcome on their return to New York. A special medal was awarded by Congress. 1940s: After two unsuccessful military aircraft, Hughes built the H2 Hercules (otherwise known as the *Spruce Goose*), which had a wingspan of 97.69 m (320.5 ft) – the largest so far. The eight-engined all-wood aircraft flew once. It covered less than a mile at a height of 21.34 m (70 ft) at Long Beach, California, on 2 November 1947 with Hughes at the controls. The 190-ton aircraft is now preserved in California. Hughes died a recluse.

Police aviation

Many countries throughout the world use aircraft in the support of police forces but nowhere is the use as extensive as in the United States. For a country which can claim to be the leader in many aspects of aviation this may come as no surprise but some other countries which cover a large geographical area also have thriving police aviation units.

For many years, the Royal Canadian Mounted Police have used aircraft to help 'get their man', floatplanes and skiplanes being included in its fleet. Other countries as widely separated as Malawi, Oman and Japan use fixed-wing aircraft and helicopters to provide police services in remote areas. Many European countries make use of aircraft for a variety of law enforcement duties but there is no common pattern either in the use of aircraft among police forces in Europe or in their operation. In Britain, the Metropolitan Police and the Devon and Cornwall Constabulary use helicopters flown by commercial pilots but the State and City police helicopter units in Federal Germany are operated by policemen trained as pilots.

MBB BO 105 and other helicopters used by German police forces have become a familiar sight over motorways, and many accident victims owe their lives to the speedy aid brought by the green and white aircraft.

Although a growing number of European cities now see police helicopters providing aid to traffic and crime prevention units, none can compare with the extensive aircraft fleets operated by law enforcement agencies in America. In 1970, there were 61 of these agencies operating 118 aircraft; in just ten years, 1980 saw some 1,100 aircraft in use with 335 law enforcement agencies.

State and City police forces, Sheriff departments and even Park police units tend to operate helicopters with the same familiarity as a motor car. For the most part, the helicopters are used in support of ground units to pursue criminals, patrol highways and rescue accident victims.

However, the unique flying qualities of the helicopter have resulted in some remarkable rescue successes, notably the evacuation of 150 people from the roof of a burning hotel in Las Vegas and the hoisting of the survivors of an airliner which crashed into the Potomac River in Washington. No other type of aircraft could have achieved these rescues but other less spectacular, life-saving events occur almost every day.

A child lost in the desert; a mountain climber with a twisted ankle; a motorist far from anywhere and out of petrol – the examples are legion but police helicopters are successful in finding criminals too, by night as well as by day. Powerful searchlights mounted on a helicopter can lift the veil of darkness which cloaks the activity of the thief, mugger or housebreaker.

Although helicopters large and small are used extensively by American law enforcement agencies, fixed-wing aircraft are widely used too, patrolling borders, seeking deer poachers or simply catching speeding motorists. Three years after the New Hampshire State Police began to use Cessna Skylane aircraft to patrol its highways, a total of some 20,000 arrests had been made. By the use of a special device, it is possible to check the speed of a motor car from the air and many a motorist has been quite unaware that there is a highway patrol in the sky – the bear in the air.

While the level of police aviation is constantly increasing, the high cost of this type of operation has tended to discourage its introduction by smaller police forces. However, the Police Department in the City of Downey, California, has tackled the problem by acquiring an Eipper Quicksilver microlight aircraft. A very basic class of aircraft powered by a 25 kW (33 hp) engine, the Downey microlight weighs only 172 kg (380 lb) but has proved to be a valuable aid in the fight against crime.

The microlight is not the solution to all police aviation requirements but it opens interesting possibilities which could bring this type of aid within the reach of more law enforcement agencies.

Brian Walters

The helicopter has in recent years proved invaluable to airborne law enforcement. Many nations now use a variety of types like the Bell 222, one of seven in service in Japan. Helicopters also allow police agencies to perform medical evacuations.

Although some police forces fly microlight aeroplanes, their main user is the sports enthusiast (left); sports enthusiasts have also taken to the hang-glider (above right), which is reminiscent of early German experiments, and to the home-built ultralight aeroplane (right), which can be folded away in a simple trailer. These simple designs make flying for fun cheaper than using more conventional aircraft, even from flying clubs or flying groups. There are now several national design authorities for this type of flying and many holiday resorts around the world offer courses in the building and flying of them. A microlight has even crossed the Mediterranean non-stop.

certain risks. The added weight increased the take-off speed to near the maximum running speed of the pilot. If the pilot had full power on and tripped over, the propeller could do considerable damage to the pilot's feet.

Undercarriages needed

It was soon realised that some form of undercarriage was needed to increase the safety of such aircraft. This was really the beginning of a new breed of aircraft known as ultralights in the USA and microlights in the UK. (In the UK, the term ultralight was already assigned to homebuilt aircraft.)

Two forms of microlight aircraft have developed – weightshift and three-axis. In general both types have maintained the Dacron-and-tube structure of hang-gliders, but the weightshift types have maintained the traditional form of hang-glider control whereas three-axis types use the stick-and-rudder system found in conventional aircraft.

Weightshift microlights have two basic parts – the wing and the trike. The wing is often very much like a hang-glider and in many cases it can be used as such. The trike is usually a three-wheeled tubular structure which houses the engine turning a pusher propeller. It is attached to the wing by a joint which allows motion both fore and aft and left and right. The pilot has a crossbar which he uses to move the trike relative to the wing. By pulling on the crossbar the trike's weight is moved forward and the aircraft pitches nose down. By moving the crossbar to the right, the trike moves to the left, so the aircraft turns left. These movements are essentially the reverse of those on a stick-and-rudder aircraft.

Three-axis microlights have become by far the most popular type and as a result have become increasingly complex and sophisticated. The early microlights were very simple

and cheap, but some more modern ones approach the complexity of conventional light aircraft and have a price which reflects this. Some people in the microlight movement say that these expensive aircraft are moving away from the original idea behind microlights which was to bring flying within the pocket of most people.

Filling in the gaps

Some microlight manufacturers have moved well away from the Dacron-and-tube designs towards advanced materials such as CFRP and GRP. Resulting aircraft are often sold in kit form as homebuilt aircraft have been for many

years. The divisions between light aircraft, homebuilt aircraft and microlights are blurring and aircraft of all levels of cost and sophistication will soon be available.

Another reason for the development of microlight aircraft was the lack of legislation. To begin with there were very few controls on microlights in most countries. In many cases, the pilot did not need a licence and the aircraft did not have to be certificated or registered. But when microlight flying became more popular in the early 1980s, many civil aviation authorities took more interest. Some banned microlights completely and others introduced legislation, some of which was inappropriate.

In the USA, microlights had to be foot launchable although an undercarriage could be fitted. The idea was to keep the aircraft's size and weight down so that the result of an accident would be less severe. In the UK, specific rules were introduced governing

EIPPER QUICKSILVER MX

Country of origin: USA.
Role: Microlight.
Wing span: 9.75 m (32 ft).
Length: 5.51 m (18.08 ft).
Max weight: 238 kg (525 lb).
Engine: 1 × Rotax 377 piston engine, 25 kW (33.5 hp).
Max speed: 85 km/h (53 mph).
Range: 178 km (111 mi).
Crew: 1.
Passengers: None.

Aerobatics

Among the greatest of pleasures in aviation is performing aerobatics. Airshow attenders will be well familiar with the spectacular displays performed by the world's aerobatic teams and individual pilots. Few spectators realise that such manoeuvres could be well within their own capabilities albeit at greater altitude and with somewhat less noise and showmanship. Of course, it is first necessary to gain a pilot's licence but many PPL holders are wary of trying aerobatics when they might have considerable talent for and gain much pleasure from such flying.

Safety first

Most flying clubs have an aircraft which is either semi- or fully aerobatic. A typical example is the Cessna 152 Aerobat. While not being a competitive aircraft, the Aerobat is ideal for an introduction to aerobatic technique. Safeguards are extremely important in aerobatics. There is no formal aerobatic qualification in most countries but pilots are strongly advised to seek some instruction from an experienced aerobatic pilot or instructor. Manoeuvres should be performed high up, certainly not below 900 m (3,000 ft) until experience has been gained. A good lookout should be kept at all times and if possible aerobatics should be performed in an area where few other aircraft are likely to be, such as a corner bounded by controlled airspace. Smooth, clear air is needed and published checks should be made beforehand.

Each manoeuvre should be studied on the ground first, then demonstrated by the experienced pilot before the novice tries his hand. About 1 hour of practice on each manoeuvre should be sufficient for the beginner to try it solo. Typically three manoeuvres will be learned, such as a loop, barrel roll and stall turn. Once mastered, other things can be added, for example a slow roll. Later, more complicated aerobatics can be attempted and sequences put together.

Ian Parker

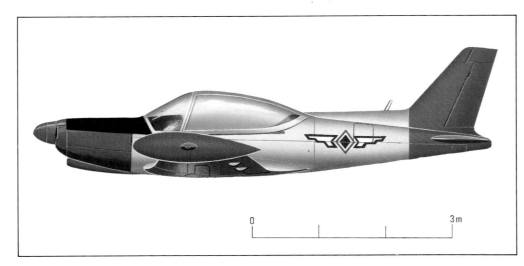

The classic aerobatic training aeroplane is the Rheims-Cessna 152 Aerobat (*top*), found at many clubs in Europe and North America. For the more specialised, the Pitts Special (*above*) is also a favourite display aeroplane and has been used by several air force academies. Pilots who fly such aircraft need special skills, which can be taught, but also a certain innate flair, which cannot. Several sponsored aerobatic teams, flying the Pitts, are found in Europe.

Left: the SIAI Marchetti SF 260 trainer has good aerobatic features and is used by many civil and military operators, like the Philippines Air Force (illustrated). In some countries, airline pilots are trained on the type and it has been used for military operations and paramilitary purposes.

of Cayley's machine was built about 10 years ago and flown to demonstrate the validity of his design. The fuselage had a boat-like structure with the pilot sitting sideways holding a tiller to which the tail was connected. This gave the rudimentary elevator and rudder controls which are found on modern aircraft. A film of the replica's test flights was made showing what Cayley's exploits must have looked like over 130 years ago.

Within a few years of Cayley's flights other experimenters began flying rudimentary gliders, exploring various methods of control. Many of them resembled today's hang-gliders, including probably the best known example built by German Otto Lilienthal. He had a conical hill built specially so that he could fly whatever the wind direction. Most contemporary gliders were foot-launched and were flown off hills, platforms and sometimes dropped from balloons.

Later gliders used wing warping for control. Instead of having hinged control surfaces

maximum weight and minimum wing area to keep the kinetic energy down. Pilot licences were also introduced, as were aircraft registration requirements. Although these rules were designed to protect microlight pilots and people on the ground, in many cases the rules have stifled the sport.

One of the original desires of microlight pilots was to get away from rules and regulations, but in many cases the sport has been virtually regulated out of existence. Accident rates have been unacceptably high, but the same situation occurred in hang-gliding in the 1970s, when civil aviation

authorities negotiated with the sport's governing body to improve training standards and reduce accident rates. The training standards are monitored by the hang-gliding associations and not by the government authorities.

Gliding

A more traditional method of flying cheaply is gliding. Sir George Cayley built a glider which was flown by his coachman in 1853. It was towed into the air like modern gliders and although it made a crash landing the structure and aerodynamics were quite sound. A replica

Gliding has its roots in the last century and was pioneered in Europe by such as Otto Lilienthal (*above left*).

Today, modern high-performance machines are called sailplanes (*below*). Some of the best of these are produced in Eastern Europe, but many of the best pilots come from the small clubs in the southern United States and Great Britain. Gliders are often owned by syndicates to reduce operating costs, whilst others prefer to fly club machines.

operated by a mechanical system of pulleys and wires, wing warping adjusts the tension in the aircraft's external rigging wires so bending the wing. The method sounds crude, but has advantages. There are no joints to cause turbulence and a good aerodynamic shape is maintained. Some recent manpowered aircraft have used the concept.

The development of powered aircraft and their rapid improvement during the First World War almost halted glider development. It was not until the 1920s that gliding really started to become a sport and Germany took the lead, a position which the country still holds today. Modern gliding techniques began to develop.

A glider pilot faces two basic problems – how to get the glider into the air and how to keep it there. Essentially there are four methods of getting a glider into the air – by bungee launch, winch, aerotow, and autotow. Bungee launching is not very common because a wind-facing hill is needed.

A rubber rope is attached to a hook on the glider's nose with the aircraft on top of the hill facing into wind. Several people pull the rope down the hill applying and maintaining tension. Another person holds the glider's wingtip while a final person holds the glider's tail fastened to the ground. On a signal from the pilot the tail is released and the glider accelerates quickly, becoming airborne in a few seconds.

Wind coming up the hillside is forced to climb and the glider is launched into this rising air, which keeps it up. The glider can turn left or right along the hillside, flying backwards and forwards gaining height. This was the first kind of lift to be used by glider pilots. It was known that birds used thermals to stay airborne effortlessly, but these columns of warm rising air were of little use to glider pilots until the variometer was invented.

The variometer is essentially an instrument which detects very small changes in pressure. Air pressure falls with increasing height so such an instrument can detect the rising or

falling of anything to which it is attached. If a pilot has a variometer in his glider, he can tell if he is going up or down. The variometer was invented towards the end of the 1920s, allowing glider pilots to leave hills and fly cross-country. The large fluffy white clouds that form on a summer day are created by thermals and therefore mark their position, but the variometer is needed when such natural signposts are not present.

Winching

The ability to use thermals to provide lift meant that gliders could be launched from sites other than on hills. But this needed a more powerful method of launching to give the glider a bigger initial start. So winching was developed. This method is not as quick as

Aerotowing (using a light aeroplane as a tug) is considered the best, if most expensive method of launching a glider. This method is especially common during gliding competitions.

bungee launching, but it is still fairly rapid and reasonably cheap. The winch has a powerful engine driving drums wrapped with steel cables. There are usually two drums and a car pulls them out across the airfield to a distance of about 900 m (3,000 ft). One drum is selected and the glider is attached to the cable. When the pilot is ready he gives a signal to take up the slack. This signal is usually transmitted to the winch driver by a flashing light. He starts to pull the cable in slowly. When the slack has been taken up the pilot signals 'all out' and the winch driver accelerates the drum, bringing the glider up to flying speed. The pilot holds the glider level until climbing speed has been reached, at which point he raises the nose. The glider climbs quite rapidly because while it is on the cable it has the power of a large engine but not the weight.

The glider usually gets up to about 300 m (1,000 ft) before the cable is released. Sometimes the cable breaks and training for such an event forms an important part of learning to glide. The driver is surrounded by a cage to protect him from a whiplashing cable. He also has a guillotine so that he can cut the glider away if the pilot finds that he is unable to release the cable. Winching is probably the most common method of launching in the UK. It is often performed from a runway, but only a few hundred metres of smooth ground are needed for the glider's ground run.

Autotow

A similar method which does need a runway or similar smooth surface is autotow. The simplest method involves a powerful vehicle pulling the glider behind it. This requires the driver to keep looking backwards to check on the glider's progress, but a better method is reverse pulley autotowing. The cable runs round a pulley so that the car faces the glider, making it much easier for the driver to see how the glider is getting on. Both these methods require quite a high level of skill from the driver.

He has to pull the glider at the correct airspeed. The groundspeed of the autotow car or the cable speed of the winch will have to be adjusted to take account of the wind. Sometimes the wind blows more strongly higher up, so the driver will have to ease off as the glider nears the top of the launch.

The pilot can signal the driver by waggling the wings (too slow) or yawing the nose from left to right (too fast). It is very important that the launch speed is correct, otherwise the glider may approach a stall, or be in danger of overstressing its airframe. Even though the correct launch speed may seem slow on the airspeed indicator, the stress on the airframe is equivalent to the aircraft being in a tight turn at high speed.

Aerotowing

The most civilised method of getting airborne in a glider is by aerotow. A light aircraft tows the glider up with a cable about 45 m (150 ft) long. This method of launching requires more skill from the glider pilot. As the tug and glider accelerate, the glider will reach flying speed before the tug does because the tug needs to go faster to take off. If the glider immediately started to climb, it would pull the tail of the tug up, making its propeller hit the ground. So the glider pilot has to keep his aircraft just off the ground while the tug builds up enough speed to take off.

Once airborne, the glider can fly in one of two positions (high or low) behind the tug to stay out of its wash. Maintaining station behind the tug takes practice and inexperienced pilots tend to over-control when trying to correct. One of the big advantages of an aerotow is that it can take the glider higher than any other method of launching and place it where lift is known to be. For example, the glider pilot may ask to be taken to an altitude of several thousand feet and released under a nice big cumulus cloud. Of course, this is the most expensive method of launch.

Learning to glide

Many people are keen to try gliding and quite a large number experience the delights of this form of aviation by taking a trial flight. But it takes quite a lot of enthusiasm to become a glider pilot because most of the time on the airfield will be spent waiting or ground-handling aircraft. It is not usually possible to book a glider and most clubs make up a flying list on a first come first served basis. This means that an early arrival is recommended, but even after a person has flown he or she will normally be expected to remain helping to launch the gliders.

A student pilot usually makes one or two flights per day. If a winch is being used and there is no lift about, the flights may only last five minutes. In the summer, gliders may be flown from 8 am to 8 pm, so that as a rough

GROB G 109B

Country of origin: Federal Germany.
Role: Motor-glider.
Wing span: 17.4 m (57.08 ft).
Length: 8.1 m (26.58 ft).
Max weight: 850 kg (1,874 lb).
Engine: 1 × Grob 2500 piston engine, 67 kW (90 hp).
Max speed: 240 km/h (149 mph).
Range: 1,500 km (932 mi).
Crew: 1 pilot.
Passengers: 1.

guide a student pilot will spend an hour on the ground waiting or helping for every minute spent airborne. This sorts out the really keen students from those just interested in 'having a go'.

One of the attractions of gliding is that with some experience it is possible for a group of people (say four) to buy and operate their own glider. A good one can be bought for a few thousand pounds and maintenance costs are usually low. Motorgliders are an option which offers some of the advantages of both powered flying and gliding. Such aircraft have been refined in recent years so that their gliding performance approaches that of the less-sophisticated ordinary gliders. But motorgliders are expensive.

They usually have the long slender wings seen on conventional gliders with airbrakes which are not normally fitted on powered light aircraft. A small engine turns a propeller, which will often be featherable. If the motorglider pilot finds good lift he can shut down the engine and turn the propeller edge on to the airflow to cut down the drag. In powered cruise, modern motorgliders are often very

Piper's Cub was used as a wartime artillery spotting aeroplane, but many found their way into gliding and flying clubs after the war, where it proved very effective. This machine is in its original US Army colours.

much more efficient than conventional light aircraft, using half as much fuel at a given speed. Storage problems are overcome by making the wings foldable, sometimes in just a few minutes.

One of the best ways of getting into gliding is to do a course lasting one or two weeks. By the end of a two-week course most beginners will have reached or be near the standard required to go solo, assuming that the weather has been reasonably good. Most clubs usually guarantee a minimum number of flights so that if the weather is bad for a long period, students may have to return during following weekends to finish the course.

The training in gliding is broadly similar to that for powered flying and there is usually no problem in changing from one to the other. Powered aircraft pilots may take a minute or two to settle down to the fact that they cannot fly with the nose above the horizon, since a glider needs to be constantly sinking with respect to the air. Perhaps the biggest difference between gliding and powered flying is circuit planning. When a powered aircraft flies a circuit it will normally turn over predeter-

mined points. Students are often shown landmarks over which to make turns. In gliding the circuit may have to be modified at any time to take account of changing wind strength, lift or sink. A glider has no second chance at making an approach so this part of the flight receives careful attention.

Airbrakes

To give gliders some flexibility on the approach they are equipped with airbrakes. The pilot will aim to be at the right height so that he can make the final approach with about half airbrake deployed. If the glider is getting too low he can close the airbrakes slightly to reduce the rate of descent. If he is staying too high, he can open the airbrakes to increase the rate of descent. On approach, the airbrakes of a glider perform the same function as the throttle of a powered aircraft. Perhaps surprisingly the airbrakes of a glider do not slow it down very much. They are used to control the rate of descent.

Once a student has mastered glider handling and circuit flying he can move on to what the sport is really all about – soaring.

A series of international badges can be gained for achieving flights of a given duration and altitude gain. Soaring is vital if these badges are to be attempted. The simplest form of soaring is ridge soaring. When a strong wind hits a hill the updraught is sufficient to keep a glider airborne. The lift occurs in a known place so that if no other kind of lift is found the glider cannot go very far.

Most long-distance flights are made by using thermals – currents of warm rising air. This warm air may come off a particular kind of crop in a field and it often forms into a cumulus cloud. The glider pilot will often search for lift under such a cloud. On some days no such clouds form, making thermal finding much more difficult.

Much of the glider pilot's skill is in finding thermals, putting his aircraft into a tight turn in the centre of the thermal, and then

Many of the modern high-performance sailplanes have retractable wheels, thus streamlining the fuselage and allowing the craft better long-distance capability. The elimination of drag is vital in such unpowered flying machines.

The motorised glider with a T-tail has aerodynamic advantage with or without engine power. Most of the designs are from Europe but are expensive to buy and maintain; nevertheless more and more are appearing on club registers.

searching out the next one. But the greatest altitudes are reached by flying in standing wave. This form of lift is rare and it is caused by wind blowing on to hills or mountains. It is a bit like ridge lift, but it can often be found at great height and well downwind of the mountain line. Wave occurs only when the upper air conditions are right, and if the wind is strong enough it is possible to point the glider head to wind and climb vertically with little or no ground speed.

Wave systems can take gliders to altitudes of 14,000 m (46,000 ft) or more, but the up-going areas have corresponding down-going areas and immediately behind the ridge of mountains there may be an area of rotating air (called rotor) which is best avoided.

High performance
The most important measure of glider performance is not the cruise speed but the glide ratio. High-performance gliders can have a glide ratio of around 50:1 which means that the glider travels 50 m (165 ft) forward for every metre it sinks. A glider designer has to try to make the aircraft able to fly slowly so that it can use weak lift, but also capable of flying quickly to get from one area of lift to another with minimum height loss. Some competition gliders carry water ballast. This enables them to fly quickly during the middle of the day when thermals are strong, but the water ballast can be dumped towards the end of the day so that better use can be made of the weaker thermals found then. Sometimes spectators can have a damp surprise if such a glider reaches its home airfield with water left, because the pilot will dump it just before landing.

Low drag
Another important consideration for the glider designer is keeping drag to a minimum. This is why gliders have long slender wings and narrow fuselages. In some gliders the pilot virtually lies down. Other ways of keeping the drag low are to have retractable wheels and a very smooth finish. Composite materials, such as glassfibre reinforced plastic, are often used. The first glider to be made of this material was built in 1957 by a German university.

Some gliders have flaps which produce the same effects as they do on powered aircraft – increasing lift and drag. The pilot will usually deploy the flaps for landing, but he can also use them for thermalling when he needs to turn tightly at low speed. Occasionally a glider pilot will run out of lift well away from his home airfield, so he will have to make a field landing. The flaps will enable him to land slowly on a surface which may be far from ideal. Some gliders have the tailplane at the top of the fin. This has some aerodynamic advantages, and keeps the tail out of harm's way.

Balloons and airships

One popular misconception is that balloons and airships are things of the past. The first untethered manned balloon flight was made in 1783 and since then lighter-than-air aviation has contributed much in both peacetime and war. Hot-air ballooning continues to increase in popularity as a sport and the world's armed forces are once again turning their attention to airships for such duties as submarine hunting, and very long-duration surveillance. Advertisers use both balloons and airships to carry their messages across the sky and commercial air transport operators are looking at airships for both freight and passenger services.

Etienne and Joseph Montgolfier are credited with building the first manned balloons. On 15 October 1783 a 15 m (49 ft) diameter Montgolfier balloon carried 26-year-old François Pilatre de Rozier to a height of 26 m (85 ft) for 4.5 minutes. This was a tethered flight, but on 21 November 1783 de Rozier and the Marquis d'Arlandes took off from the garden of Château La Muette in the Bois de Boulogne,

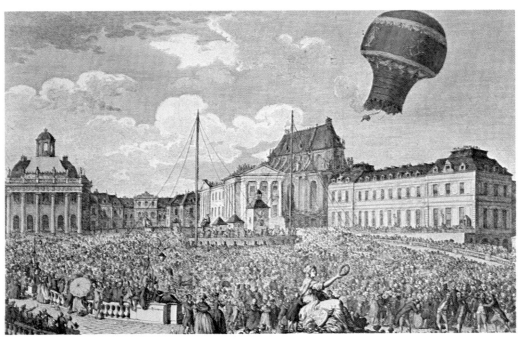

Above: the Montgolfier brothers are the best-known of the early balloon-makers, and on 19 September 1783 one of their hot air craft flew over Versailles with farm animals on board. Manned flight soon followed.

Below: airships were the rage of the inter-war years, with hydrogen- and helium-filled designs considered as the way forward for travel. However, this R101 crashed at Beauvais in France in 1930, putting a stop to developments in Britain for several decades.

Paris, and flew 8.5 km (5.3 mi) in 25 minutes – the first free manned balloon flight.

There is evidence that a Brazilian priest, Bartolomeu de Gusmao, demonstrated a model hot-air balloon at the court of King John V of Portugal in 1709, but the early development of manned hot-air ballooning was due largely to the Montgolfier brothers. They were paper makers by trade and this was the material from which their balloons were made. The first unmanned Montgolfier flew from Annonay near Paris in 1783, and later that year animals were flown before the first manned flight was made.

Hydrogen balloons

In 1766, the British scientist Henry Cavendish isolated hydrogen. He called the gas Phlogiston. Hydrogen is the lightest chemical element and therefore it has a very low density – some 14 times lighter than that of air at standard temperature and pressure. This makes it an ideal substance with which to fill a balloon. Hot-air balloons work on the same principle of having a less dense substance inside the balloon than outside, but this density difference has to be maintained with a heat source. Once filled with hydrogen, a balloon can be sealed up and it will provide lift for a long time. Unfortunately, hydrogen molecules permeate substances easily and the gas is gradually lost.

In the same year that the Montgolfiers were flying their hot-air balloons (1783), Professor Jacques Charles flew hydrogen balloons. The first was unmanned and was released from the Champs-de-Mars, Paris, and landed about 25 km (15.5 mi) away at Gonesse. The envelope was made of a rubberised silk. Later that year, Charles built and flew a manned balloon with Marie-Noël Robert, who with his brother had invented the rubberisation process. They took off from the Tuileries Gardens, in Paris, in front of a large crowd and flew 43.5 km (27 mi) in two hours.

As with any development in technology it was not long before the balloon was examined for its military potential. Many of the early balloonists remarked on the view from on high and it was not long before it was realised that this would be of immense value during a battle. France's artillery service formed a balloon division initially with four aerostats. On 26 June 1794 the balloon division's commander, Captain Coutelle, ascended in the balloon *Entreprenant* during the battle of Fleurus in Belgium, and was able to signal information which probably played an important part in the defeat of the Austrians. This was the first use of an aircraft in war.

Ballooning spread quickly and on 7 January 1785, the English Channel was crossed by Jean-Pierre Blanchard of France and Dr John Jeffries of the USA. François de Rozier and

The first man-carrying free balloon flight takes Pilatre de Rozier and the Marquis d'Arlandes into the air near Paris, France, in November 1783. It was another example of the use of hot air for lift generation.

Jules Romain were killed on 15 June 1785 while trying to cross the Channel northwards. It is believed that the accident was caused by the venting of hydrogen which was ignited probably by the flame which kept the air hot, for this was a hybrid balloon using both hot air and hydrogen to keep it aloft – a rather dangerous combination. Blanchard and Jeffries's southward crossing had not been without incident, however. Their balloon had very little margin of lift and the occupants had to throw out their belongings, including most of their clothes, to remain airborne. Since the flight was made on a very cold day this must have made it a very uncomfortable experience.

It was about eight years later that Blanchard took ballooning to the USA and a historic flight was made from Philadelphia in 1793 before five men who all subsequently became President of the USA – Washington, Adams, Jefferson, Madison, and Monroe. Ballooning soon became popular in the USA and in 1859 a flight was made from St Louis, Missouri, to Henderson, New York, a distance of 1,770 km (1,100 mi). This flight turned balloonists' thoughts towards crossing the Atlantic, a feat which was achieved for the first time only a few years ago.

The Americans were slower to see the military potential of balloons, but with the outbreak of the American Civil War in 1861, several balloonists joined the Union Army. One of them was Thaddeus Lowe, who demonstrated a captive balloon for military reconnaissance to President Lincoln in 1861. Later that year, the President instructed Lowe

to build a large balloon, which was used to observe the fall of artillery fire and transmit corrections. The results were very good so Lowe was ordered to build four more balloons and train crews. In May 1863, Lowe resigned and the Balloon Corps lost its impetus and was disbanded.

Mass-produced balloons

During the Siege of Paris, which started in September 1870, many balloons were used to carry messages and important people out of the city. The Prussian Army, which had surrounded Paris, believed that the city would soon surrender, but a military balloon company was formed and many balloons were built. At the beginning of the siege there were about six balloons in Paris, but a production line was set up at the Gare d'Orléans and at a derelict music hall in the Elysées-Montmarte.

Development of balloons continued and they were successfully used during the siege of Paris, in the Franco-Prussian War, when the world's first air mail service was started – operating, however, only one-way, out of the city.

Circus acrobats and sailors were trained to fly the balloons out of Paris, presumably because they had experience of heights. By the end of the seige on 28 January 1871, some 64 balloons had carried 155 passengers and crew with about three million letters. This prompted the development of the anti-aircraft gun.

Barrage balloons

A free-flying balloon usually gives a very smooth ride, but tethered balloons are far less stable. Kite balloons were developed to overcome this problem. A tail fin kept the balloon pointing into the wind, providing a much more stable observation platform. Later two other fins were added on some types, improving stability still further. This type was used extensively during the First World War for observation on both land and sea. During the Second World War similar balloons were used as barrage balloons to deter low-level bombing. The wires that restrained barrage balloons were a great threat to all aircraft.

During the Second World War, the Japanese bombed North America with automatic balloons. High-altitude, high-speed winds called jetstreams blow in many areas, particularly east across the Pacific. The Japanese launched their automatic balloons into these jetstreams, height being controlled

Barrage balloons were particularly successful during both World Wars in deterring enemy aircraft from making low-level attacks on high-value targets. These were photographed in 1940 in southern England during the Battle of Britain.

with a barometric device at between 9,000 and 11,500 m (30,000 ft and 38,000 ft). Gas was vented to bring the balloon down and ballast was dropped to allow it to rise. When all the ballast was gone and the balloon descended, the weapons were dropped. These consisted of two incendiary bombs and a 15 kg (33 lb) anti-personnel device. About 9,000 balloons were launched and some 12 per cent got through, but the likelihood of hitting a significant target was small and the main result was fire in open areas. A news blackout denied the Japanese knowledge of how effective the balloons were and the programme was stopped.

Modern ballooning

The two main uses for balloons today are science and pleasure. Weather data has been collected by balloons for many years and scientific instruments have been flown to great altitudes by balloons. Hot-air ballooning has also become a popular sport and many companies have realised the advertising value

of such aircraft. Although most balloons are approximately spherical there is no reason why they have to be this shape and it is possible to build them to virtually any design. Companies often have well recognised objects in their advertisements and balloons are frequently made to the same shape. For example, a brewery could have a balloon made in the shape of a beer bottle or a construction company could have a balloon made in the shape of a house.

Ballooning as an aerial sport is increasing in popularity, although it is a good deal more expensive than most people imagine. The hourly charge on a hot-air balloon is often more than that on a light aircraft. All aviation sports are weather-dependent, ballooning more so than most. Balloons cannot be inflated if it is windy and visibility has to be fairly good.

During summer, balloon meets are held and these can be some of the most colourful events in aviation. Balloons of all shapes, sizes and hues are flow together, often in a race. The concept of a balloon race may seem a little odd, but by controlling the balloon's altitude carefully it is possible to make best use of the wind. Although any balloon must travel with the wind, the wind direction may vary with altitude, allowing some directional control.

As with any flying machine, balloons are

In recent years there has been a powerful resurgence in hot-air ballooning helped by the use of the medium for advertising, as well as recreation and sport. This is a meeting in northern France in the summer of 1985.

sharing airspace with all sorts of other aircraft such as airliners, military planes, and light aircraft, so they must be flown by responsible people who have been trained. In many countries a balloon pilot's licence is needed and apart from the handling of the aircraft, the course leading to this qualification covers much the same areas as does that for any other form of pilot's licence – navigation, meteorology and air law.

Airships

From the beginning, it was realised that balloons would be far more useful if they could be steered. To do this a powerplant was necessary and the balloon had to be tubular rather than spherical. Early airships were powered by electric motors, steam engines, and coal gas motors, but in most cases the powerplants provided so little power that the airship had a tiny forward speed, making it barely controllable. The development of the petrol engine gave airships the power/weight

ratio they needed to be truly steerable.

Even Count Ferdinand von Zeppelin's first airship was underpowered. The LZ1 was 128 m (420 ft) long and it contained 11,328 cu m (400,000 cu ft) of gas which was held in 17 separate cells – a new feature. Power was supplied by two 15 hp Daimler engines, which together weighed 771 kg (1,700 lb). The LZ1 first flew on 2 July 1900. It was expected to reach 45 km/h (28 mph) but could manage only 26 km/h (16 mph).

Airships were used extensively during the First World War, particularly by the Germans. Zeppelins bombed Britain, although their success and reputation have been said to have resulted from the lack of defences in Britain rather than the effectiveness of the airships. Zeppelin attacks on Britain ended with the loss of the L70, which was shot down by a de Havilland DH4 on 5 August 1918.

After the war, airships continued to be developed for civil uses including passenger transport. Britain's Barnes Wallis designed the R100 and the R101 but the latter crashed and was destroyed. In March 1936, Zeppelin completed the *Hindenburg*, which was the world's largest rigid airship with a length of 245 m (803.8 ft). It was powered by four 1,000 hp Daimler engines, which gave it a top speed of 130 km/h (81 mph). On 6 May 1937,

the *Hindenburg* docked in Lakehurst, New Jersey, and burst into flames. Because it contained hydrogen, it was consumed very quickly, but 62 of the 97 people on board survived. This was probably because the hydrogen, being very much lighter than air, ascended away from the gondola containing the passengers and crew.

This marked the end of major airship operations, although some have remained in service for duties such as publicity flying. The best known in this sphere are probably the Goodyear airships, of which there have been over 300 built for various purposes.

Airships reborn

With the coming of modern materials and technology, interest in airships has sprung up again for both civil and military purposes. In the UK, Airship Industries has developed the Skyship 500 and 600, which will employ such advanced technology as fly-by-light control systems. Digital messages from the pilot's controls will be transmitted to the flying surfaces by pulses of light which will travel along fibre optic conductors.

Skyship 500 is one-fifth of the length of the *Hindenburg* and about a third of the diameter, enclosing one-fortieth of the gas – in this case helium, which is not flammable. Although it is

Munich is the scene for this Zeppelin airship flight in 1909 (*above*); in later years, these craft were used to bomb London. Today, there is interest in using modern airships, like the British-designed Skyship 500 (*right*) for civil, commercial, paramilitary and naval patrol purposes. Modern airships are filled with non-flammable helium rather than hydrogen, as were the early Zeppelins, but helium could not be obtained for later Zeppelins because of American government embargoes.

a non-rigid airship, the Skyship 500, like others, becomes extremely rigid when inflated. Rigid airships, which usually have metal structures inside the envelope, are not favoured today.

As the altitude of an airship increases, the external pressure decreases and the envelope gas would try to expand, straining the skin. The gas could be vented off, but helium is very expensive, so ballonets are fitted to control internal pressure. The ballonets are air pouches inside the envelope which can be filled by air from the propeller wake or by electric fans. On take-off the ballonets are usually partially filled to an extent governed by the amount of helium needed for lift-off. As the altitude increases the ballonets empty, allowing the helium to expand. The maximum ceiling (or pressure altitude) of the airship is therefore governed by the amount of gas needed for take-off, because once the ballonets empty the airship cannot go higher without venting gas. As the airship descends the ballonets are filled again to maintain a pressure in the envelope slightly above atmospheric, usually by about 5 millibars maximum.

The front and rear ballonets can be inflated to different amounts to give fore and aft trim. A full ballonet holds 600 m³ (21,186 cu ft) of air which weighs 530 kg (1,168 lb). This acts over a moment arm of 16 m (52 ft). The laminated polyester fabric and Saran polyurethane film of the envelope leaks helium at 1 per cent per month. Titanium dioxide white polyurethane gives the outside a reflective coating to reduce the effects of solar heating.

Zip fastener

The Skyship 500 has a handle outside the captain's window, connected to a rip line which opens the envelope when pulled. This is for use in emergencies (when the airship is on the ground) since it costs many thousands of pounds to fill the envelope with helium. Spring clips prevent inadvertent operation. The propellers on Skyships can be swivelled upward, forward, or downward, giving positive vertical control as well as thrust.

Ten passengers can be carried in the gondola of the 500, which has large windows allowing in a great deal of daylight compared with most aircraft. There are no rudder pedals, yaw being controlled by the control wheel in the manner that normally moves the ailerons on a winged aircraft. In addition to the normal flight instruments the panel has gauges showing the envelope pressure and the thrust angle of the propellers, which are ducted.

Above the captain's head are four handles on heavy vertical shafts which operate the air valves for emptying the ballonets. Additional instruments on the panel above the windscreen include those which show helium temperature and water ballast contents. The temperature of the helium is important in lifting performance and calculation of the load sheet. It is related to the outside air temperature expected at the destination so that the buoyancy (either positive or negative) at the destination can be predicted. The maximum allowable 'lightness' (positive buoyancy) is 200 kg (440 lb) and if this is exceeded then precious helium has to be vented off.

The Skyship 500 has 350 kg (770 lb) of water ballast which can be jettisoned quickly to stop a rapid descent. Two Porsche 930/10 engines drive ducted propellers, which have four pitch settings. The blade pitch is set electrically and powerful springs return the blades to coarse pitch if there is a failure. Engine rpm is shown on vertical LED strips to the right of the captain's instrument panel.

Nominally seven 10 kg (22 lb) ballast bags are taken off by the ground crew for each person on board. Fore and aft trim cannot be felt on the ground because the nose is attached to the mast, so ballonet inflation has to be read off.

An 'astrodome' in the gondola roof allows

the general condition of the envelope interior to be seen. The ballonets can be observed to see how full they are and for fore-aft trim to be estimated. Typically the airship will not be neutrally buoyant, so that up or down thrust will be needed to hold height. When ready to take off the airship is released from the mast and handled by the ground crew, four men holding each of two bow lines. With the pilot assisting by controlling power and rudder, the ground crew can manoeuvre the airship quite easily.

A typical take-off would involve the thrust being vectored to 45° up and the engines being throttled up near their 205 hp maximum output. The thrust vectoring gives a much greater degree of control than on airships without the facility. The minimum effective control speed is 19 km/h (10 kt), but on many occasions the wind speed will be greater than this, so full control will be available from the moment of release. A take-off speed of 39 km/h (20 kt) is desirable to be able to handle gusts.

A climb rate in excess of 6 m/sec (1,200 ft/ min) would mean that the ballonet valves would be fully open and the envelope pressure would need monitoring. But the Skyship 500 is easily capable of going up at 7.6 m/sec (1,500 ft/min). Airships are a good deal less responsive than aeroplanes to levels of control and their pilots have to make allowances for

SKYSHIP 500

Country of origin: UK.
Role: Transport airship.
Width: 14 m (45.92 ft).
Length: 52 m (170.58 ft).
Load: 1,600 kg (3,527 lb).
Engines: 2 × Porsche piston engines, 152 kW (204 hp).
Max speed: 101 km/h (63 mph).
Range: 870 km (541 mi).
Crew: 2 pilots.
Passengers: Up to 10.

The Goodyear tyre company maintains a number of airships for publicity and commercial purposes. They have become familiar sights, particularly at sporting events, where they are used for filming. One was used as camera platform for the Los Angeles Olympic Games.

the fact that they must respond to aerial disturbances by extensive control movements.

Mixed fluids

Controlling an airship is effectively handling a sack of mixed fluids. The gondola is well below the centre of buoyancy (7 m (23 ft) below in the case of Skyship 500), giving a pendulum effect, and the rear-mounted control surfaces give the impression that all disturbances come from the back. An airship is not banked into a turn like an aeroplane. The control wheel applies rudder and the centrifugal effect causes the gondola to swing out. Large rudder deflections will cause the rudder to lock into the airflow. This is caused by the tail having a high sideways speed, the airflow holding the rudder at full deflection. It is a common airship characteristic and the condition is fairly stable and not dangerous.

Because the engines are on the gondola, the airship tends to pitch nose-up when power is added, and nose-down when power is taken off. Maximum operating speed of the Skyship 500 is 102 km/h (55 kt) and as this speed is approached the controls become more effective but more force is needed to move them – a characteristic of a simple mechanical control system.

In an airship, power controls speed, with attitude making little difference. If power is set at 75 km/h (40 kt) in level flight, say, the nose can then be pushed down into a fairly steep dive and the airspeed will remain the same. This is because in buoyant flight weight and lift remain in near balance at all angles. At 30° nose-down pitch and 75 km/h (40 kt) a descent rate of over 10 m/sec (2,000 ft/min) is possible.

8

TECHNICAL TERMS AND EXPLANATIONS

There are many terms in aviation which can give the layman trouble. This section has gathered these together for easy explanation. It also includes those terms which will become commoner as new materials, methods and systems are introduced in the quest by nations to use aviation in order to develop their technological bases. The importance of composite construction, digital flight controls, instrument landing systems and microchips are explained with illustrations and examples are given of many of them.

Wings are already far more efficient than those of even a decade ago – thanks to the use of computers in design and to modern materials. However, engineers are within reach of further improvements which will substantially change the appearance of aircraft.

The wings of the latest high-speed combat aircraft are as much inclined planes with near symmetrical section as wings, and develop up-thrust by reaction of the air they push downwards due to their **angle of attack** (AoA). However, the classic wing develops **lift** by suction – reduced pressure over the upper surface caused by air having to travel a greater distance, and therefore faster, over the wing than under it. Wings also have **drag** – made up of **induced** or lift-related drag, and **profile drag** due to shape and **skin friction**. The degrees of lift and drag depend on the wing's shape, particularly its **camber** – the overall curvature by which it deflects airflow slightly downwards – and its angle to the airflow (angle of attack, as mentioned, but also known as **angle of incidence**).

A wing's efficiency is sensitive to forward

Flying in restricted airspace, such as the mountains of Switzerland, calls for aircraft like the STOL Pilatus Turbo-Porter with specially designed wings which generate extra lift thus allowing low landing and take-off speeds.

Combining the speed and fuel efficiency of a turboprop with the Vertical Take-Off and Landing (VTOL) performance of a helicopter, the V-22 Osprey is the first modern tilt-rotor design to approach production; it is due to fly in 1987/88.

speed, and different wing shapes suit different speed regimes. A high-lift wing, as used on gliders or **short take-off and landing** (STOL) aircraft such as the Twin Otter, is characterised by pronounced camber, and a high **aspect ratio**; that is, its **wing span** (tip to tip dimension) is large compared with its **chord**, or fore-and-aft dimension. For the slowest speeds, as during take-off and landing, camber is often accentuated with **high-lift devices**. **Leading edge slots** are common; these allow air to pass through the wing to reinforce the layer of air next to the surface, or **boundary layer**, above the wing. Leading edge **slats** and trailing edge (the wing's after edge) **flaps** extend the cambered wing forward and back. **Slotted flaps** can be extended in sections to produce a marked increase in wing area and camber. These are a type of **Fowler flap**; other types are **simple** (hinged) or **split flaps**. **Blown flaps** have air blown over them to augment the airflow and resulting lift.

For high-speed flight a wing needs to be thinner and flatter, as any excess camber increases drag markedly. In the **transonic** region (transition from **subsonic** to **supersonic** – faster than sound), shock waves build up over straight wings causing **buffet** and adverse control effects. Angling or **sweeping** the wings back can delay the onset of these **compressibility effects** and associated **buffet stall** – breakdown of wing lift. A **delta** wing (as on the Mirage) combines sweepback with a long trailing edge for control surfaces, and high strength. A **variable geometry** aircraft like the F-111 or Tornado can vary its wing sweep to suit different speed regimes. **Supercritical** wings are thin and have a relatively flat

Modern airliners, like this Airbus A300, have various lift aids and airflow regulators on the main wings which are not only designed for safety but also efficiency. An important requirement of modern international airlines is fuel economy.

top surface to achieve a weak shock wave at transonic speeds, compared with a strong one on conventional wings. This pushes the rise in drag to much higher airspeeds.

Enhancing performance

Efforts to improve wing efficiency are focused on determining the best **profiles** (shapes in section), with the help of computers; using composite materials to reproduce these in actual wings – possibly embodying several variations in section in a given wing; and controlling the airflow over the wing. Future **mission adaptive wings** (MAW) will be able to change wing profile in flight.

A major contributor to drag is disturbed airflow, including rotating **vortices** (small whirlpools of air) caused by discontinuities such as projections and wing tips. One recent development, made practical by improved materials technology, is to add vertical **winglets** – like 'turned-up' wing tips – to capture some of the wing-tip vortex energy and turn it into thrust.

Another possibility which has tempted engineers for years is promotion of **laminar flow** – smooth, low-friction airflow – by controlling the **boundary layer** of air next to the wing's skin. Normally, laminar flow occurs at the wing's leading edge but breaks up a short proportion of the chord aft into 'dirtier' **turbulent flow**. If this disturbed flow could be smoothed out, wings would become more efficient, with less frictional drag and more lift. Some experimental aircraft achieve controlled flow by suction, applied through many tiny holes in the wing's surface. A clean, smooth surface is needed and keeping the wing wet with a chemical-water mixture can keep it free of squashed insects and other build-up which could be enough to disturb the flow. Laminar flow can also be encouraged by suitable design of wing section. Good laminar flow can provide a 20% reduction in **fuel burn** on future airliners.

New shapes

New concepts are putting into question the traditional relationship of the wing to the rest of the aircraft. Most dramatic is the **forward swept wing**, flown experimentally on the

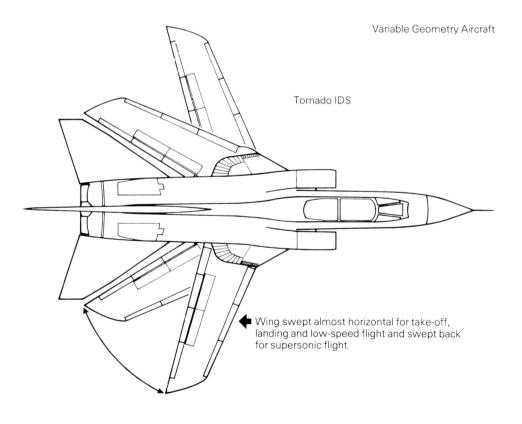

Variable Geometry Aircraft

Tornado IDS

Wing swept almost horizontal for take-off, landing and low-speed flight and swept back for supersonic flight.

Based in the United Kingdom, the F-111 nuclear-capable fighter was the first successful swing-wing (variable geometry) aircraft to enter regular service with the US Air Force. It is used both tactically and for strategic duties.

207

Grumman's X-29 is a remarkable reverse-wing concept developed originally in the 1940s; this modern version combines great manoeuvrability with good lift efficiency. The aircraft may well enter US Air Force service by 1998.

Grumman X-29 and providing lift efficiency and great **agility** (manoeuvrability). The strains on such a wing in combat are enormous, and only **composite materials** (see helicopter section) are making this a candidate for advanced tactical fighters. Composites enable designers to put strength precisely where it is needed and to tailor the **aeroelasticity** of the wing – its flexibility related to aerodynamic requirements. Because they are essentially plastic moulded materials, the shape or section of the wing can be varied, making possible features such as **discrete variable camber** (several camber sections incorporated in the same wing).

An essential adjunct to such advanced concepts is a **fly-by-wire** control system, in which **electrical signalling** of control surfaces from a computer takes the place of mechanical control runs. The advantage of fly-by-wire is that it provides effective control where a human pilot cannot, in conditions of **relaxed**

Giving new life to old designs, the Dassault-Breguet Mirage IIING (New Generation) uses a canard foreplane to improve flight characteristics, especially at low speed, important for a fighter operating in modern combat conditions.

static stability.

A conventional aircraft is stable in the pitch axis, any oscillation produced by a momentary disturbance rapidly dying away. In a relaxed or negative stability design (centre of gravity at or behind the centre of lift) the oscillations would grow were it not for the electronics. Such designs can, however, respond more quickly to control movements, promoting tactical superiority. Because of their dependence on advanced digital flight control systems, such aircraft are sometimes described as **control configured vehicles** (CCVs).

A related development is the **canard** or **foreplane** located ahead of the **mainplane** (wing) as seen on the Mirage NG. A canard is relatively small, providing less than a quarter of total lift. If it provides more than a quarter of the lift, the surface is regarded as a **forward** or **tandem wing**. Both planes provide lift, in contrast to a conventional aircraft where the **tailplane** has to provide a down-load to counteract the fact that lift acts aft of the **centre of gravity** in order to confer positive pitch stability. Again, electronics provide **artificial stability** or else the agile relaxed stability canard aircraft could not function.

The ultimate in 'relaxed stability' is probably the forward swept wing aircraft with canard – rather resembling an arrow in backward flight!

For passenger aircraft, use of a foreplane enables the four wings to be located at the ends of the fuselage, offering superior views for passengers, and **pusher turboprops** leave clean wings with location of noise source aft.

Engines – more power, less fuel and cheaper

The three major strands of engine development today are to increase the power available from a given weight or volume of engine, to reduce **specific fuel consumption** (sfc) – consumption for a specific power – and to reduce **cost of ownership**, the costs of buying and running an engine.

Engine types

Both the **reciprocating** piston engine powered by **Avgas** (aviation gasoline) and the **jet** engine are **internal combustion** engines, gases burning inside the engine doing work by expanding rapidly. For economic reasons jets have largely superseded piston engines, except on small **general aviation** (GA) aircraft and some **commuter** airline types intended for modest speeds and altitudes.

Jet engines work on the **reaction** principle; ejecting a jet of gases from the back of the engine at high speed creates a reaction which pushes the engine forward. The jet is created by mixing pressurised air with vaporised **kerosene** (paraffin) and burning it in a **combustion chamber**. The rapid expansion involved in this process causes the gas to be expelled with great force. The burning process is continuous rather than intermittent as in a reciprocating (up and down or back and forth) engine.

Air for combustion is drawn into the engine through the front **air inlet**. The low-pressure air is pressurised by a **compressor** and pushed into the combustion chamber. The compressor is driven by a **turbine**, which is driven in turn by jet gases leaving the combustion chamber at the engine's rear or **hot end**. Jet engines are therefore also called **gas turbine** engines. The compressor and turbine together form the **core** of the engine.

Gas turbines are of various types. Transonic and supersonic aircraft are powered by **turbojets**, typified by low frontal area and high **jet velocity**. In these, the simplest and earliest form of gas turbine to be developed, all the air entering the engine passes through the engine core.

Commuter and short-haul passenger aircraft are powered by modern, fuel efficient turboprop engines which, although they may look old-fashioned to some passengers, are in fact of the latest technology and have a fine safety record.

Piston engine (Wasp)

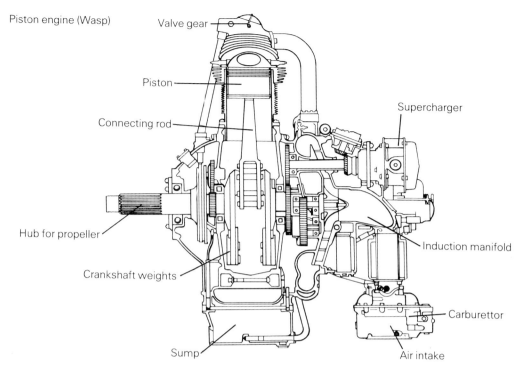

Valve gear

Piston

Connecting rod

Supercharger

Hub for propeller

Crankshaft weights

Induction manifold

Sump

Carburettor

Air intake

Propfan

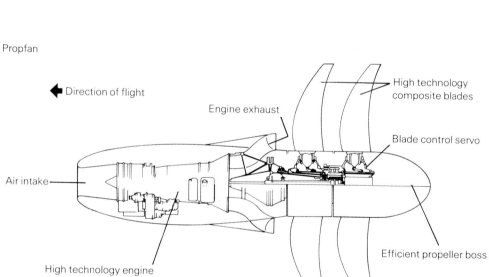

Direction of flight

Engine exhaust

High technology composite blades

Air intake

Blade control servo

High technology engine

Efficient propeller boss

Bypass turbofan

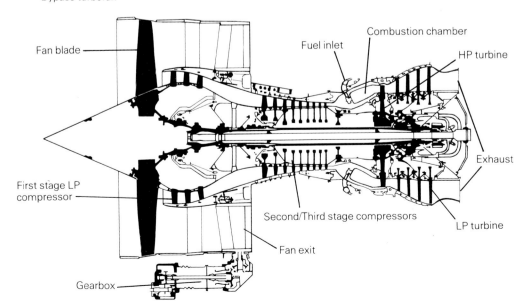

Fan blade

Fuel inlet

Combustion chamber

HP turbine

First stage LP compressor

Exhaust

Second/Third stage compressors

LP turbine

Fan exit

Gearbox

Modern high-performance aircraft require engines of equally high ability, capable of Mach 2 flight, yet giving good endurance and fuel economy, for example the General Electric F404 power plants for the Northrop F-20 Tigershark.

For high subsonic speeds – typically 965 km/h (600 mph) – a lower jet velocity is more efficient and the **turbofan** engine is used. This is a **by-pass** engine, where part of the air is compressed for combustion, while the rest is only slightly compressed and is ducted round the core. Some propulsion derives from the large ducted **fan** at the front of the engine. This is driven by the turbine, either together with the compressor, or at a reduced speed through gearing (**geared fan**). The ratio of air by-passing the core to that passing through it is termed the **by-pass ratio** (BPR). A BPR for a typical modern turbofan is eight to one.

Power is expressed in units of thrust, normally kilograms (kg), tonnes (t) or pounds (lb).

The **turboprop** is a turbojet having an extra **power turbine** which drives a **propeller** through a **reduction gear**. It is highly efficient at moderate speeds and altitudes, typically 640 km/h and 9,000 m (400 mph and 30,000 ft). Power is expressed in **kilowatts** (kW) or **shaft horsepower** (shp) – power measured at the shaft – plus the **residual thrust**.

Turboprops may be due for a new lease of life through the **propfan**, an aerodynamically advanced propeller-cum-fan, **ducted** or **unducted**, which could provide greatly reduced fuel consumption and **direct operating costs** (DOC) for high subsonic speed aircraft by the turn of the century. Propeller-driven aircraft could be flying at jet speeds.

A **turboshaft** is similar to a turboprop, but

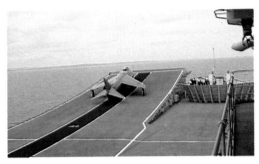

Enhancing the load-carrying capability of the Short Take-Off and Vertical Landing (STOVL) Sea Harrier, the ski-jump ramp on the forward end of an aircraft carrier's deck gives the fighter a semi-ballistic launch.

without its propeller. Instead an output shaft is used to drive something else, such as a generator (in a power station) or a helicopter rotor.

A **ramjet** is a jet without rotating compressor or turbine internals. Air is compressed by the **ram** effect of the engine's forward motion. The engine is simple, but efficient only above **Mach 1** (speed of sound).

A **rocket** is a jet engine which relies on fuels which support combustion without the need for air.

Vectored thrust is a means of changing the

Helicopters like this Westland Sea King use turboshaft engines, in this case the Rolls-Royce Gnome. Mounted on top of the cabin, the engines are easily accessible and can be replaced even in the confines of a ship's hangar.

direction of, or vectoring, the jet and is used to achieve **vertical or short take-off and landing (V/STOL)**, or variations such as **short take-off and vertical landing (STOVL)**. In the **Pegasus** turbofan used in the **Harrier** and AV-8A V/STOL aircraft, thrust is vectored by four **swivelling nozzles**, two of which carry hot jet gas and two fan by-pass air. **Plenum chamber burning** (PCB) – burning fuel in the forward nozzles – could augment thrust available in future marks. **Reversed thrust** is an extreme form of vectoring in which **clamshell doors** or **retractable ejectors** reverse the hot gas thrust completely, or **blocker doors** on turbofans

reverse the cold air flow. It is used to brake aircraft after landing.

Reheat or **afterburning** provides short-term **boost power** by burning additional fuel in the jet pipe to augment exhaust velocity and thrust.

Like blowlamps, jet engines can occasionally go out (**flame out**) in flight but can be brought back by **relight** procedures. **Surge** is unequal delivery and burning of fuel, with varying engine speed.

Improving performance and efficiency

The routes to better performance are more pressure or higher temperature – technically greater **mass flow** (of air through the engine) or higher **combustion temperature**.

Modern compressors can pressurise air up to some 30 atmospheres (**pressure ratio** 30:1). Compressors can be single or **multi-stage**. State-of-the-art engines may have several compressor stages driven independently at optimum speeds on separate **concentric shafts**, each driven by an associated turbine drive stage. The stages are known as **spools**, a three-spool engine having three stages.

Rocket motors often power modern missiles with high supersonic performance, like the US Navy's Tartar system (illustrated), used for medium-range shipborne area air defence. Other missiles have ramjet motors.

Vectored Thrust of the Harrier

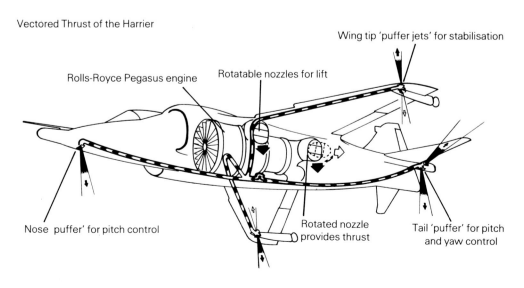

Wing tip 'puffer jets' for stabilisation

Rolls-Royce Pegasus engine

Rotatable nozzles for lift

Nose 'puffer' for pitch control

Rotated nozzle provides thrust

Tail 'puffer' for pitch and yaw control

In Europe and the United States, considerable research and development work is being carried out to find more efficient engines, and NASA is involved in the Unducted Fan Engine which will power the Japanese–American Boeing 757 programme.

For small airlines, a low ownership cost of both engines and airframes is vital to keep the airline in profit; servicing must be kept to a minimum, as with this Dash 7, which is in daily airline service around the world.

Compressor stages may be **axial** or **centrifugal**. An axial stage has alternate sets of rotating and stationary aerofoil blades (**rotors** and **stators**) impelling air through a converging duct. A centrifugal stage works by throwing air outwards against the casing. High-speed centrifugal stages are currently seen as a means of improving compressor efficiency.

Raising combustion temperature improves efficiency, but raising it significantly above 1600°C (2,912°F) calls for **exotic materials** technology. Air-cooled turbine blades, probably of titanium, are the first stage, though metals with **single crystal** or **directionally solidified** structures will be needed to remove potential fault paths inherent in the normal crystalline structure of metals.

To maintain close **tip clearances** between blades and the turbine casing (a 0.076mm (0.003in) reduction in clearance can produce a 1% rise in **stage efficiency**) demands high **thermal stability**. This explains moves to the use of materials such as silicon nitride and **ceramics** for blades – which need not then be air cooled – for casings or for both. An alternative on some engines is **active clearance control** (ACC), involving digital control of cooling air. Turbine stages will revolve on **air bearings**, since liquid lubrication media would barely tolerate the temperatures involved.

Combustion chamber **burners**, igniters and annular 'flame tubes' will also utilise advanced materials able to withstand higher temperatures and pressures.

Cost of ownership

Making engines economical to buy and run involves attention to a range of disciplines.

One factor is reducing complexity and **part count**. A major way to achieve this is with fewer compressor-turbine stages, achieved by making remaining stages highly efficient. The ideal in breaking the historical efficiency-complexity link is the single-stage compressor. An associated single-stage turbine could comprise a single unit blade and disc combination termed a 'blisk'.

Another approach is to minimise costs of design and production. Computer methods are reducing input of manpower. **Computer-assisted engineering** (CAE) embodies **computer-aided design** (CAD), **computer-aided manufacture** (CAM), with perhaps the two linked as CADCAM. Factories are being automated with machines under **computer numerical control** (CNC) and **digital numerical control** (DNC). **Flexible manufacturing systems** (FMS) apply low-manpower automation methods to **batch** (limited volume) **production**.

Engines can also be designed with higher **maintainability**. One method is to enable **first-line servicing** engineers on the spot to **repair by replacement**, simply removing a faulty engine module and replacing it with a **line replacement unit** (LRU), sending the faulty unit back for second- or third-line attention. This **modular design** philosophy results in minimum aircraft **downtime** and maximum revenue earning time. Engines are also designed in the first place for higher **reliability** and therefore less frequent maintenance.

There will be continued moves towards replacing critical components **on condition** (when sufficiently worn), rather than at the end of approved **rated lives**, which inevitably have to be conservative. This will depend on perfection of continuous self-monitoring and diagnosis, through intelligent **health and usage monitoring** (HUM) systems backed up by vibration-sensing **accelerometers**, fibre-optic **borescopes** – needle-like visual inspection probes which can be permanently installed – and **chip detectors** allied with **debris monitors**, for examining particles in lubricant, as well as the normal temperature, pressure and flow sensors. The latest engine technology combined with systems such as **engine instrumentation and crew alerting system** (EICAS) has already 'engineered out' the third crew member, the flight engineer.

Attack can also be made on **fuel burn**, through **fuel flow management**. Engine fuel can be metered and controlled digitally – comprehensive systems are called **full authority digital engine control** (FADEC). Implications extend to the flight deck. **Performance management systems** calculate efficient flight profiles for the pilot or **automatic flight control system** (AFCS) to follow. A further step combines performance and navigation management in a full **flight management system** (FMS). Such a system can optimise lateral and vertical navigation together with

As part of the UTC/Fiat rescue package for Westland, the Black Hawk helicopter is to be built under licence in the UK; in 1986 trials began to fit the RTM 322 turboshaft engine for the European and other export markets of the UH-60B version.

Self-defence tactics for modern fighters

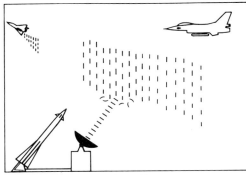

i) Ground-based radar is unable to penetrate 'window' dropped by fighter.

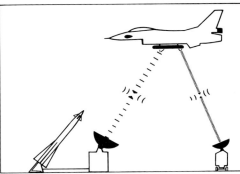

ii) Ground-based radars are unable to 'lock-on' to fighter carrying electronic counter-measures (ECM) pod.

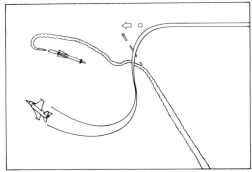

iii) Fighter avoids missile using evasive manoeuvres.

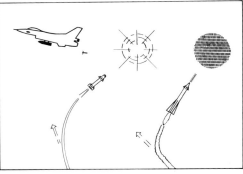

iv) Fighter drops chaff to decoy SAM and uses infra-red flares to decoy air-launched missile.

fuel flow management. In civil applications it may even control the aircraft according to a previously entered **flight plan**, right down to automatically selecting and tuning radio navaids.

Finally engine ownership costs for a given aircraft can be constrained by reducing the number of engines. Large turbofans in the 25 to 30 tonne thrust category are enabling big twins to fly routes previously flown by three- and four-engined aircraft. Sufficient **one-engine inoperative** (OEI) performance and higher reliability may persuade regulatory authorities to clear the twin-engine airliners like A320 for long-distance transoceanic operation.

Towards the invisible aircraft

Stealth – the art of not being seen or detected – brings great advantage in warfare and has been in vogue for a long time; witness **camouflage**, to make a presence merge visually into the background; **decoys** to distract attention from the real **threat** (or victim), **smoke screens** and other obscuring barriers, and **disguise**.

These means emphasise the historical pre-eminence of the visual sense but recent times have seen the principle extended into other parts of the **electromagnetic** (EM) **spectrum**. For example, aircraft designed for deep **penetration** of hostile airspace seek to avoid **radar** detection by flying very fast and very high, or conversely by **terrain following** at tree-top height, effectively hiding below radar beams or in radar **clutter** – masking reflections caused by ground features.

Such non-visual **detection** techniques have given rise to a new arena of warfare, **electronic warfare** (EW). A force will try to **jam** an opponent's radar with noise, swept frequency, spot frequency or other jamming, or **spoof** it with **false returns**. These constitute **electronic counter measures** (ECM). The simplest ECM is the time-honoured **chaff**, a cloud of airborne **dipole** (linear antenna-like) reflectors which can be used as screens or decoys, launched from aircraft or chaff-dispensing

An aspect of electronic warfare is the Boeing E-3 Sentry airborne warning and control system (AWACS), which uses the large Westinghouse airborne early warning radar atop the aircraft's fuselage. The aircraft is flown by the USAF and NATO.

For electronic surveillance of enemy signals (ESM), helicopters like the Westland Lynx are fitted with the Racal MIR-2 system. One of the antennas can be seen on the nose while others, covering 360°, are found on the fuselage.

munitions. To combat ECM the opposing force will equip its radars with high-power **burn-through** capability, **frequency agility** (hopping from one frequency to another, often in **pseudo-random** fashion), or other **electronic counter-counter measures** (ECCM).

The alert pilot's eye, which was once sufficient to detect threats, now has to be supplemented by devices which can detect **hostile emissions** from radars and other sensors over a range of frequencies. These are collectively called **electronic support measures** (ESM). In combat, ESM becomes an essential defence, warning a pilot of **locked-on** enemy radar – an indication that he has been **acquired** as a target – and probably identifying threats by comparing their emissions with a

library of stored EM signatures. On-board ECM and ESM are vital in reducing an aircraft's **vulnerability** in a **high-intensity environment** (a fierce battle).

Since remaining undetected is paramount in a **covert** (literally under-cover) **strategic surveillance** (popularly referred to as a 'spy') mission, **stealth** has become an important and organised concept since the mid-1970s, embracing several technical disciplines.

Specifically, the idea is to achieve a low **signature** (characteristic emission pattern) in any spectrum from any angle. The main signatures are **visual**, **radar**, **thermal** (infra-red), **electronic** (radiation), **acoustic** (sound) and **pressure**.

Visually the best defences are small size, non-reflecting surface, camouflage finish and very high or very low altitude.

Various means for making the aircraft quiet externally range from **baffles** and sound-**absorbent** materials to the 'high-tech' cancellation of sonic vibrations at source with artificially created vibrations which are exactly **antiphase** (in the opposite sense). **Anechoic** (non-reflecting or echoing) structures which **attenuate** (reduce the volume of) sound by bouncing waves repeatedly among acute-angled pyramidal surfaces, **absorbing** a proportion at each bounce, can be used in ducts and other cavities.

The chief means of reducing **visibility** to radar are to minimise reflective area or **radar cross-section** (RCS), and to absorb any incident energy.

Radar is easily reflected by non-acutely angled structures and vertical surfaces, espe-

212

Maritime patrol aircraft, like the Lockheed P-3 Orion, have modern sensors, including radar for surface search duties, as well as carrying the anti-submarine and anti-ship weapons with which to prosecute a target once it has been detected.

cially flat ones. The substantial side surfaces in a conventional aircraft – fuselage, nacelles and particularly fins – can be designed out in an **all-wing** form having buried engines **blended** into the wings. **Stores** – detachable munitions such as rocket bombs – are carried either internally or on **conformal** weapon pallets whose shape blends, or conforms, with that of the aircraft, avoiding angles or projections.

Flying wings (or **lifting bodies**) can be inherently unstable at high Mach numbers, particularly when vertical stabilisers are minimal – making automatic flight control systems (AFCS) essential for effective control.

Engines with their dense rotating masses and angular inlets and exhaust ducts are highly reflective, requiring special attention to **low profile**, **serpentine** (curvy) intake tunnels, streamwise baffles, shields and possibly inlets which are **flush** below the aircraft.

To the second line of defence, absorbing radar energy: metals tend to reflect, but some plastics, including ABS thermoplastic, absorb the energy, heating slightly in the process. Carbon-based plastics, such as **carbon-fibre reinforced composite** (CFC) and carbon impregnated foams and **reinforced carbon-carbon** are particularly effective, raising the prospect that 'invisible' aeroplanes will be plastic. Absorbent structures such as plastic

Using early 'stealth' technology, the Lockheed SR-71 Blackbird has reportedly penetrated the USSR, Warsaw Pact nations and the People's Republic of China to acquire intelligence for the United States national security agency. The aircraft are still active.

Many paramilitary forces now require low-cost night vision systems for surveillance duties. One such is the McDonnell Douglas MD 500 Nightfox, which has a nose-mounted Forward Looking Infra-Red to detect heat sources, like the human body.

honeycomb and radar-anechoic surfaces are useful for lining inlet and exhaust ducts.

Aircraft can also be coated with radar **ablative** finishes which, in effect, spread incident energy away from electromagnetic **hot spots**. One, known as 'iron ball' paint, apparently contains microscopic iron particles which spread energy by conduction.

Thermal signatures are tackled in a similar way, by absorbing and spreading the heat. Extended absorbent ducts are fitted, for instance, to helicopters as **IR suppressors**. An example is the Apache's 'black-hole' system.

Stealth aircraft will have exhaust systems obscuring the hottest parts of engines from as many angles as possible, and variable nozzles closable to restrict rear angle view of the engines. **Closed-loop** cooling and **environmental-control systems** (ECS) may be needed which dump heat into fuel or structure rather than disposing of it overboard. Airframe **friction heating** can be reduced with thermal barrier finishes.

Stealth also demands minimising of 'own goals' caused by emissions from one's own **active** (transmitting) sensors and electronic systems. Radars must detect targets using low-energy emissions of only short duration, leaving no more pulses than necessary available for enemy detection. Another approach is to substitute sensors which are more localised and harder to detect – **forward-looking infra-red** (FLIR) or lasers, for example. A third ploy is to go **passive**, using sensors which 'listen' only and do not transmit.

Reliance on Doppler/radar for navigation can be avoided with accurate inertial navigation (see electronics section), perhaps integrated with an automatic satellite tracking system such as **global positioning system** (GPS), or star tracking equipment.

Own antennas are, by definition, excellent reflecting surfaces. **Covert strike radars** (CSR) avoid their give-away effects by 'hiding' themselves when not in use, tilting their antennas into non-reflecting positions, or being located behind **fenestrated** radomes having radar-transparent **windows** which can be 'closed' against pulses arriving from outside. Non-reflecting electronically steered **conformal arrays** are another possibility.

Stealth aircraft already exist. The famous American U-2 'spy' plane was succeeded by the Lockheed A-12 TR-1, which, together with the SR-71 'Blackbird', was the first serious **reduced RCS** aircraft to fly. Lockheed's Mach 4 ramjet-powered D21 drone was

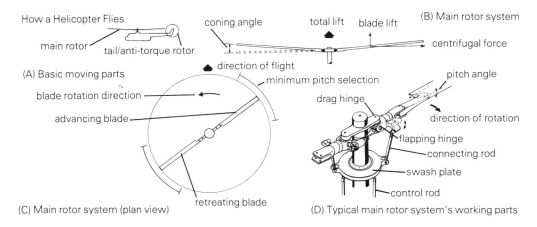

How a Helicopter Flies
main rotor — tail/anti-torque rotor
(A) Basic moving parts

coning angle — total lift — blade lift
(B) Main rotor system
centrifugal force

blade rotation direction
advancing blade
direction of flight
minimum pitch selection
retreating blade
(C) Main rotor system (plan view)

pitch angle
drag hinge
direction of rotation
flapping hinge
connecting rod
swash plate
control rod
(D) Typical main rotor system's working parts

To move forward, the pilot inclines his rotor disc forward. To do this he commands increased pitch on each blade as it sweeps the rear half of the circle of rotation, and reduced pitch as the blade passes around the front of the helicopter. Blades therefore fly higher or lower according to their cyclic position, tilting the rotor disc and providing a forward force vector as well as lift.

Rotor disc inclination is used to move the helicopter in any direction, including sideways and backwards, and is applied using the **cyclic** control. To control **heading** the pilot **yaws** the helicopter by varying tail rotor pitch using 'rudder' pedals.

The rotor disc could be tilted by using control rods to incline a universally jointed hub, but in all current designs the pilot controls a tilting **swash plate** which then transmits both cyclic and collective pitch inputs to the blades via connecting rods and **pitch arms**. The swash plate is a non-rotating anulus encircling the main drive shaft. A forward tilt of the plate causes the pitch arm connecting rods to be pulled down in the forward sector and raised in the after sector, creating an inclination of the rotor disc which mimics that of the plate. The swash plate can also be raised or lowered without tilt to provide collective pitch control. As collective is applied, an associated linkage to the engine governor ensures that more engine power is available to drive the rotor in coarser pitch.

also stealthy. Then came Lockheed's F-19 stealth fighter. The **Advanced Technology Bomber** (ATB), **Advanced Tactical Fighter** (ATF) and **Advanced Cruise Missile** (ACM) all incorporate significant stealth technology.

How a helicopter flies
A helicopter is an aircraft which uses rotating aerofoils to achieve vertical and horizontal flight, and to hover.

Lift is provided by one or more large horizontally rotating **main rotors**. Most helicopters have a single main rotor, and to counter the tendency of the helicopter itself to rotate about the main rotor drive shaft, a **counter-torque tail rotor**. The tail rotor is kept small by giving it the large moment arm of an extended fuselage or tail boom. Counter-torque options include reaction jets.

Some larger helicopters such as the Chinook have a twin main rotor configuration, providing greater lift and avoiding the need for a tail rotor. The rotors can be disposed in tandem or co-axially.

First-generation helicopters (like the Bell 47) were typically powered by a single piston engine driving the rotors through the **main**

Bell's Model 47 is a good example of a first-generation helicopter (it was the first civil type) and is powered by a single piston (reciprocating) engine. This example is used for agricultural spraying in Zimbabwe, a typical modern role.

The main rotor hub is an important part of the helicopter's dynamic flight system. This is the main rotor system on a Mil Mi-6 Hook transport helicopter, fitted with conventional titanium spar rotor blades. Modern systems are made of composites.

gearbox, reducing to low rpm for the main rotor, the **tail rotor gearbox** and **transmission** shafts; these moving parts together constituted the helicopter's **dynamic system**. With the advent of compact, high thrust-to-weight ratio **turboshaft** engines twin engines mounted over the cabin with the gearbox became popular for safety and reliability, and now even triple-engined layouts are specified, as for the Anglo-Italian EH 101.

Rotor dynamics
The main rotor comprises several individual aerofoils or **blades** connected to a **hub**. They are connected through flexible elements or through articulating **hinges** so as to allow the blades to move relative to the hub in several axes; blades can **flap** up and down, **lead** or **lag** their 'standard' positions as they move through the air, or vary their **pitch** relative to the airflow.

Pilots use blade pitch and rotor speed as the primary means to control the helicopter. To rise vertically, the pitch of all the blades is increased together to present an increased angle of attack to the airflow, using the **collective** control. In this case the plane of the **rotor disc**, the circular area swept by the blades, is horizontal over the helicopter.

Despite its modern appearance, cockpit systems and engines, the Bell 222UT still has the original teetering main rotor system of the 1940s to provide lift. However considerable research has been undertaken into the blade design and size.

Rotor disc tilt is also used to counteract the basic asymmetry of a helicopter's lift dynamics. When a helicopter is moving forward, its **advancing blade** – the one moving forward into the airstream – would create more lift than the **retreating blade** were this effect not compensated for. To equalise lift, the advancing blade is allowed to flap higher, losing effective lift, while the retreating blade inclines downwards. Once the limits of flapping movement are reached, the helicopter cannot fly forward faster without becoming unstable, and modern light helicopters like the Jet Ranger are typically limited to a forward speed of some 278 km/h (150 kt).

A related asymmetry is a tendency of the blades to speed up slightly in the advancing sector of the disc and slow down again in the retreating sector. Drag hinges are used to accommodate the resulting lead/lag movement and to suppress the vibration which would otherwise occur.

Helicopters hover more easily near the ground, where air pushed down by the rotor is prevented from dissipating and helps 'cushion' the aircraft. This phenomenon is called **ground effect**. In the event of engine failure a helicopter can descend with the rotor windmilling and, by careful pitch control, make a gentle landing. This is called **autorotating**.

Rotor construction
Conventionally, rotors have been fully **articulated** – that is mechanical **hinges** have been used to allow for flap and lead/lag movements. The development of **composite materials** which can both flex repeatedly and bear heavy loads has given designers improved means of

Federal Germany's MBB BO 105P anti-tank helicopter is designed with a rigid rotor system made of composite materials. It is extremely agile and highly manoeuvrable; a display helicopter of the type has been looped and rolled.

accommodating the required movements and superseding complex mechanical hinges. **Rigid rotors** (so called) have no articulating hinges and **semi-rigid** rotors have flapping hinges only. Articulating designs themselves are becoming simpler and easier to maintain as designers introduce **elastomeric** hinge elements.

Modern materials are also revolutionising the design of rotor blades. Composites can be moulded to any required shape, so blades are being produced where sections are varied progressively from hub to tip to suit different airflow conditions. Compared with conventional metal blades, which are of constant section, the new blades with their varied profile, **camber** and twist, together with **swept tips** to delay the effects of **rotor stall** on the retreating blade, will improve aerodynamic efficiency by a third.

Composite materials comprise glass, Kevlar (aramid) or carbon strands reinforcing a plastic matrix. They combine lightness and strength with high tolerance to fatigue and damage. Composites are also coming into use for **primary** (main) and **secondary** airframe structure and helicopters with all-composite airframes have been built in America by Bell and Sikorsky. Composites are also coming into use for fixed-wing aircraft.

Electronics
The capability of a modern helicopter depends heavily on its electronic systems. Due to their lack of inherent stability many helicopters have a **stability augmentation system** (SAS) to make them easier to fly. Civil helicopters are increasingly equipped for **instrument flight rule** (IFR) operation, single or dual pilot. Day/night and all-weather battlefield operation calls for **night vision systems** (NVS) and, for low-level **nap-of-the-earth** (NoE) tactics, precise control assisted by an advanced **automatic flight control system** (AFCS). For minimum hull exposure, weapon sights are mounted over the cabin, or over the main rotor hub in a **mast-mounted sight**. The AFCS plays an essential role in **anti-submarine warfare** (ASW) and **search and rescue** (SAR) missions where highly precise flying is called for. In both battlefield and maritime roles, the newest of fire-and-forget missiles are replacing wire- and radar-guided missiles, which leave helicopters 'hull up' and open to detection. **Cockpit management systems** are easing crew workload by automatically managing communication, navaid and some mission control and selection functions.

In civil operation, **electronic flight instrumentation systems** (EFIS) are using colour cathode-ray tubes (CRTs) to improve the display of flight and mission information to the flight crew. In the 'glass' cockpit, conventional

Top: half-way between the modern electronic cockpit and the older analogue style, the Westland Lynx-3, a private venture design, is fitted with a Racal RAMS mission management system (TV screen left centre) to reduce aircrew workload in action.

Above: in many of the larger civil helicopters, electronic flight instrumentation systems (EFIS) are now becoming a standard fit, especially if they operate in crowded airspace as for example the Westland 30s of Pan Am, seen here over New York.

controls may give way to **sidestick controllers** for ergonomic reasons. Electronic **health and usage monitoring** (HUM) is seen as a way of increasing reliability. Helicopter vibration will be alleviated in flight by using **higher harmonic control** (HHC) – making small cyclic rotor blade adjustments to cancel out the effects of vibration sources. Other on-board sensors will monitor blade path height or **tracking** as an aid to subsequent balance and pitch adjustments.

The future
Future helicopters will be faster, safer and more reliable. Cockpit electronics will be better integrated, based on computers communicating on two-way digital data highways – or **databuses**. Helicopters will be self-monitored to ensure efficient operation and improved maintainability. Mechanical control linkages will give way to electrically signalled **fly-by-wire** or **fly-by-light** systems incorporat-

ing intelligent actuators.

The helicopter forward speed barrier will be broken by concepts such as the **tilt rotor**, now going ahead in America as the V-22 Osprey programme, and the **advancing blade concept** (ABC), under which lift symmetry at speed is maintained by having two coaxially contra-rotating main rotors.

The growing role of avionics/electronics

Avionics (electronic systems in aircraft) account for a third to a half of the purchase price of sophisticated modern aircraft and further advances in operational capability depend on their continued evolution.

Digital technology

Most of the technological expectations for the next two decades are underpinned by **digital** technology. Unlike earlier **analogue** electronics, which represent continuously variable quantities in like or analogous fashion, a digital system breaks up a variable into a number of discrete steps, each of which can be represented numerically. The numerical system used is the **binary**, in which quantities are expressed in terms of noughts and ones – representable by the clean-cut ON/OFF states of electronic switches. Binary digits, or **bits**, can be stored and processed electronically, and **logical** manipulation of **data** by electronic computers is fast, accurate and versatile.

Originally the large numbers of electronic switches required were realised as **discrete** components – initially valves, then **transistors**, which actively switch – and **passive** resistors, capacitors and inductors. Nowadays circuits are **monolithic**, literally etched out of

The high-technology 'glass' cockpit of the Airbus A320 airliner uses television-type screens and very few conventional instruments, thus reducing pilot workload by presenting only the information actually required at any one time.

blocks of **semiconductor** – usually **silicon** (occasionally germanium). The smaller they are the faster they switch binary state and the more compact 'black boxes' become. Currently thousands of individual transistor circuits can be **integrated** on tiny slices, or **chips**, of silicon.

The electronics generation currently in service is typified by numerous individual **integrated circuits** (ICs), mounted on **printed circuit boards** (PCBs), normally installed on frames and inserted as plug-in **cards** into standard **cases** (best known are the various ARINC aviation standard ATR cases). The ability to integrate more and more active circuits on to each individual chip has led to the computer-on-a-chip and through **large-scale integration** (LSI) to **very large-scale integration** (VLSI).

As well as more processor power for a given volume, processor manufacturers constantly seek higher speeds, with switching times measured in **nanoseconds** (thousand-millionths of a second), and **very high-speed integrated circuits** (VHSIC) will enable processors to work at rates up to several **million instructions per second** (MIPS). High speeds are needed for the complex **real-time** processing characteristic of aircraft systems, where, for example, a control system must act to counter a disturbance immediately it happens.

Several different semiconductor and logic technologies have been used to achieve required combinations of speed, capacity, power supply levels and resistance to **electromagnetic interference** (EMI). These include **polar**, **bipolar**, transistor-transistor logic (**TTL**), emitter coupled logic (**ECL**), metal oxide-silicon (**MOS**) and the currently popular **CMOS** (complementary metal oxide-silicon). Processing power has become comparatively cheap and no longer limits system capability. A trend towards **distributed processing** is seeing **microprocessor** intelligence

distributed around the aircraft and communicating on a 'ring main' digital **databus**, so enhancing survivability. On the other hand, strides in VLSI miniaturization may eventually favour **central processing** based on powerful **multiprocessors** or **array processors**, in which elements work in parallel to execute large processing loads fast.

When not being processed, data is stored in **memory**. Capacity is expressed in **bits**, bit groups called **bytes** (they can be thought of as syllables) or digital **words**; the most common measure is **kilobytes** (Kb) or **megabytes** (Mb). Word length in avionic computers has generally been eight bits, but needs for greater precision and resolution are prompting **16-bit** and **32-bit** system development.

A semiconductor device holding data long-term for reference rather than processing is a **read-only memory** (ROM) chip. Where the stored data can be erased and replaced under programme control, the chip is referred to as **erasable programmable memory** (EPROM) or **electronically erasable programmable memory** (EEPROM). **Random access memory** (RAM) provides virtually instant access to data, no matter what its **location**, unlike **serial** memory – as in magnetic tape or discs – where **access time** depends on the data's position on the magnetic medium. The latter is often used as **back-up** memory, the most frequently accessed data being held in **main store**. Memory is also either **volatile**, data 'evaporating' whenever the power is disconnected, or conversely **non-volatile** if storage survives disconnection.

A computer's intelligence lies largely outside its **hardware** in **software** – the coded instructions or **programs** on which its operation depends. Code can be written in a **low-level language** such as Assembler, something near the **machine code** which is directly understandable by the computer. Alternatively it can be written at **high level**, using English-type statements which the computer translates into machine code under instructions from a specialist set of software called a **compiler**. The high-level approach therefore demands more memory and processing power, but speeds software preparation. The trend is towards the use of high-level, real-time orientated languages such as CORAL and PASCAL, although system producers are also looking ahead to ADA, which is being promoted hard by the United States Department of Defense.

Where digital and analogue systems come together (many sensors are still analogue), **analogue-to-digital** and **digital-to-analogue converters** are needed at the interfaces. Computers combining analogue and digital elements are termed **hybrid computers**.

In transmitting digital data, signals can be

interleaved and combined or **multiplexed** onto a single line as a **serial** data stream, a converse de-multiplexing process returning a processed stream back to parallel signal format. In **synchronous** transmission or processing, a **system clock** is used to synchronise data trains. In **asynchronous** systems, operations follow each other sequentially without the need for a clock.

The Glass cockpit

This term reflects the tendency to replace conventional **dedicated** (single-purpose) electromechanical flight instruments by TV-type **cathode-ray tube** (crt) **multi-function** displays and, in military aircraft, additional **head-up** or **head-down** displays (HUD or HDD). The aim is to present an ever-growing volume of information more rationally, avoiding cockpit clutter and crew overload or confusion. The best way of doing this is subject to much research into the physical aspects of the **human factor** (ergonomics).

Current **electronic flight instrumentation systems** (EFIS) have two or more crts – typically six – of 127 to 200 mm (5 to 8 in) size (diagonal), on the main instrument panel, presenting normal flight data (altitude, indicated air speed, heading, aircraft attitude), but much more besides. Choosing from a displayed **menu**, or list of options, a pilot could select displays of aircraft system **status** – engines, hydraulics, electrics, for instance – navigational displays, automatic flight guidance, flight management, displays from sensors such as radar or infra-red, or tactical situation displays. Computers manage the display system and ensure that aircrew are presented with all essential 'need-to-know' information on a **priority** basis. Fault or alarm conditions automatically produce a crew alert (e.g. from an **engine instrumentation crew**

CRTs can give useful information, such as special search patterns for the SA 365N Dauphin II rescue helicopters of the Irish Air Corps. This Sperry system is part of the standard EFIS package offered for other shipborne helicopters.

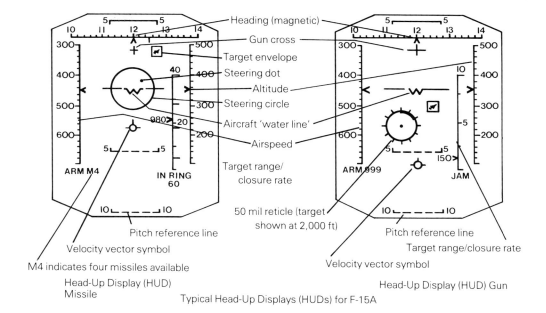

Typical Head-Up Displays (HUDs) for F-15A

Head-Up Display (HUD) Missile — labels: Heading (magnetic), Gun cross, Target envelope, Steering dot, Altitude, Steering circle, Aircraft 'water line', Airspeed, Target range/closure rate, Pitch reference line, Velocity vector symbol, M4 indicates four missiles available, ARM M4, IN RING 60

Head-Up Display (HUD) Gun — labels: 50 mil reticle (target shown at 2,000 ft), Pitch reference line, Target range/closure rate, Velocity vector symbol, ARM 999, JAM

alerting system, EICAS).

Shadow-mask three-gun colour crts have supplemented earlier **monochromatic** (black and white) displays. The recent development of a **beam index display** shows promise that a single electron gun can be used to create colour, avoiding the definition and brightness pitfalls of a three-gun system under stress of vibration or 'g' loads. Avionic displays can have simultaneous **raster** scan – the repeated line scanning and flyback system familiar to TV viewers – and **calligraphic** or direct **stroke-writing** mode (also known as **cursive**). The former is better for area painting, as in a radar or mapping display, the latter for **alpha-numeric** (alphabet and numeral) **symbology** (symbols).

Development emphasis is on larger, clearer conventional displays, reducing the heating inherent in shadow-mask technology, and developing **plasma**, **liquid crystal** and other **flat panel displays** to avionic standards of brightness and robust reliability. New, innovative display formats are displacing transitional solutions which mimic conventional **horizontal situation indicator** (hsi), and **attitude and direction indicator** (adi) presentations. An American manufacturer has recently proposed a single large full-panel display on which all data would be called up as required.

Head-up displays enable military pilots to assimilate flight and mission data without interrupting their view of the outside world. Symbology is presented on a transparent panel between the pilot's eyes and the windscreen, **collimated** for infinity so that he can see the symbols clearly when focused on the outside scene. Weapons aiming and firing symbols can be presented having a 1:1 correspondence (in scale with) the actual target area. **Holographic** or **diffractive** display optics will meet current

The US Army's advanced attack helicopter, the McDonnell Douglas Apache, is equipped with a pilot's night vision system and target acquisition and data system, mounted on the nose, to enable it to be flown and fought at night.

needs for wider **field of view** (FOV). **Image intensified** or **forward-looking infra-red** (FLIR) scenes can be presented on HUDs, together with navigation or mission symbology **overlay**, to assist night and other operations.

The overlay technique is also used in head-down displays superimposing, for example, symbology over an electronic map, or over a colour radar image in ground avoidance mode. This technique is also employed now to superimpose symbology over television and forward-looking infra-red images.

Information can also be presented to a pilot with a **helmet-mounted display**, via one of his eyes (**monocular** display). Such helmets can be useful in weapon aiming, with weapons being **slaved** to the pilot's **line of sight** so that he only has to look at the target to direct his weapons.

Protruding from the long nose of the Sikorsky S-76B helicopter are the pilot tubes used to provide pressure-related information for flight instruments. Other helicopters have these sensors mounted above the rotor hub or cockpit as needed.

Sensors and guidance

Electronics have greatly supplemented the data available to a pilot flying 'blind' under **instrument flight rules** (IFR).

Once he had only the humble **pitot-static** (air pressure) and **gyroscopic** sensors. The gyroscope has become partly electronic, as in the **tuned rotor gyro**, or completely 'high-tech' as in the **ring laser gyro** – a sensor measuring rate of movement in an axis by comparing the time taken for two laser beams to travel a triangular or square (not a ring) path in opposite directions. Gyros rely on their **inertia** to hold orientation with respect to the

aircraft and are the basis of **inertial navigation systems** (INS). In some, the gyros 'move' relatively in the traditional way, but in other systems they are restrained in **strap-down** mode and provide signals related to aircraft orientation and rate of movement. Associated **accelerometers** sense accelerations. Modern INS incorporating ring laser gyros can be astoundingly accurate.

Other **navaids** include **automatic radio direction finding** (ADF) based on radio beacons, **hyperbolic** radio aids which relate aircraft position to lattices of curved lines (hyperbolae) on charts – **Omega**, **Decca** and **Loran** are such; vhf-omnidirectional radio range (VOR), a method of navigation using selected VOR **radials** or radio tracks; **distance measuring equipment** (DME) used mainly during approaches; and **area navigation** (R Nav) which permits direct flight paths between programmed **waypoints** – selected positions *en route* to a destination. A related military system is the Tactical Air Navigation System (TANS), made by Racal Avionics.

In airport terminal areas, **instrument landing systems** (ILS) guide aircraft to safe landings. Some, to **Category IIIb** standard, are cleared for use with aircraft landing in **zero/zero** conditions (zero visibility, zero decision height). Radio beams provide vertical guidance (**glideslope,** GS) and lateral guidance (**localiser,** LOC). More versatile

systems operating at higher frequency, suitable for fixed-wing or rotary-wing aircraft guidance, are **microwave landing systems** (MLS).

Collisions with the ground have become fewer thanks to **radio** (radar) **altimeters** allied with **ground proximity warning systems** (GPWS). Military pilots have **terrain following radar**, optimised for ground-hugging operation.

Radar (radio direction finding and ranging) has long proved an effective, versatile sensor. Transmissions at **microwave** frequencies can be **pulsed**, to provide **range** measurement, or may be **continuous wave** (cw). Generally high frequencies and short pulses provide good **resolution** (distinguishing of target objects one from another) and high **definition**, with high **pulse repetition frequencies** (prf) favoured for good target **illumination**. Radars project a **beam**, shaped with a curved **antenna** (analogous to a searchlight reflector), though some radars **steer** their beams, at least in part, electronically using flat or **planar** arrays. The beam can be used to **scan** in search mode or steered to follow or **track** a target. Modern radars in combat aircraft can do both in **track-while-scan** mode, some with **multiple target** capability. Suitable choice of basic frequency, prf, **pulse length** and other parameters permit specialised ground **mapping**, **weather** and other modes, and some combat **multi-mode** radars have a capability of up to 14 different modes. Target illumination for weapons systems is another radar function, and there are dedicated **collision avoidance** radars to prevent collisions in flight. Powerful airborne radars such as the Thorn EMI search water carried aloft to extend the scanned horizon for naval fleets provide an invaluable **airborne early warning** (AEW) function against low-flying aircraft or missile threats.

Doppler radar provides accurate measurement of relative motion, utilising the Doppler effect well known at audio frequencies. **Secondary surveillance radar** (SSR) can obtain for ground controllers aircraft data such as identification and height; it depends on each aircraft being fitted with a **transponder** which transmits when triggered by the ground **primary radar**. The military equivalent, vital for distinguishing friendly from opposing forces, is **Identification Friend or Foe** (IFF).

Radars have means of cancelling returns from precipitation – rain, snow, etc – and from ground and sea **clutter** to provide good **sub-clutter visibility**.

Instrument Landing System Approach

ILS localiser

Aircraft too high/right of centre line

Autoland system check-point

Aircraft on track

ILS transmitter

Inner marker

Aircraft too low/left of centre line

Instrument Landing System
(flight deck indicators shown in the three boxes)

Outer marker

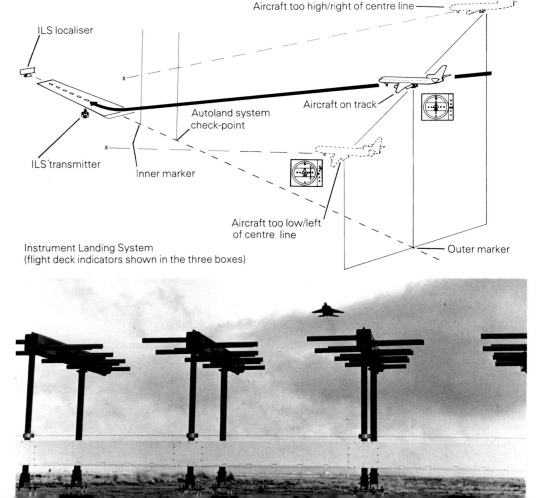

Modern military airfields and many civil airports are equipped with instrument landing systems (ILS) for precision approaches in all weather conditions by appropriately equipped aircraft. Landing here is the McDonnell Douglas F-15 Eagle.

Fokker's F27M Maritime (here in Thai colours) can be equipped with FLIR, surveillance and mapping radars, IFF and Electronic Warfare equipment, and life-saving gear such as rafts and flares. Bombs and missiles can also be carried.

Progress in radar design is achieving high capability from very compact systems. **Solid-state** (semiconductor) transmitters, for example, take up much less room than previous **travelling wave tube** or tuned oscillator **klystron** or **magnetron** transmitters, and progress here has been complemented by development of **microwave integrated circuits** (MICs), the microwave equivalent of electronic ICs.

Other areas of the **electromagnetic spectrum** exploited by sensors are the infra-red, well utilised in **forward-looking infra-red** (FLIR), and thermal emissions, used in **thermal imaging**. Extremely faint light images can be effectively amplified by **image intensifiers**, used in **low-light television** and **pilots' night vision systems** (PNVS), examples of which are found on the PAH-2 and Apache helicopters and in aviation **night vision goggles** (NVGs).

Cockpit automation

Electronic options to ease the flying task range from **stability augmentation** devices, through a **flight director** or **flight path controller** – which displays symbols to tell the pilot which way to fly – to a full **automatic flight control system** (AFCS). This up-to-date autopilot leaves the pilot out of the **control loop** and couples **actuators**, which move control sur-

faces and throttles, to **sensors** via a computer which generates the control laws or **algorithms**. The AFCS can be coupled to ILS and MLS, control height, speed and heading, and carry out advanced control procedures such as **terrain following** (TF) for low-level strike aircraft, or **automatic hover** for helicopters.

Several means of controlling the growing profusion of cockpit controls include **common control units** (ccu) based on multifunction programmable **keyboards** with associated **control display units** (cdu) to label key functions; **cockpit management systems** (CMS) to handle control of communication and navigation systems from data entered initially; and **voice input** devices in which connected-word **speech recognisers** compare pilot voiced commands with stored voice **templates** and initiate appropriate responses.

Communication between the many distributed units of machine intelligence will be provided by **databus**, a 'ring main' data highway such as the military MIL STD 1553 B. A civil near-equivalent, though less truly a full databus, is ARINC 429. Buses can be **synchronous** or **asynchronous**, **uni-** or **bi-directional** (one- or two-way data traffic). 1553 is a synchronous bi-directional bus.

Communication of data between aircraft or from aircraft to ground or seaborne units relies on an external radio **datalink**. Some of these, such as that included in the American **Joint Tactical Information Distribution System** (JTIDS) are **secure**, providing a high resistance against hostile eavesdropping (passive) or active **jamming.** The system will be applied to the Sea Harrier FRS 2.

Control

Fly-by-wire and **fly-by-light** systems may gradually supersede mechanical control runs in some aircraft categories, control actuators being signalled electrically or by light pulses transmitted through **fibre-optics**. This will

reduce weight, provide more control options, and be less vulnerable to damage. Systems will be closely integrated with flight and mission computers. Conventional control columns will give way to **sidestick** or **sidearm** controllers, providing fingertip control, including multiple switch selections. Alongside the pilot, perhaps in his seat arms, these will be of the **positive displacement** (moving) type or **force feel** (non-moving, pressure sensitive) type. This revolution will enable designers to improve cockpit layout, ergonomics and the overall **man-machine interface** (MMI) greatly.

System integrity – the capacity a system has for delivering its required performance and not letting the aircrew or mission down, especially during **critical** flight phases – depends on **reliability** and on **fail safe** or **fail operational** characteristics and effective **self-monitoring.**

Reliability implies high average or **mean time between failures** (MTBF) and high system reliability can be achieved with means such as high **circuit integration** in avionics to minimise interconnections, operating at **conservative ratings** and well within environmental limits, and using **parallel redundancy** techniques.

Redundancy means having a back-up **channel** or **lane** available in case of failure in the first lane. The channel can be **active**, executing out-of-loop processing, or available on **hot-standby**, warmed up ready for use but not actually working.

To **fail operational** a system must tolerate a single failure without operation being affected; a back-up channel takes over from the first channel if a fault develops, and initiates a crew alert and channel failure indication. A **fail soft** characteristic means that a system responds to a fault, for example a **runaway** (rogue input signal), 'softly' – with limited effect on the control surfaces so that there are no great perturbations in the aircraft's flight path.

Systems can be **simplex** – single channel non-redundant, **duplex** – twin channel with crew-operated changeover, **triplex** – two channels voting out a faulty channel on a majority basis, or even **quadruplex** – a four-channel majority-voting system which can tolerate two faults.

Continuing progress in the miniaturization, speed and '**smartness**' of electronics will expand the contribution the technology can make to future sensing, communication and flight control.

Future avionics may not simply ease pilot workload; avionics have already enabled single pilots to fly and land blind (single-pilot IFR) and will ultimately have the power to control pilotless aircraft.

For military helicopter aircrew, night vision goggles allow safe flying operations to be conducted at night, especially at low level and in confined spaces. The goggles use image intensifiers but require some ambient starlight.

For nearly 25 years, airliners have been using auto-land systems which bring them through fog and low visibility for safe approaches to a set height above the airport runways. Today, such routine operations are taken for granted.

The publishers wish to thank the following for their kind permission to reproduce the photographs in this book:

Robin Adshead 1, 2-3, 26 left, 52 centre, 56 above, 57 below, 61 above, 70 centre, 79 below right, 81 above, 104 below left, 134, 144 below left and right, 155, 156 centre, 179 above, 199, 203, 211 above left and below, 215 above; Aerophoto 80 below, 154 below, 191 below, 218 below; Airbus Industrie 210 above; Air Portraits Colour Library 20 above, 27 centre, 30 below, 49 bottom right, 55 above, 75 above, 125 below left, 182, 187 below, 192 below, 193 below, 194 below, 204 below, 207 above, 208 below left; Alitalia 113 below; Ansett 114 above; Basil Arkell 146 left; Aviation Photographs International 54 above; Avions Marcel Dassault 85 above and centre; Paul Beaver 8, 11, (The Science Museum, London) 14 below, 23 below, 24 above, 36 centre, 48 below, 50, 51 centre, 52 above, 54 below, 61 below right, (Aérospatiale) 71 below, 73, 77 centre, 79 above, 80 above, 82 above, 84 above, 123 centre, 126 above, 130 centre, 135 above right, 137 below, 139 above, 140 above and below left, 142 below, 143

below, 144 above, 145, (John Charleville) 146 above right, (Aérospatiale) 146 below right, 147 below, 152 below, (British Aerospace) 156 above and below, (Euromissile) 157 above, (British Aerospace) 157 below, (Aérospatiale) 159, 160 above, 162, (Novosti) 164, 164-5, (NASA) 165, (Aérospatiale) 166 left, 172 below, 176, 176-7, (British Aerospace) 177, 179 below, 180, 181, (Avions Marcel Dassault) 184 above, 184 below, 190, 206, 208 above, 210 above left and right, (Charles Bryant/De Havilland Aircraft) 211 above right, (Westland Helicopters Ltd) 212 centre, (McDonnell Douglas) 213 centre, 214 below left and right, 215 centre, (Euromissile) 215 below, (Airbus Industrie) 216, 217 centre and below, 218 above, 219 above and below left; Bell Helicopters 4-5, 192 above; Andrew Besley 122-3; Bildarchiv, P.K. 195 above, 200 above, 201; Boeing Aerospace Company 59 above, 79 below left, 100 above; Brasilia Airways 124 above; Brian Mackenzie Service 118 above; British Aerospace 57 centre, 66 below, 74 above, 152 above, 208 below right; Cathay Pacific 129 below; de Havilland (Terry Shwetz) 83 and 125 above left, 133; Delta Air Lines 99; Bob Downey 41 above right, 44 above, 212 below; Embraer, Brasil 84 below; European Space Agency 174 above; Fokker B.V. Holland 112 above; Kenneth Gatland 160 below right, 161 above; General Dynamics 123 above; James Gilbert 191 above; G. R. Photography 124 centre, 188 below, 194 above, 195 below, 196, 198; Grumman Aerospace 9; Hawker Siddeley Aviation 207 above; Mike Hooks 35 above, 41 below; Angelo Hornak 150; Hughes Aircraft Company 153; Leslie Hunt 45 centre, 67 below; Imperial War Museum 23 above, 28, 30 above; Japan Airlines endpapers; Lockheed-California Company 74 below, 75 below, 77 above, 79 above, 101 above, 213 above; Lufthansa Bildarchiv 88 below, 94 below, 106-7, 129 above, 130 below; McDonnell Douglas Ltd 7 below, 53 below, 62 above, 67 above, 70 below, 121 below, 131, 148, 209 right, 219 below right; Messerschmitt Bolkokow-Blohn 37 centre; Military Archive and Research Services 10, (RAF Museum, Hendon) 12 above, (Science Museum) 12 below, (Imperial War Museum) 31 below, (K Niska) 51 above, (Department of Defence) 65 centre, (US Air Force) 69, (Douglas Aircraft Co.) 91 above, (Imperial War Museum) 151 above, (Shorts Ltd, Belfast) 158 below, (General Dynamics) 210 below; NASA 161 below, 167, 168, 169 above, 170, 171, 174 below, 175; Northrop Corporation 56 centre, 147 above; Novosti Press Agency 166 above and below right; Pan American Airlines 108 below, 127 above; Photo News 140 below right; Popperfoto 14 above, 20-1, 22 above, 89 below, 187 above; Qantas Airways 132 above; Quadrant Picture Library (Flight) 6, (Flight) 16 above, (Aeroplane) 16 below, (Flight) 18 above, 18 below, (Aeroplane) 29 below, (Aeroplane) 38, (Flight) 40, (Flight) 46 below, (Flight) 51 below, 60 below, 72 centre, (Flight) 88 above, (Flight) 102 above, 115 below, 142 above, 135 above left, 151 below, (Flight) 182-3, (Flight) 200 below; Betty Rawlings 193 above; Herbert

Rittmayer 139 centre; Rockwell International 172 above, 173 above, Singapore Airlines 126 below; Smithsonian Institute 13; South African Airways 100 below right, 103 above; Space Information Centre, Belgium 173 below; John Stroud 7 above, 17 above, 22 below, 24 below, 27 above, 34 below, 35 below, 36 below, 41 above left, 48 above, 49 bottom left, 62 below, 63 centre, 86, 87, 89 above, (The Aeroplane) 90, 92, 95 above, 96, 97, 98, 100 below left, 101 below, 102 below, 104 above and below right, 105 above and centre, 107 left and right, 108 above, 109, 110 above, 111 centre, 114 centre, 115 above, 116 below, 117, 127 below, 136 left, 138 left, 178, 189, 205, 214 centre; Suddeutscher Verlag 186 below, 202, 204 above; Swiss Air 114 below, 125 below right, 128; J W R Taylor 15, 17 below, 25, 26 right, 29 above, 31 above, 32 below, 37 below, 39, 49 top and above, 56 below, 59 below, 61 below left, 76 above, 78, (Boeing Company) 91 below, 92-3, 95 below, 111 above and below, 132 below, 138 right, 154 above, 158 above, 160 below left, 163, 213 below; Michael Taylor (Lockheed-California Co.) 46 above, 47, 49 below, 63 above, 68, 76 below; Topham Picture Library 103 below, 136 right; United Airlines 113 above.

The publishers would also like to thank James Goulding and R. L. Ward for producing illustrations – James Goulding: 19 top right, 19 centre, 20 bottom left, 27 bottom right, 94 top, 105 bottom, 110 centre, 112 bottom, 116 centre, 118 centre, 121 top, 121 centre, 124 bottom left, 125 top right, 135 bottom right, 137 centre right, 139 bottom right, 141 right, 142 centre, 143 top right, 184 top left, 184 bottom left, 185 centre, 185 bottom, 188 top right, 188 centre, 194 centre left, 197 bottom right; R. L. Ward (Modeldecal): 19, 27, 42, 43, 53, 55, 62, 120, 153, 169, 186, 207, 209, 210, 212, 214, 217, 218.

Acknowledgements

It is very rare that a book is a single person's effort, if only because it takes the collective talents of many people to get a book into print. This is certainly true of the *Encyclopedia of Aviation*, which has taken the combined expertise of several writers to provide both the chapters and the features: Philip Birtles, Bob Downey, Mike Gething, George Marsh, Graham Mottram, Dr J. Royse Murphy, Ian Parker, Don Parry, Mark Pulsford and Brian Walters wrote most of the main chapters and supporting sections. Thanks also go to Lyn Greenwood who compiled the index; to Richard Leask Ward and James Goulding who prepared the majority of the technical and colour side-view illustrations; and to Robin Adshead, Gordon Roberts, Michael Taylor, John Stroud, MARS, Quadrant Picture Library, Defence Helicopter World, Commuter World, SPACE, Air Portraits, McDonnell Douglas, Boeing, Airbus Industrie, Lieutenant Mark Watson RN, Bell Helicopter Textron, Martin Brodie, Embraer, Northrop, Westland, Sikorsky Aircraft, the Fleet Air Arm Museum and others who provided photographs.
Michael Gilliat, Philip Wilkinson, Stephen Adamson, Jeremy Bratt, Laurence Bradbury, Celina Dunlop

and Sara Hunt are to be thanked for their editorial direction, design skills, picture research and production control. In addition, my wife and researcher, Ann Beaver, contributed immeasurably by keeping the office straight when pandemonium broke out.
Most of all, I would like to pay tribute to my friend and colleague, Bob Downey, who was sadly killed in a motoring accident in October 1985 during the preparation of this book – but not before he had completed his share of the special insert sections which support the main narrative. All the contributors knew and respected Bob's work and it is perhaps fitting to dedicate this book to his memory.
Paul Beaver
General Editor
June 1986